About the Artist/Author

Vincent (Vince) Perez, world-renowned and award-winning medical illustrator, has dedicated his life to perfecting his work. The hundreds of illustrations that grace this anatomy atlas attest to his passion for the human form and his commitment to its accurate representation. For decades, Vince's work has illuminated the study of anatomy for millions of students and health-care professionals worldwide. In particular, beginning with the first anatomy guide he created for BarCharts in 1994, he has produced guides of the full body as well as of each system in detail, rounding out their extensive *QuickStudy* academic and medical product line with his instructive illustrations. His meticulous rendering of the human body—whether in broad view of the form as a whole or in minute focus of isolated parts and cells—showcases its features with unsurpassed intricacy of detail.

Vince's client list is as impressive as it is diverse, spanning the globe and encompassing the industries of health care, biotechnology, pharmaceuticals, mass media, telecommunications and transportation: ABC TV; British Airways, Ciba-Geigy, Ltd.; Cutter Laboratories; Disney; Lucasfilm, Ltd.; Pacific Bell; Potlatch Corporation; Simon & Schuster; Sterling-Winthrop; Syntex Corporation; Time, Inc.; Wright Medical Technologies; and many more. In addition to the commercial illustrations he produces for his numerous clients, Vince's multimedia art forms have enjoyed—and continue to enjoy—audiences in virtually every corner of the world, from the United States to Europe to Asia. In addition to being featured at the Osaka World's Fair in Japan, his work is showcased in the permanent collections of several world-class institutions, including the Vatican Museum (Vatican City, Italy), the San Francisco Museum of Modern Art and the California Legion of Honor (San Francisco, California), and the National Portrait Gallery (Washington, D.C.).

Always, his art awes and inspires spectators. Awards and accolades galore attest to this fact: Gold Award from the Western Art Director's Club; Award of Excellence from the Rx Club (New York); CA Award of Excellence, Design Annual; Award of Excellence, Mead Show, National Medical Enterprises, Inc.; Fourth Annual LULU Award of Excellence; San Francisco Society of Illustrators Technical Illustration Gold and Silver Awards; and the crowning achievement, two pieces chosen by the Society of Illustrators (New York) for selection among 100 of the best medical illustrations ever produced in the United States.

Born and raised on the East Coast, Vince has lived in California since 1964. Today he divides his time amongst his many beloved pursuits: first and foremost, his art; then his teaching of future artists and his volunteerism in the Alameda, California, community he calls home. A full-tenured professor at the California College of the Arts (CCA), where he has chaired the illustration and drawing programs that he helped develop, Vince instructs students in illustration, life drawing, printmaking, and of course, anatomy. He has also taught anatomy and illustration at the University of California at Berkeley (UCB). He holds a B.F.A. in graphic arts and illustration from Pratt Institute in New York, and an M.F.A. in painting from the California College of Arts & Crafts (now CCA). He also completed graduate work in fine arts at the University of the Americas in Mexico City (Mexico).

VINCE PEREZ

An artist whose talent truly runs the gamut—he is equally at home creating intricate medical illustrations, cutting-edge graphic designs and whimsical woodcuts—the genius of Vince Perez comes to life on every page of this atlas, and his at-once informative and innovative anatomical renderings are sure to enrich your medical experience at whatever point you are in your education—beginning student awed by the medical universe, seasoned professional seeking further knowledge, or curious layperson expanding a home-health library. No matter, you are sure to find the Perez-illustrated journey to be as instructive as it is unforgettable.

Let Vince's unparalleled representation of the human body illuminate your foray into the world of medicine! (To see a sampling of his work, please visit *www.perezstudio.com*.)

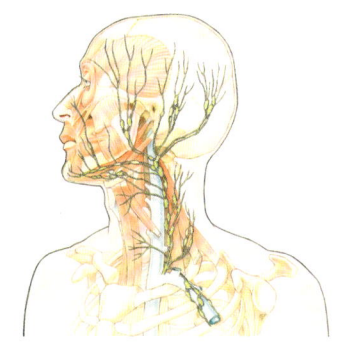

DISCLAIMER: This *QuickStudy® Atlas of Human Anatomy* is intended for *informational purposes only*. It is not intended for the diagnosis, treatment or cure of any medical condition or illness, and should not be used as a substitute for professional medical care. BarCharts, Inc., its writers and editors and designers, are not responsible or liable for the use or misuse of the information contained in this guide.

All rights reserved. No part of this publication may be reproduced or transmitted in any form or by any means, electronic or mechanical, including photocopy, recording, or any information storage and retrieval system, without written permission from the publisher.

©2006 BarCharts, Inc.
ISBN 13: 9781423251415
ISBN 10: 1423251415

BarCharts and QuickStudy are registered trademarks of BarCharts, Inc.

Publisher: BarCharts, Inc.
6000 Park of Commerce Boulevard, Suite D
Boca Raton, FL 33487
www.quickstudy.com

Artist: Vincent Perez
Images © Vincent Perez / *www.perezstudio.com*

Editing:	Lisa Drucker
Proofreading:	Kaaren Ashley
	Mona Moskowitz
	Peter Miller
Art Direction:	Rich Marino
Design/Layout:	Andre Brisson
	Andrea Hutchinson
	Dale Nibbe
	Latoya Danford

Printed in Thailand

ATLAS *of* HUMAN ANATOMY

Vincent Perez

BarCharts, Inc.®

Boca Raton, Florida

Contents

| About the Artist/Author | i |

1
Surface Anatomy — 1
Full Body: Multiple Views
Head: Anterior & Lateral Views
Eye & Ear
Mouth & Nose
Right Arm & Hand: Lateral & Medial Views
Hand: Dorsal & Palmar Views
Axilla & Breast
Hips: Female & Male Views
Leg & Foot: Medial View/Dorsal & Plantar Views

2
Skeletal System — 13
Female & Male Skeletons: Anterior Views
Skeleton: Lateral & Posterior Views
Vertebral Column
Lumbar Vertebra
Cervical & Lumbar Vertebrae: Posterior Views
Skull: Multiple Views
Anterior Skull & Cut Cervical Vertebrae: Posterior Views
Skull & Cervical Vertebrae: Posterior/Anterior Views
Clavicle: Superior & Inferior Views
Hyoid Bone: Multiple Views
Scapula: Anterior & Posterior Views
Thoracic Bones
Left Hip Bone: Anterior & Posterior Views
Elbows: Anterior & Posterior Views
Hands: Dorsal & Palmar Views
Knees: Anterior & Posterior Views
Feet: Dorsal & Plantar Views
Bone Structure

3
Joints & Ligaments — 33
Spine
Temporomandibular & Hyoid
Temporomandibular Joint
Craniocervical
Sternoclavicular & Shoulder
Elbow: Anterior & Lateral Views
Wrist & Hand: Dorsal/Palmar Views
Finger: Medial View
Lumbar Spine
Connective Components of Pelvis
Hip Ligaments
Pelvis: Superior & Posterior Views
Knee Ligaments: Anterior & Posterior Views
Right Foot: Multiple Views

4
Origins & Insertions — 47
Head & Trunk: Anterior & Posterior Views
Arm: Anterior & Posterior Views
Clavicle: Superior & Inferior Views
Hand: Dorsal & Palmar Views
Leg & Foot: Anterior & Posterior Views
Foot: Dorsal & Plantar Views
Base of Skull
Hyoid Bone: Superior View

5
Muscular System — 59
Surface Muscles (Layers I, IA & II): Multiple Views
Deep Muscles (Layers II, III, IV, V, VI & VII): Multiple Views
Head Muscles
Muscles of the Eye – Extrinsic Eye Muscles: Multiple Views
Deep Neck Muscle: Lateral View
Muscles of Respiration
Components of the Hand: Dorsal View
Components of the Finger: Cross Section
Arm & Hand Muscles: Multiple Views
Palmar Hand: Layers I, II, III, IV & V
Dorsal Hand: Layers I, II & III
Lateral & Medial Hand
Leg & Foot Surface Muscles: Multiple Views
Dorsal Foot: Layers I, II & III
Plantar Foot: Layers I, II, III, IV & V
Lateral Foot
Medial Foot
Muscle Microstructure: Extensor Indicis

6
Nervous System — 101
Nervous System: Anterior & Posterior Views
Cutaneous Innervation – Dermatomes & Peripheral Nerve Distributions:
  Anterior & Posterior Views
Cervicobrachial Plexus: Posterior View
Lumbosacral Plexus: Posterior View
Spinal Cord
Sciatic Nerve
Trigeminal Nerve
Nerves of the Face & Head
Nerve Structure

7
The Brain — 113
Brain in Place
Brain: Multiple Views
Brain: Frontal Section
Brain: Horizontal Section
Brain: Ventricles
Brain: Arteries

8 The Senses — 125
Head: Eye, Ear, Nose & Mouth
Seeing
Hearing
Smell
Taste
Touch

9 Digestive System — 133
Digestive System: Anterior & Posterior Views
Mouth & Salivary Glands
Tongue
Stomach
Bile & Pancreatic Duct
Small Intestine (Schematic)
Large Intestine
Ileocecal Sphincter & Appendix
Rectum

10 Respiratory System — 145
Respiratory System: Anterior & Posterior Views
Nasal & Oral Cavity
Nasal Septum
Paranasal Sinuses
Larynx
Bronchial Tree
Muscles of Respiration
Alveoli Cluster
Oxygenation of Alveoli Cluster

11 Circulatory System — 155
Circulatory System: Anterior & Posterior Views
Venous & Arterial Systems
Head & Neck (Schematic)
Skull & Arteries
Arteries of Brain: Inferior View
Circle of Willis
Brain & Neck
Blood Circuits
Hepatic Portal Veins
Blood Vessels

12 The Heart — 169
Heart: Anterior, Posterior & Interior Views
Coronary Arteries & Cardiac Veins
Circulation
Nerves & Arteries
Heart in Diastole
Heart in Systole
Beginning & End of Diastole
Beginning & End of Systole

13 Lymphatic System — 181
Lymphatic System
Head & Neck
Arm Axilla & Thorax
Heart & Lungs
Thoracic Duct
Deep Abdominal & Inguinal Nodes
Stomach & Pancreas
Large Intestine
Nodes & Vessels

14 Urogenital System — 191
Male Urogenital System: Anterior & Lateral Views
Male Urinary System: Anterior View
Female Urogenital System: Anterior & Lateral Views
Right Kidney
Renal Corpuscle
Nephron

15 Reproductive System — 201
Male Reproductive System
Female Reproductive System
Stages of Sperm & Ovum
Uterine Cycle
Fetal Circulation
Full-Term Baby: Prior to Delivery & Being Delivered
Sexual Intercourse

Index — 212

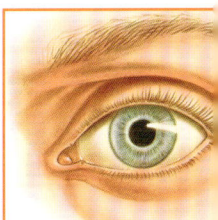

1

SURFACE ANATOMY

SURFACE ANATOMY

FULL BODY

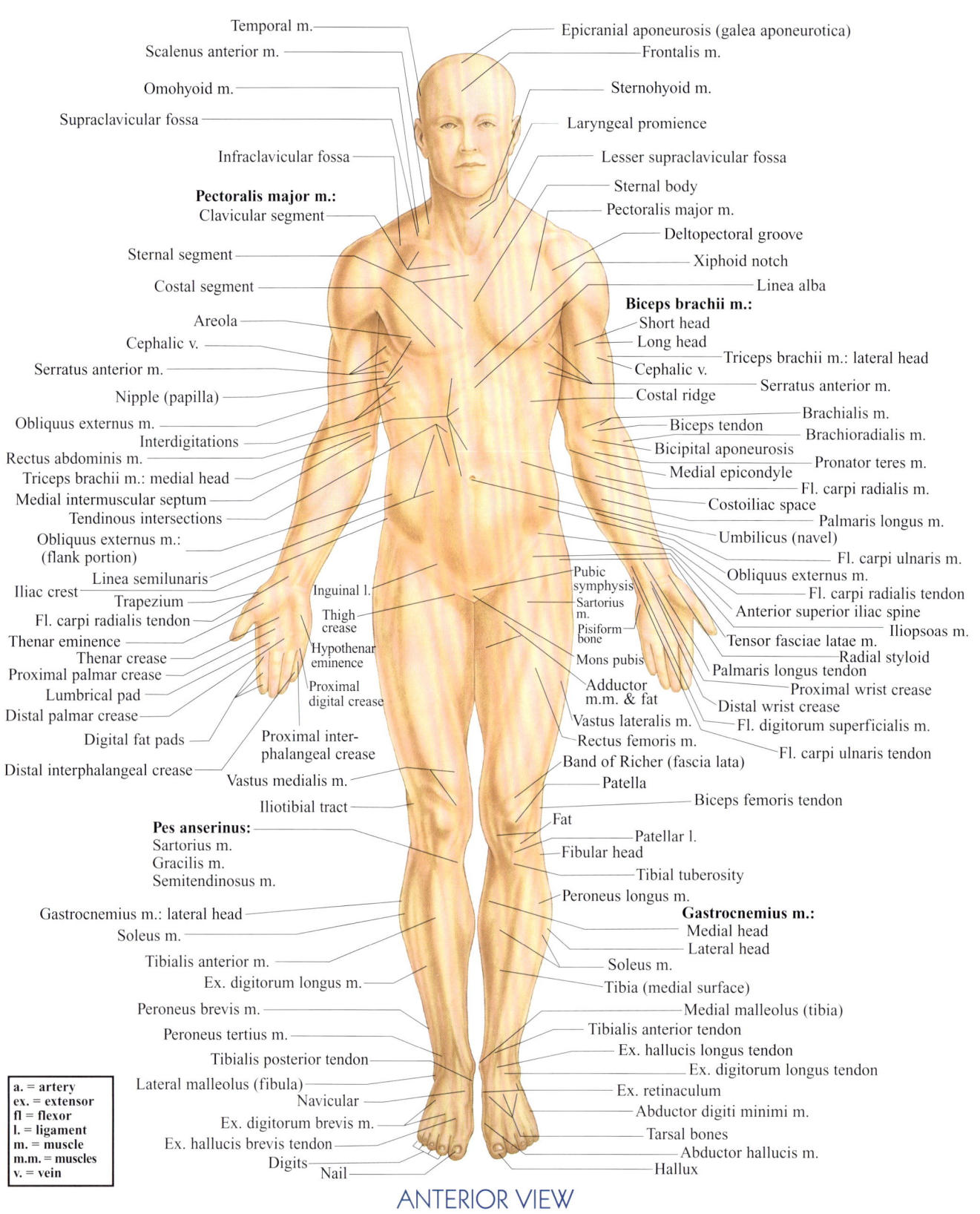

ANTERIOR VIEW

a. = artery
ex. = extensor
fl = flexor
l. = ligament
m. = muscle
m.m. = muscles
v. = vein

FULL BODY

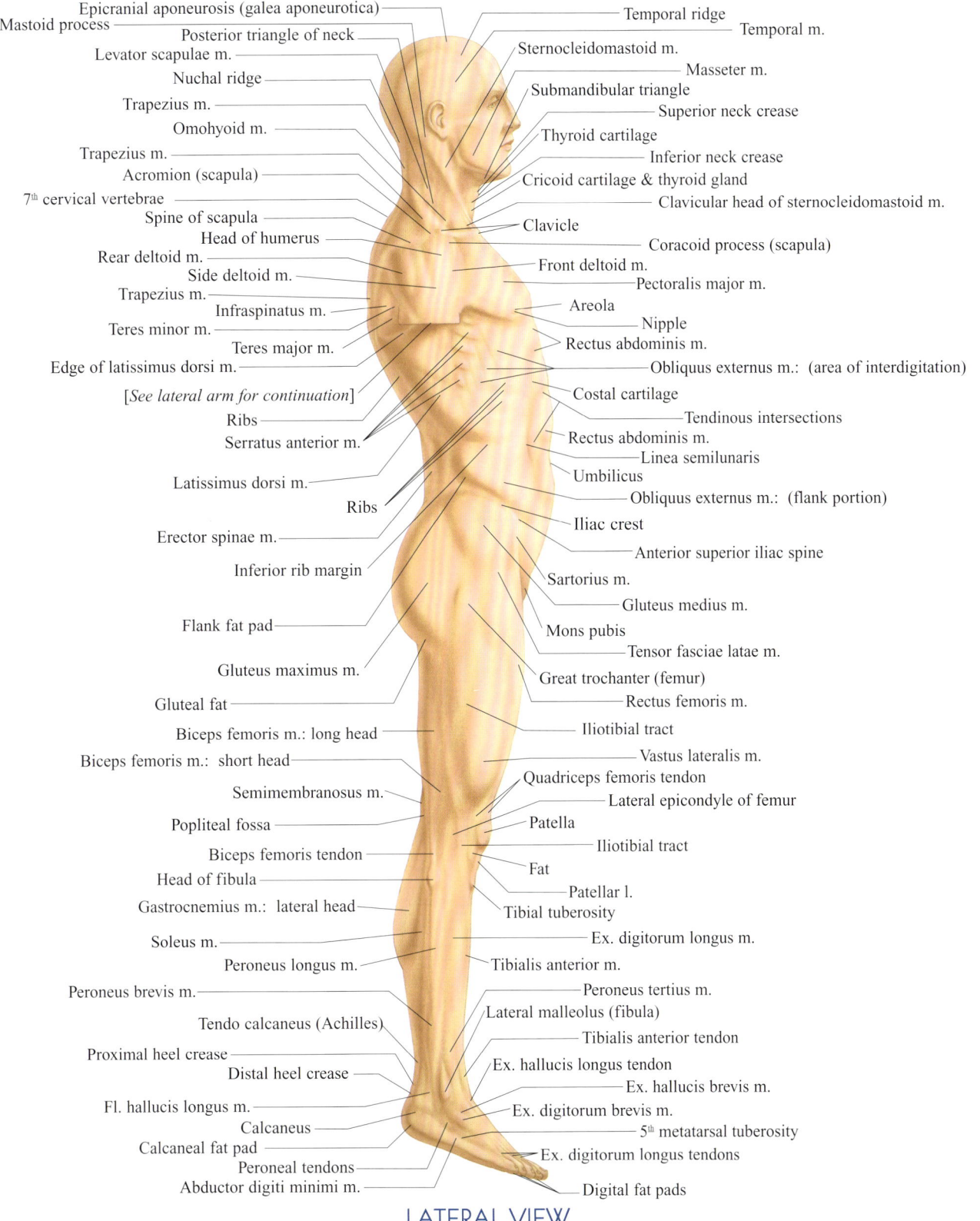

LATERAL VIEW

SURFACE ANATOMY

FULL BODY

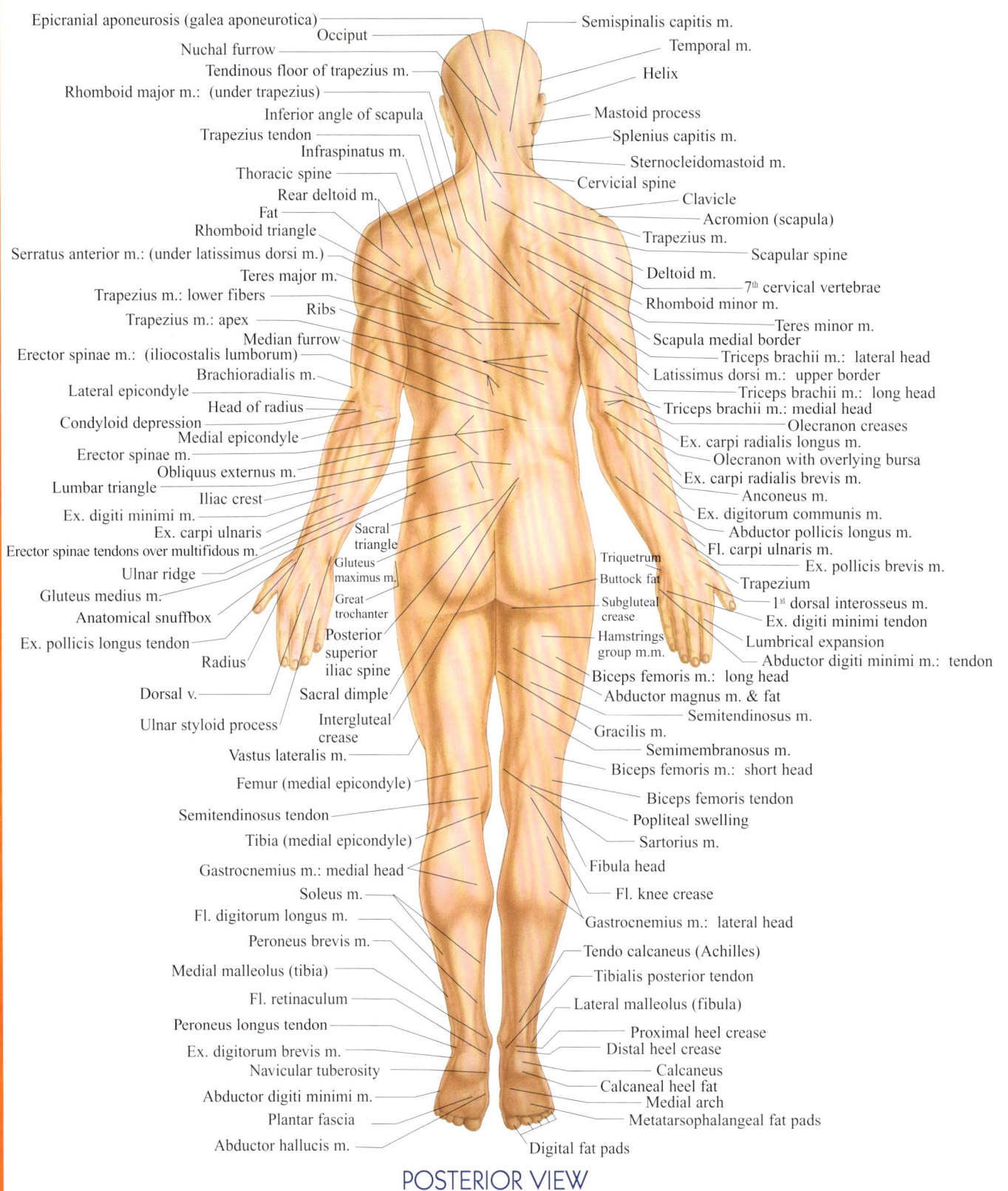

POSTERIOR VIEW

SURFACE ANATOMY

HEAD

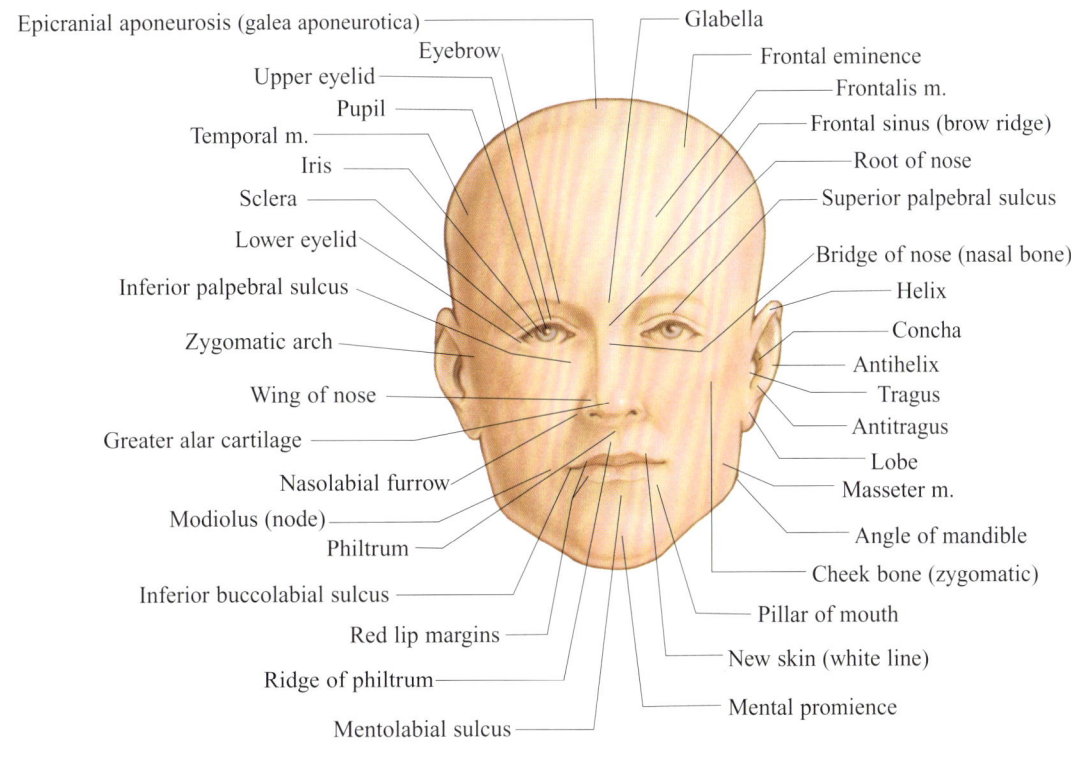

ANTERIOR VIEW

EYE

EAR

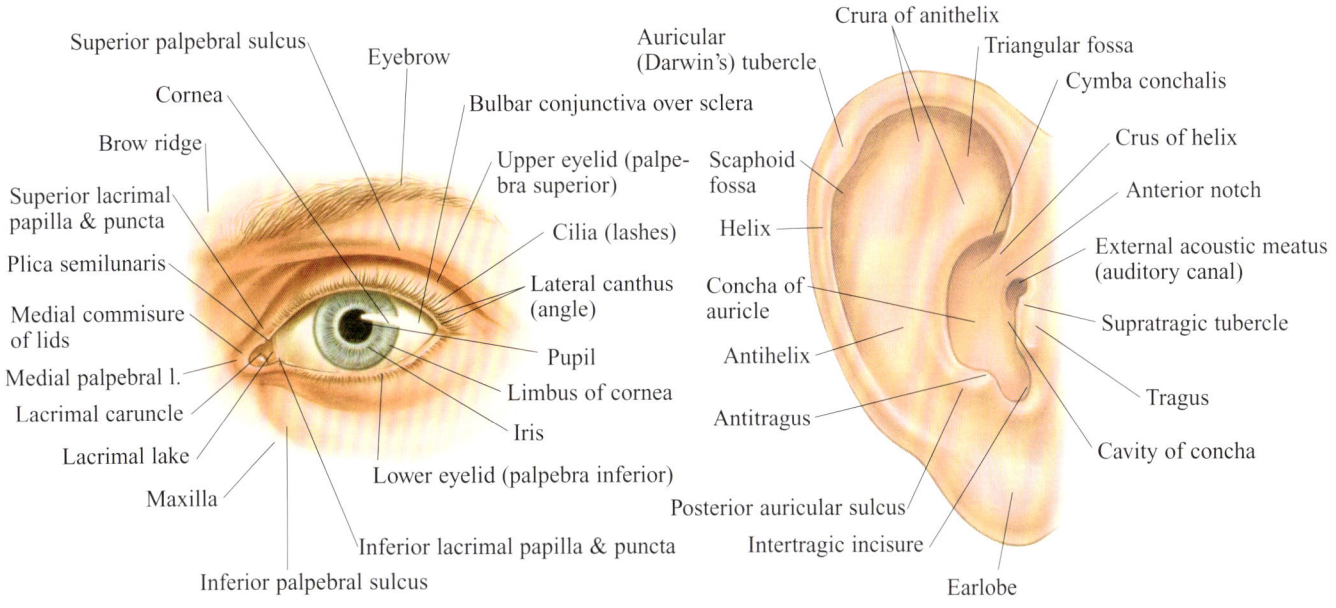

SURFACE ANATOMY

HEAD

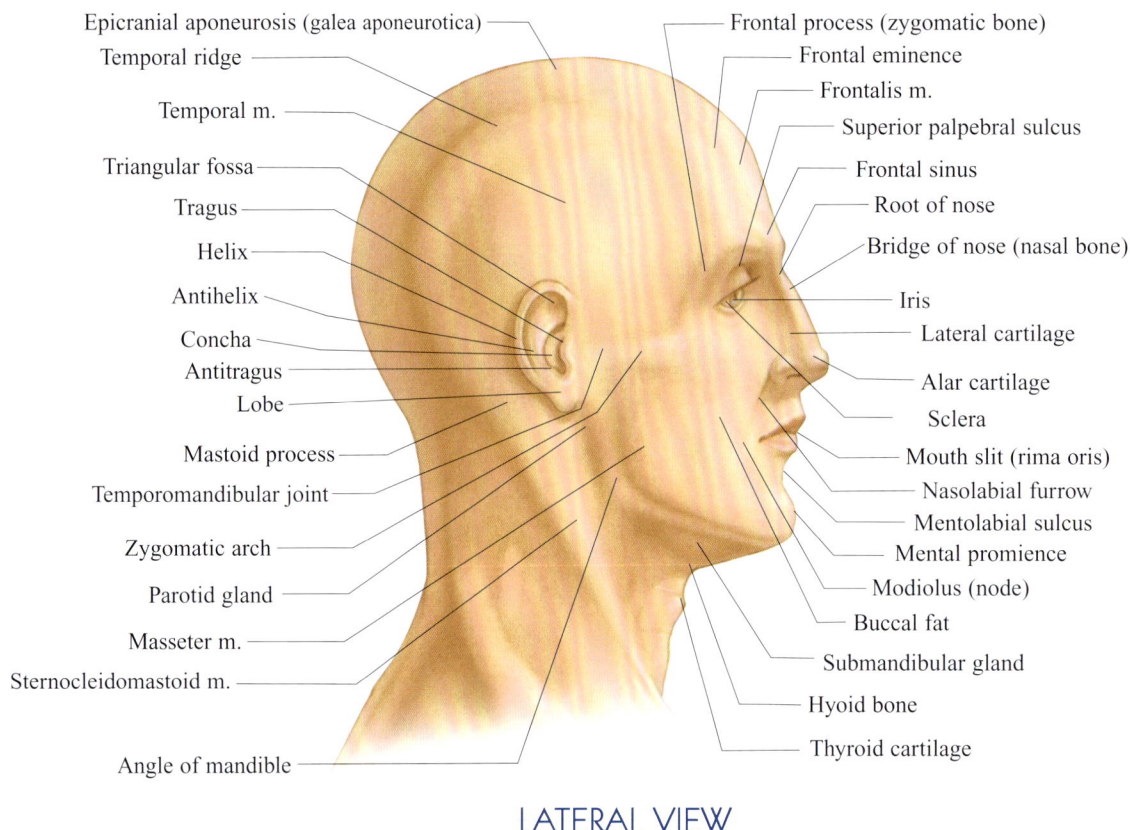

LATERAL VIEW

MOUTH & NOSE

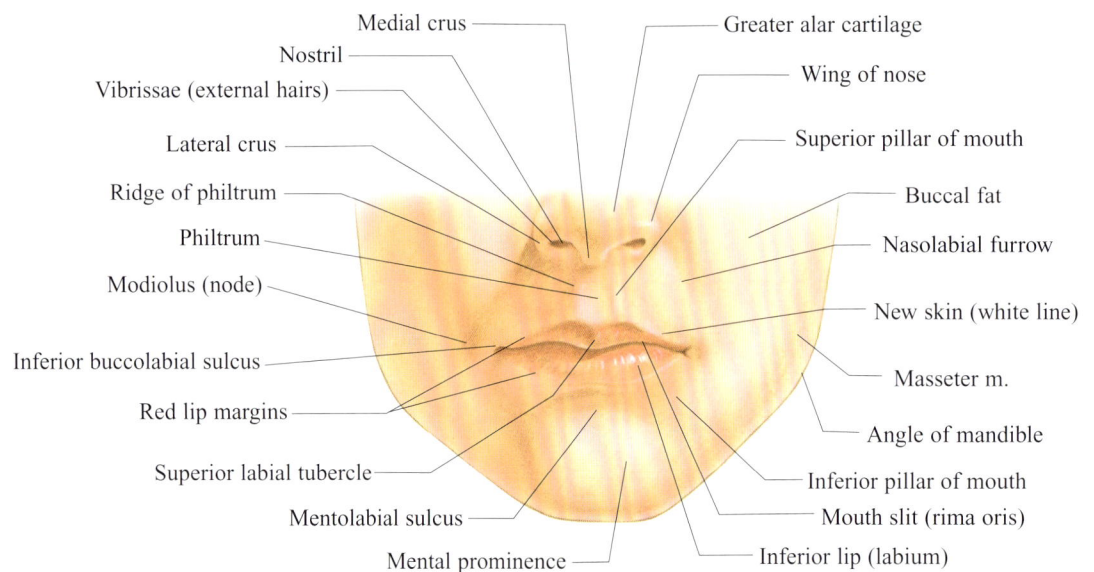

SURFACE ANATOMY

RIGHT ARM & HAND

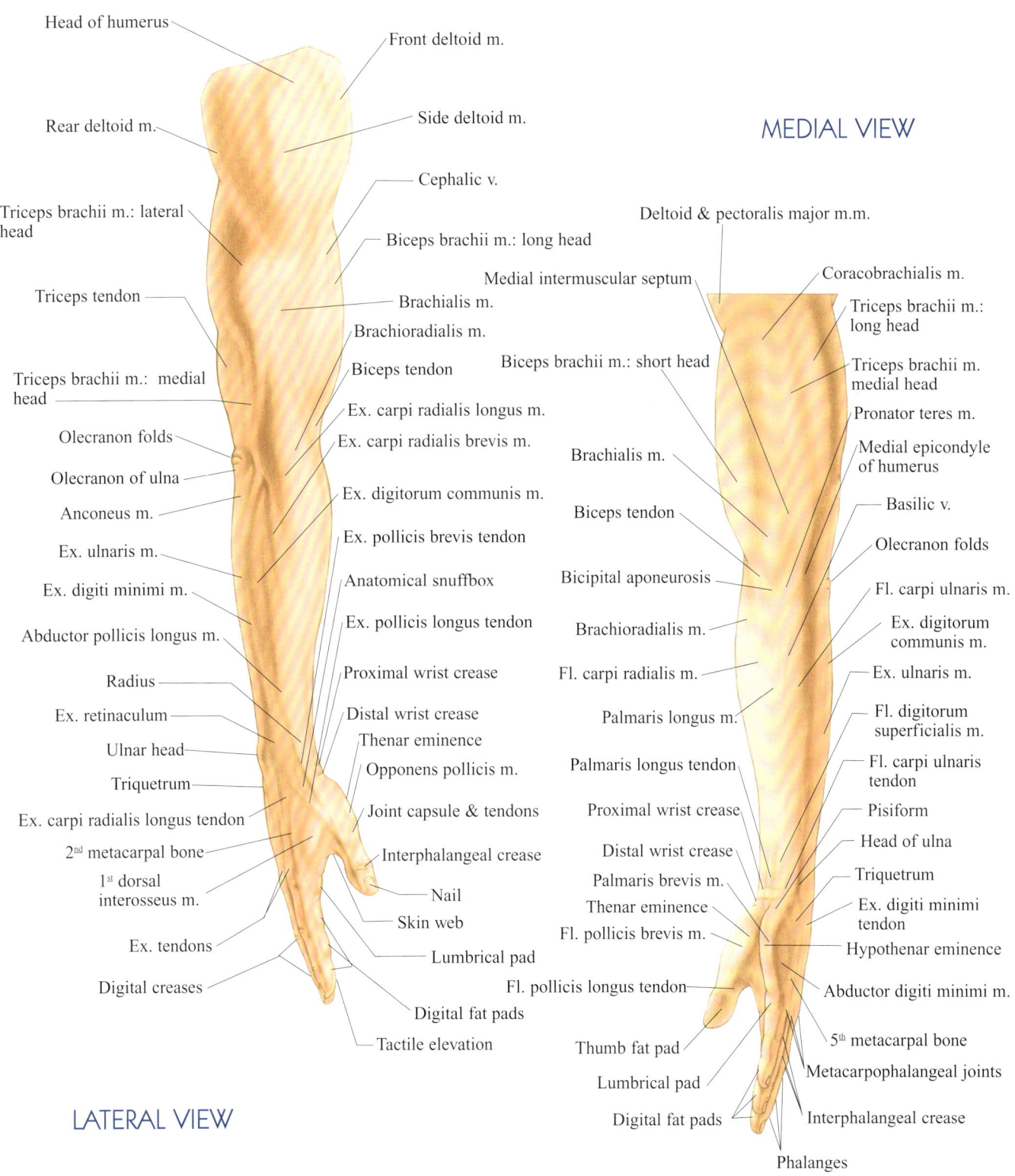

LATERAL VIEW

MEDIAL VIEW

7

SURFACE ANATOMY

HAND

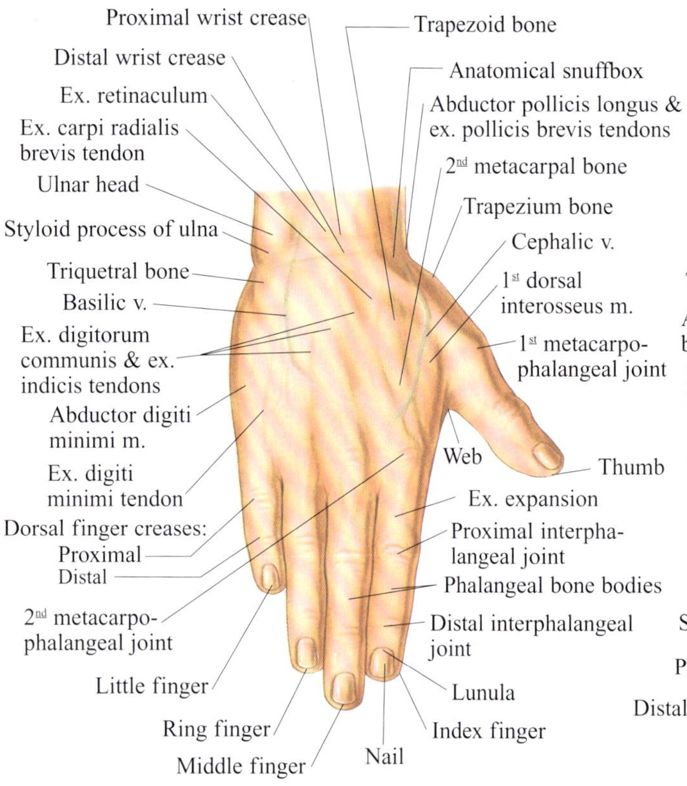

DORSAL VIEW

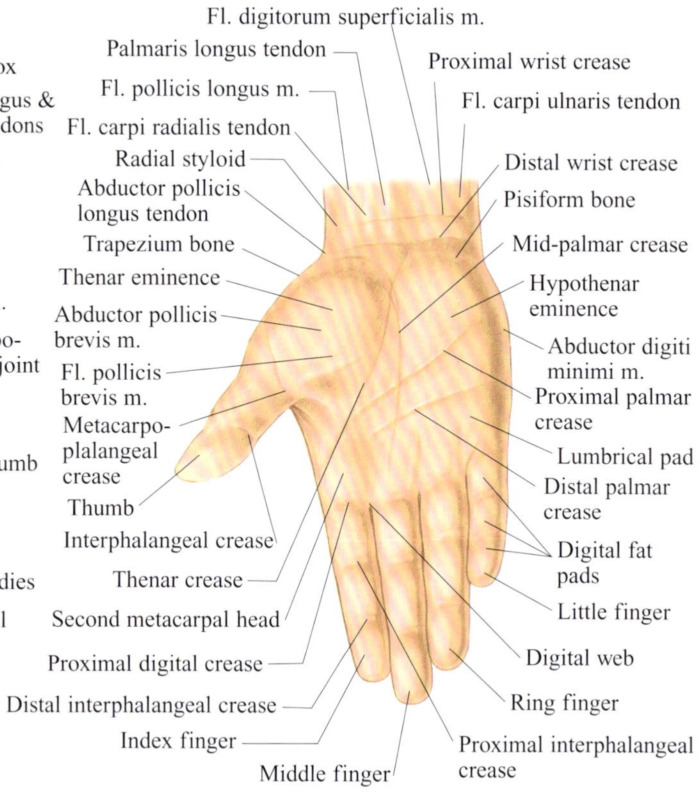

- Proximal wrist crease
- Distal wrist crease
- Ex. retinaculum
- Ex. carpi radialis brevis tendon
- Ulnar head
- Styloid process of ulna
- Triquetral bone
- Basilic v.
- Ex. digitorum communis & ex. indicis tendons
- Abductor digiti minimi m.
- Ex. digiti minimi tendon
- Dorsal finger creases:
  - Proximal
  - Distal
- 2nd metacarpo-phalangeal joint
- Little finger
- Ring finger
- Middle finger
- Trapezoid bone
- Anatomical snuffbox
- Abductor pollicis longus & ex. pollicis brevis tendons
- 2nd metacarpal bone
- Trapezium bone
- Cephalic v.
- 1st dorsal interosseus m.
- 1st metacarpo-phalangeal joint
- Web
- Thumb
- Ex. expansion
- Proximal interphalangeal joint
- Phalangeal bone bodies
- Distal interphalangeal joint
- Lunula
- Index finger
- Nail

PALMAR VIEW

- Fl. digitorum superficialis m.
- Palmaris longus tendon
- Fl. pollicis longus m.
- Fl. carpi radialis tendon
- Radial styloid
- Abductor pollicis longus tendon
- Trapezium bone
- Thenar eminence
- Abductor pollicis brevis m.
- Fl. pollicis brevis m.
- Metacarpo-phalangeal crease
- Thumb
- Interphalangeal crease
- Thenar crease
- Second metacarpal head
- Proximal digital crease
- Distal interphalangeal crease
- Index finger
- Middle finger
- Proximal wrist crease
- Fl. carpi ulnaris tendon
- Distal wrist crease
- Pisiform bone
- Mid-palmar crease
- Hypothenar eminence
- Abductor digiti minimi m.
- Proximal palmar crease
- Lumbrical pad
- Distal palmar crease
- Digital fat pads
- Little finger
- Digital web
- Ring finger
- Proximal interphalangeal crease

AXILLA & BREAST

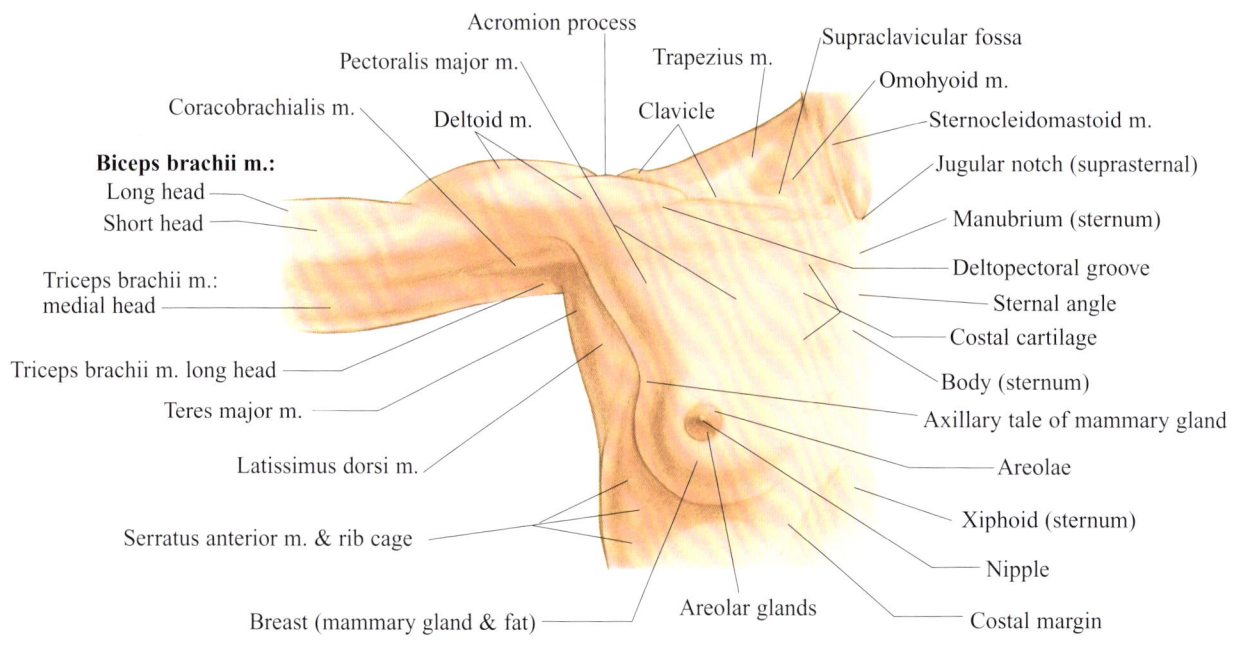

- Acromion process
- Pectoralis major m.
- Coracobrachialis m.
- Deltoid m.
- **Biceps brachii m.:**
  - Long head
  - Short head
- Triceps brachii m.: medial head
- Triceps brachii m. long head
- Teres major m.
- Latissimus dorsi m.
- Serratus anterior m. & rib cage
- Breast (mammary gland & fat)
- Trapezius m.
- Clavicle
- Supraclavicular fossa
- Omohyoid m.
- Sternocleidomastoid m.
- Jugular notch (suprasternal)
- Manubrium (sternum)
- Deltopectoral groove
- Sternal angle
- Costal cartilage
- Body (sternum)
- Axillary tale of mammary gland
- Areolae
- Xiphoid (sternum)
- Nipple
- Costal margin
- Areolar glands

8

SURFACE ANATOMY

HIPS

FEMALE VIEW

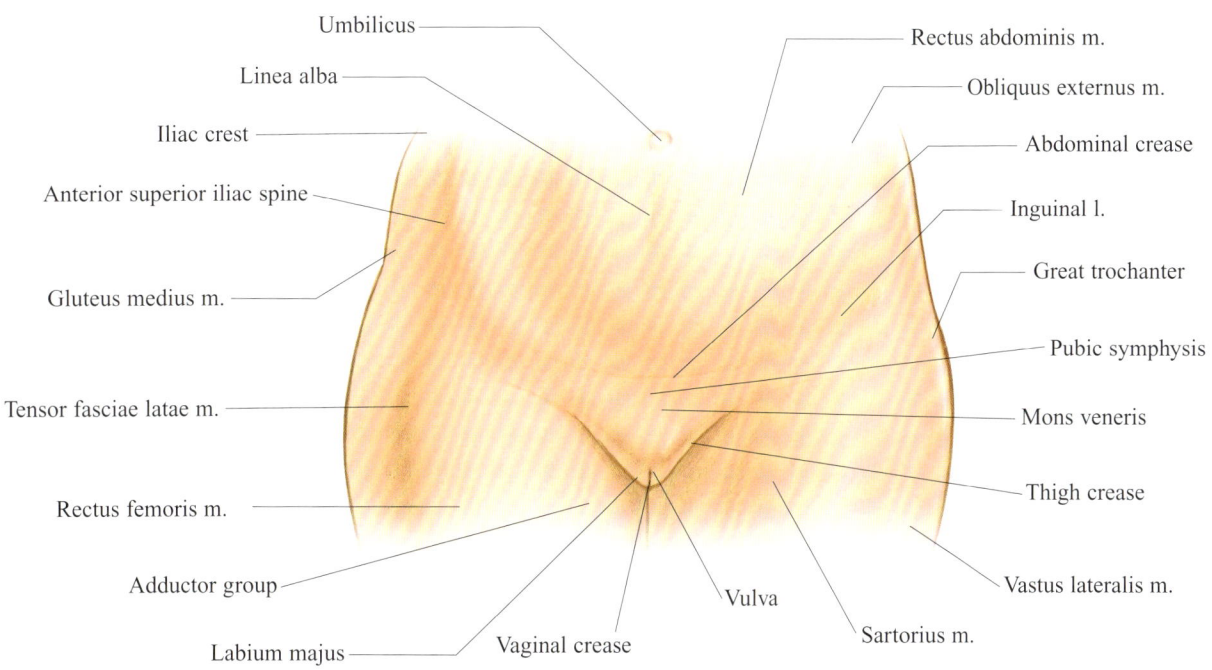

MALE VIEW

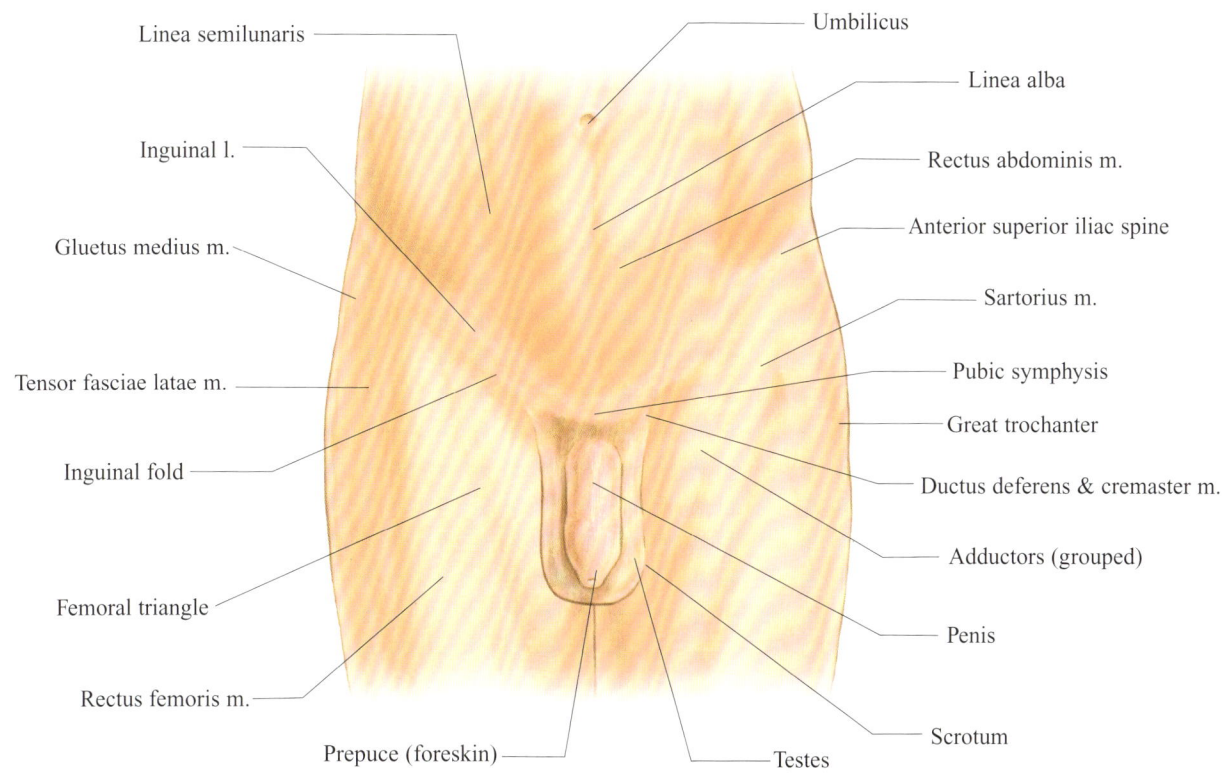

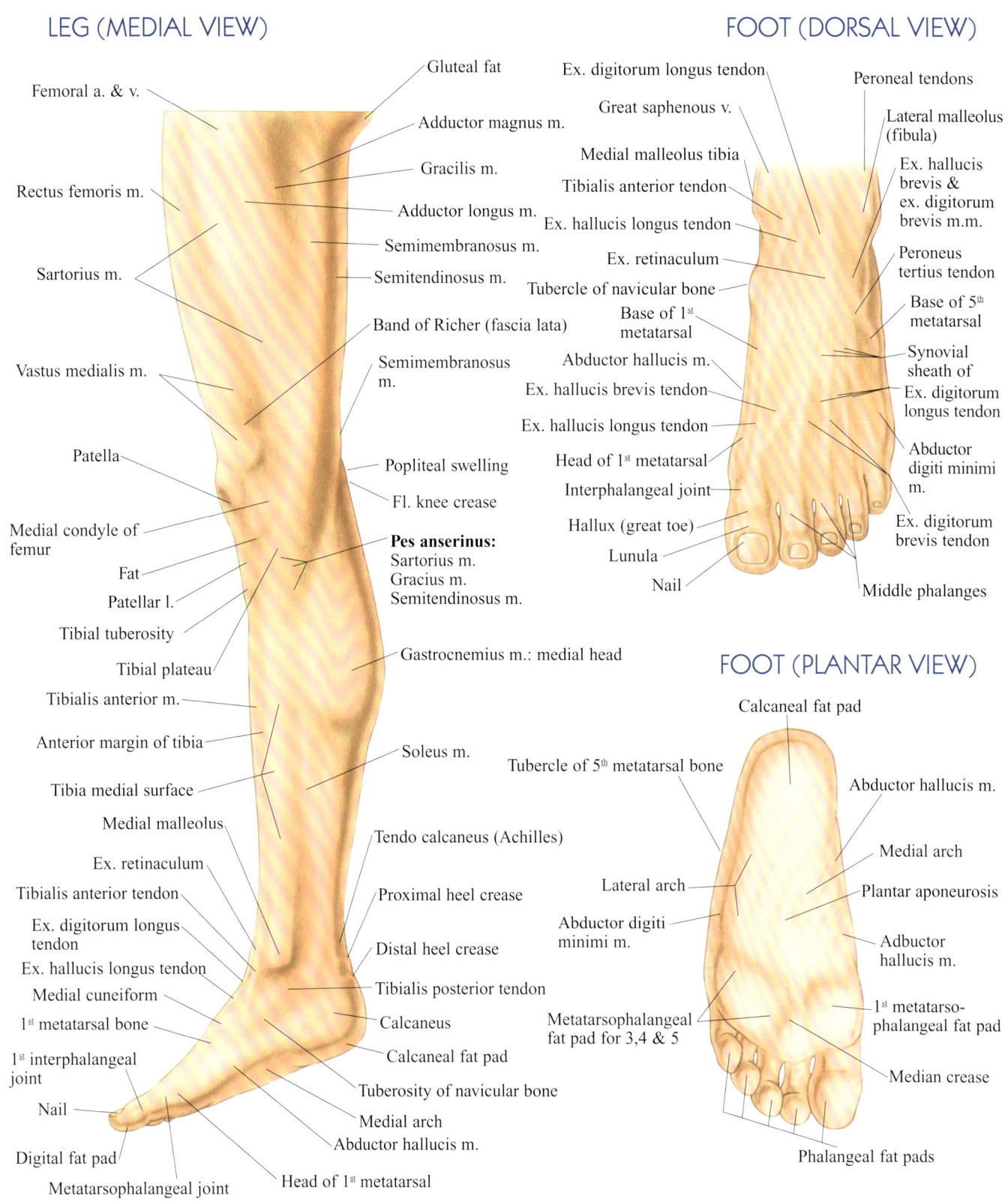

SURFACE ANATOMY

NOTES

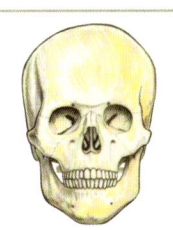

2

SKELETAL SYSTEM

SKELETAL SYSTEM

FEMALE SKELETON
ANTERIOR VIEW

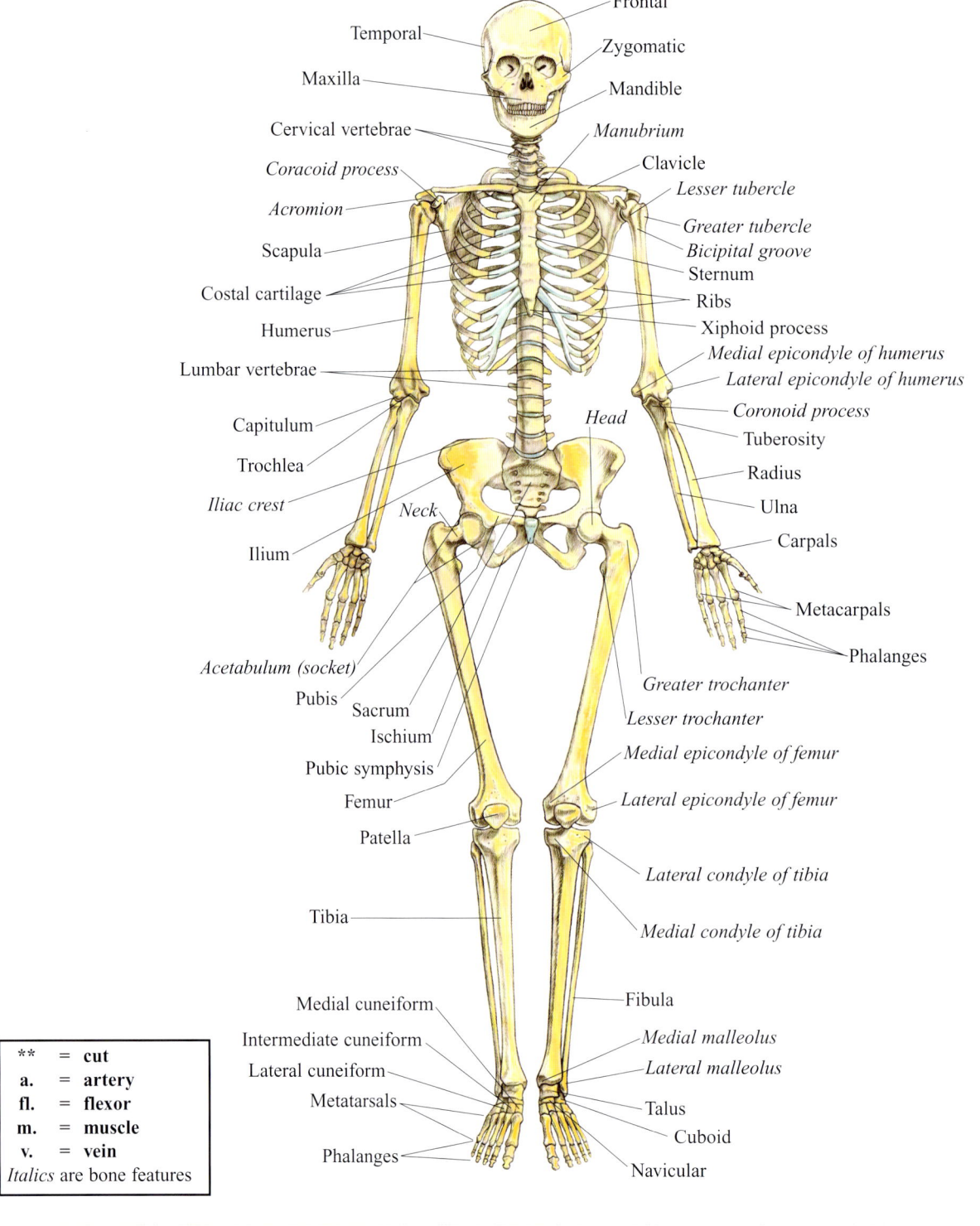

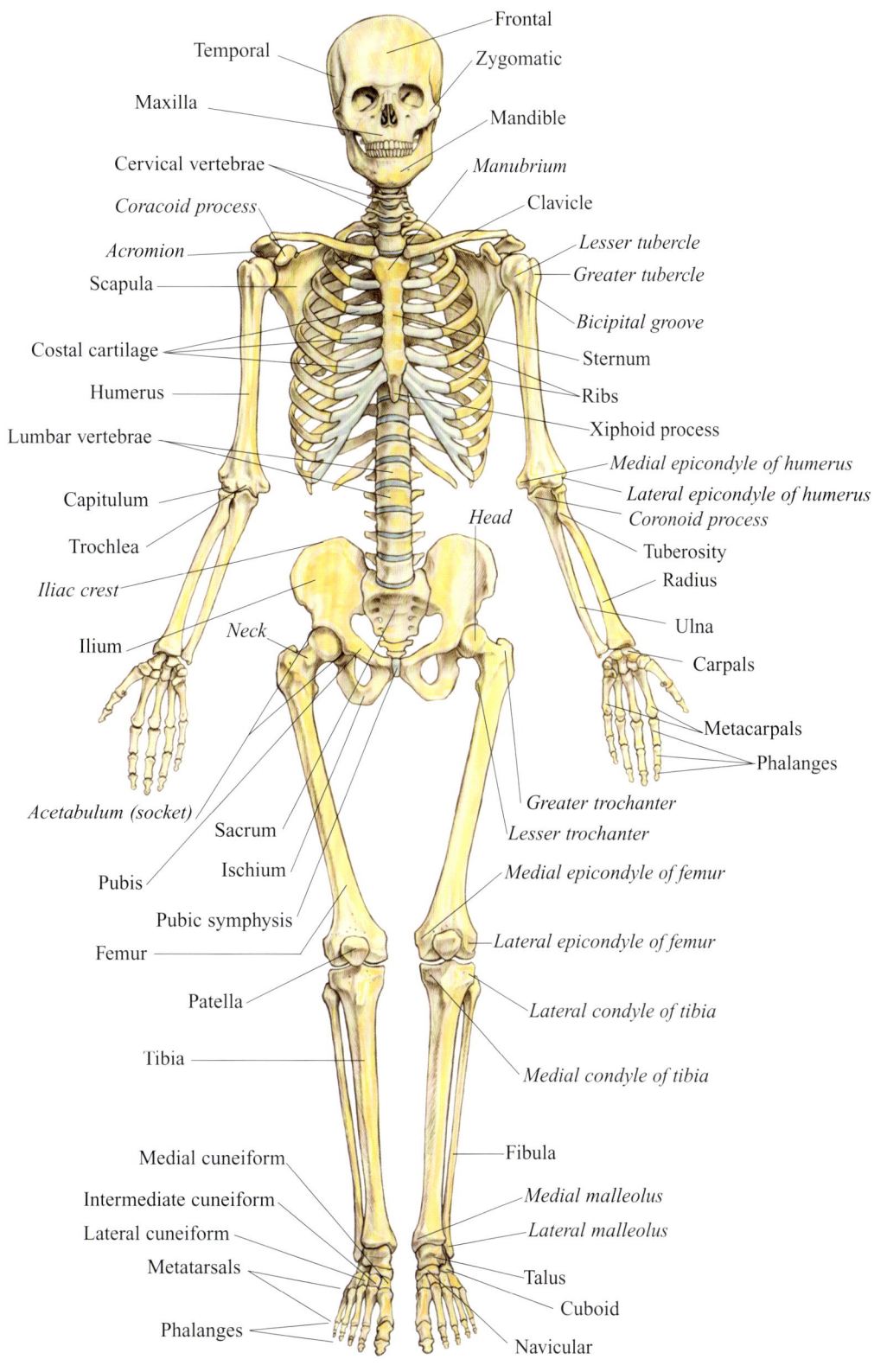

SKELETAL SYSTEM

SKELETON
LATERAL VIEW

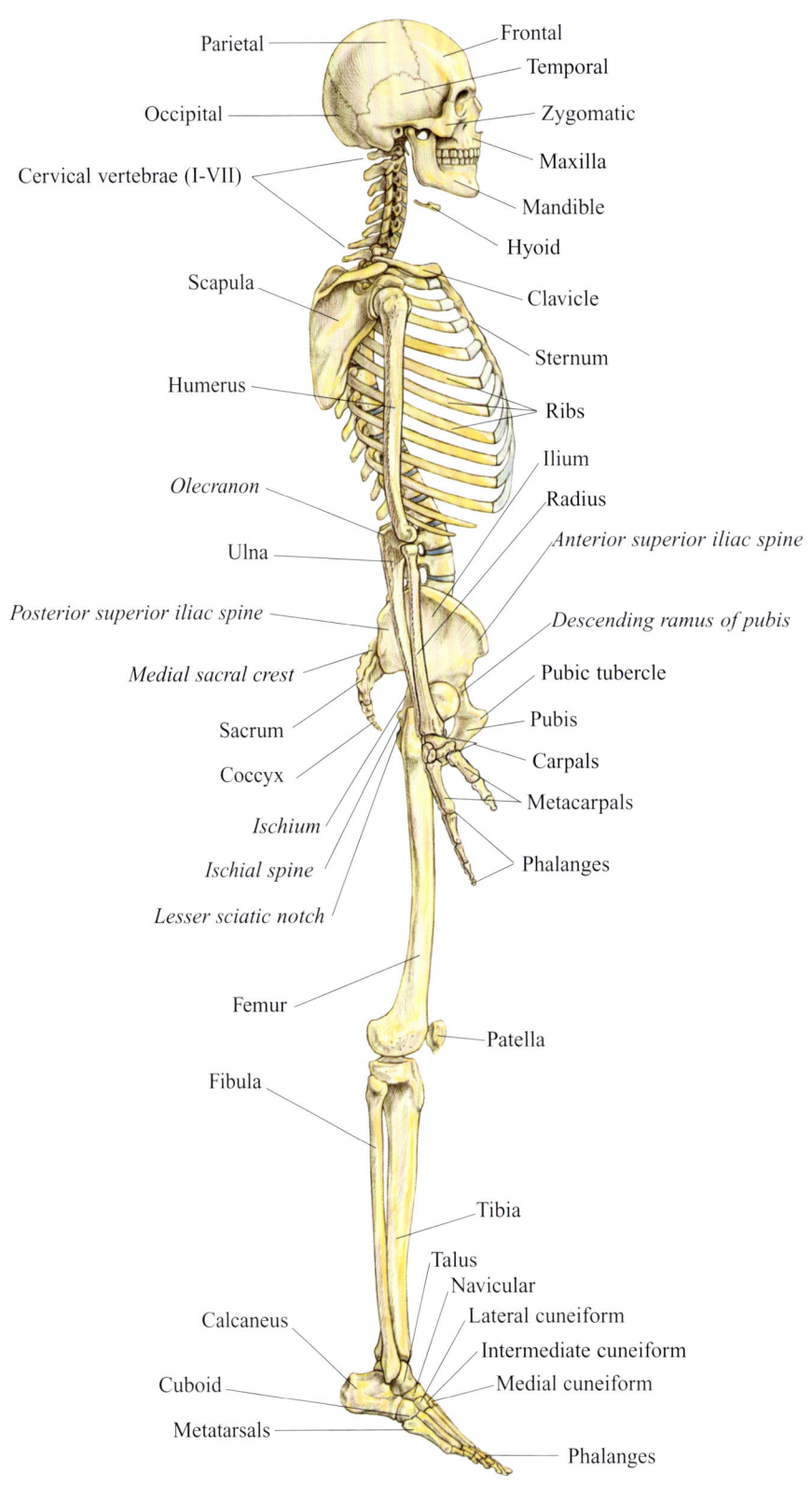

SKELETAL SYSTEM

SKELETON
POSTERIOR VIEW

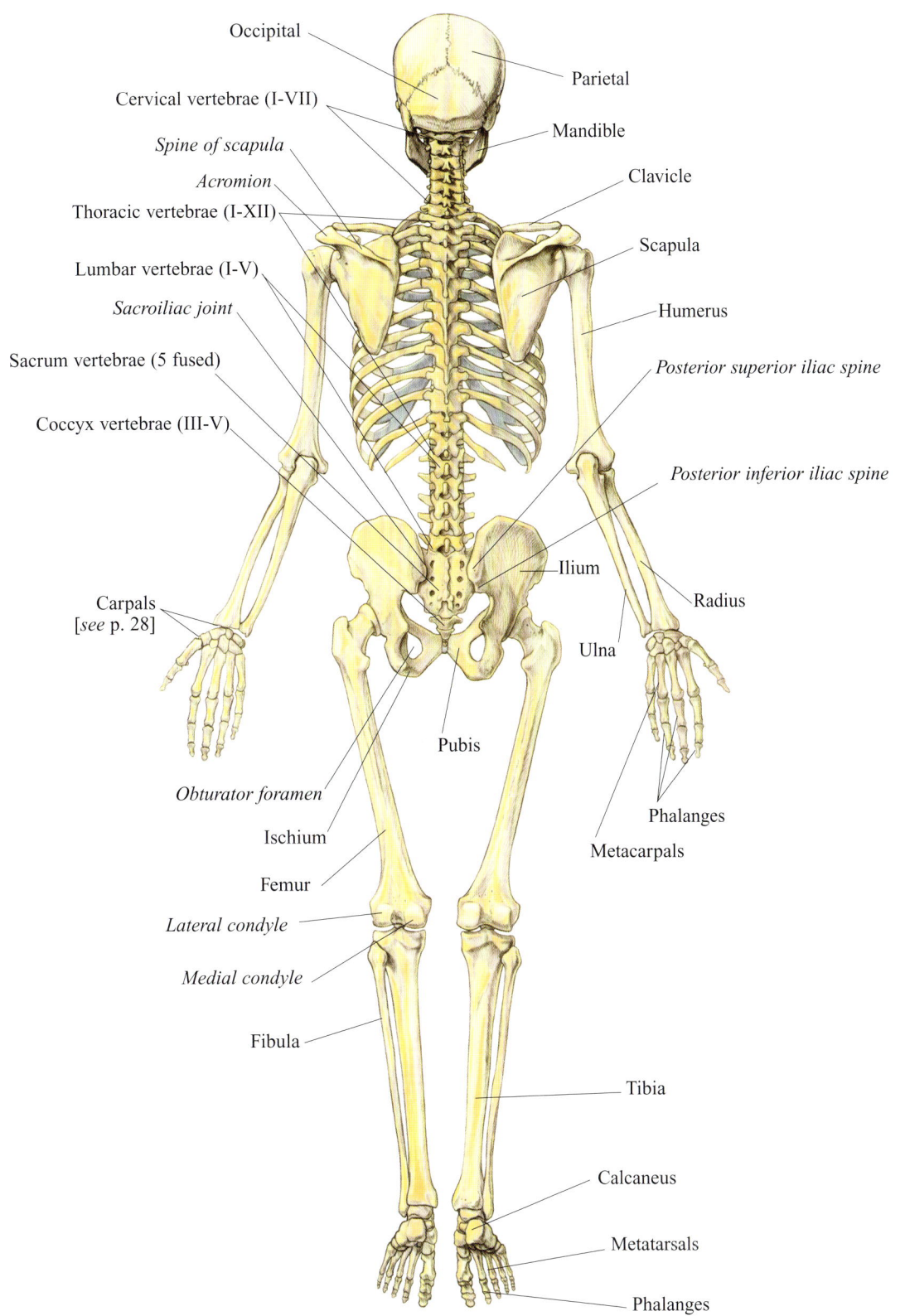

SKELETAL SYSTEM

VERTEBRAL COLUMN
LATERAL VIEW

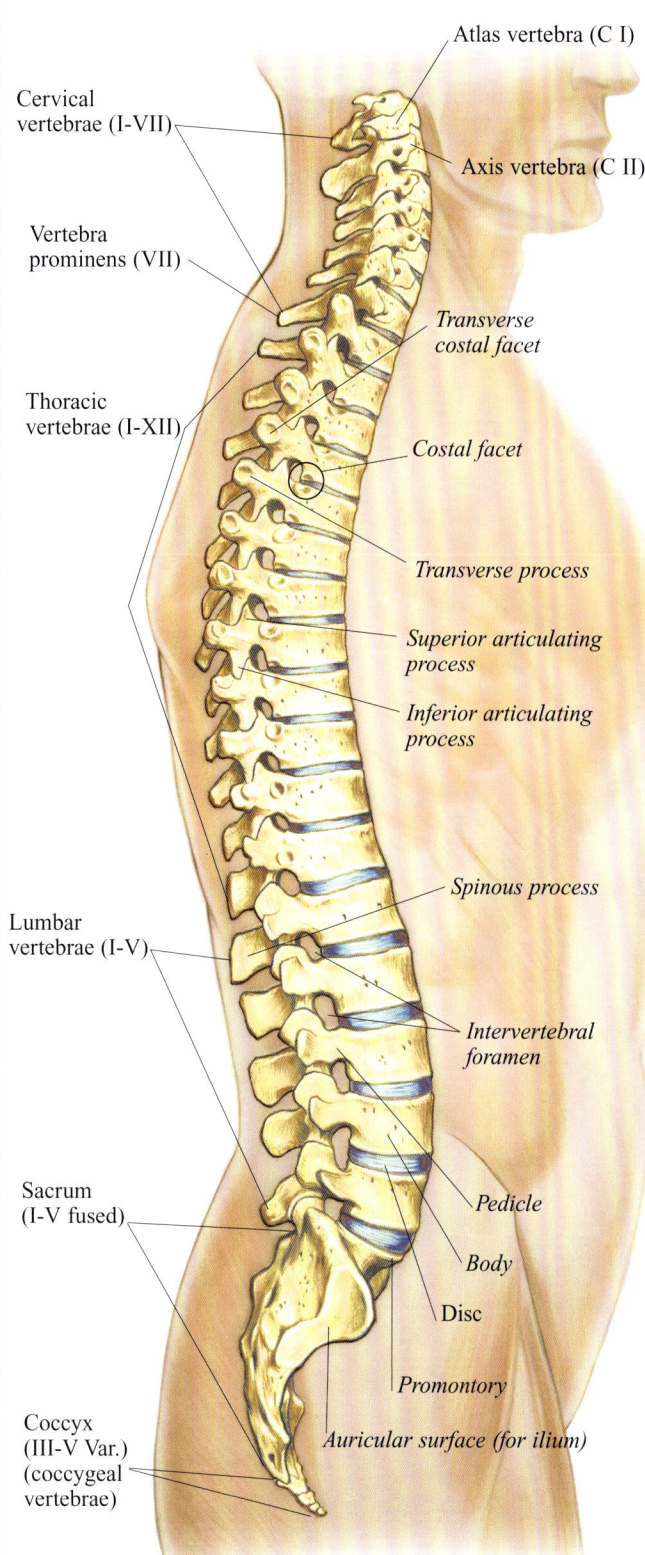

LUMBAR VERTEBRA
SUPERIOR VIEW

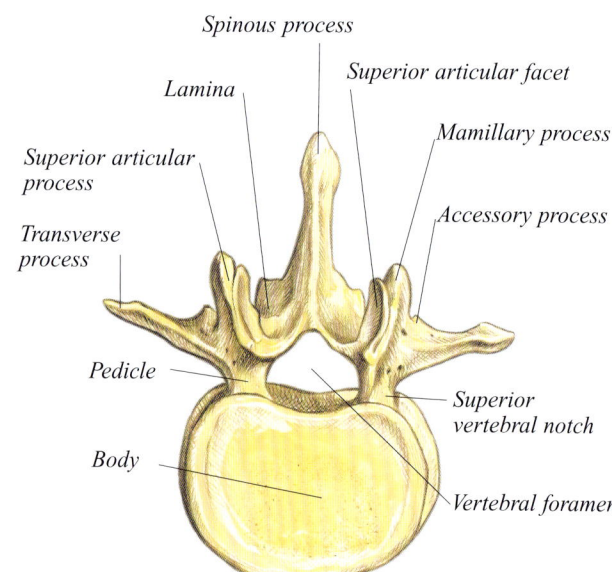

CERVICAL VERTEBRAE
POSTERIOR VIEW

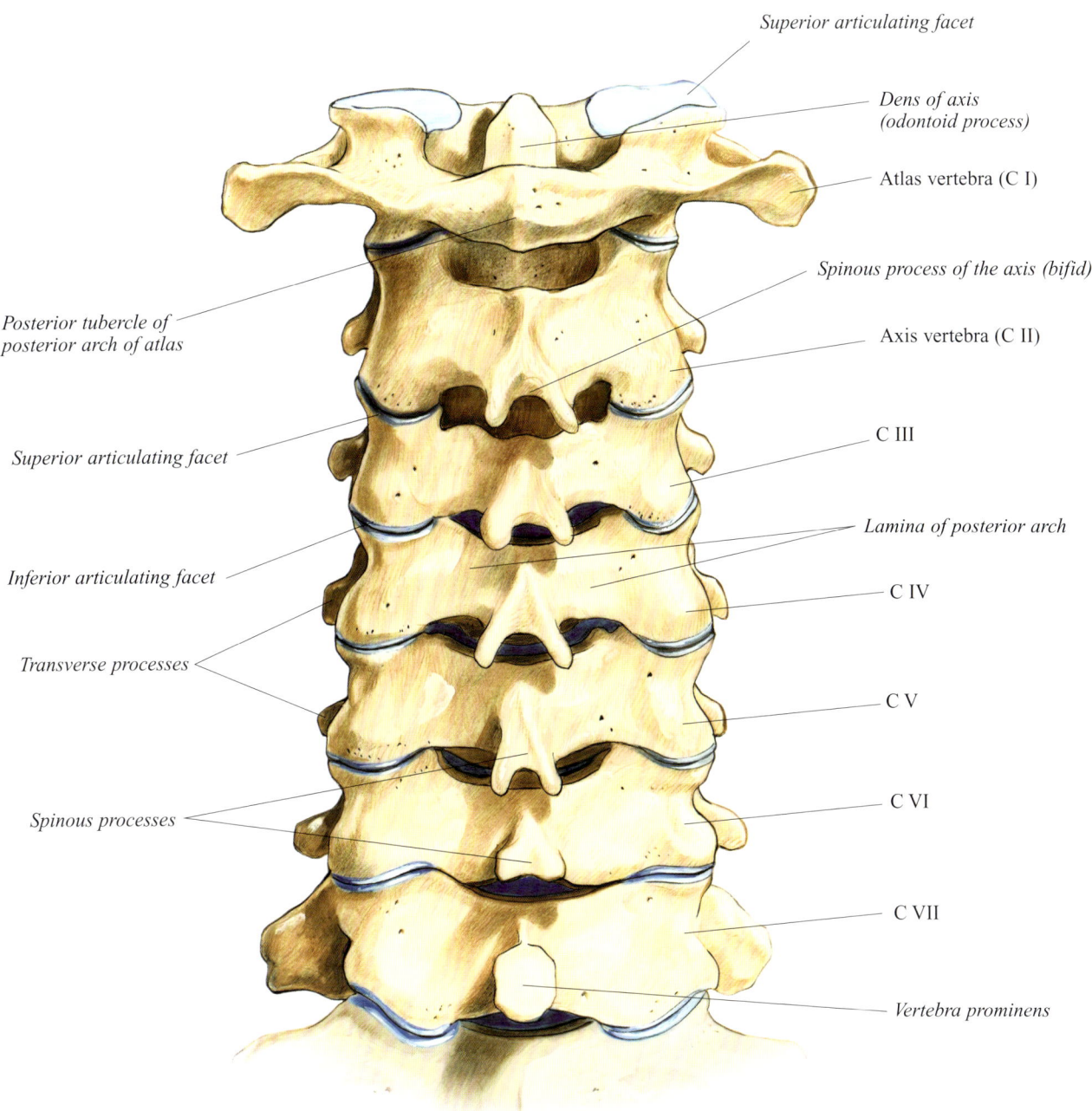

LUMBAR VERTEBRAE
POSTERIOR VIEW

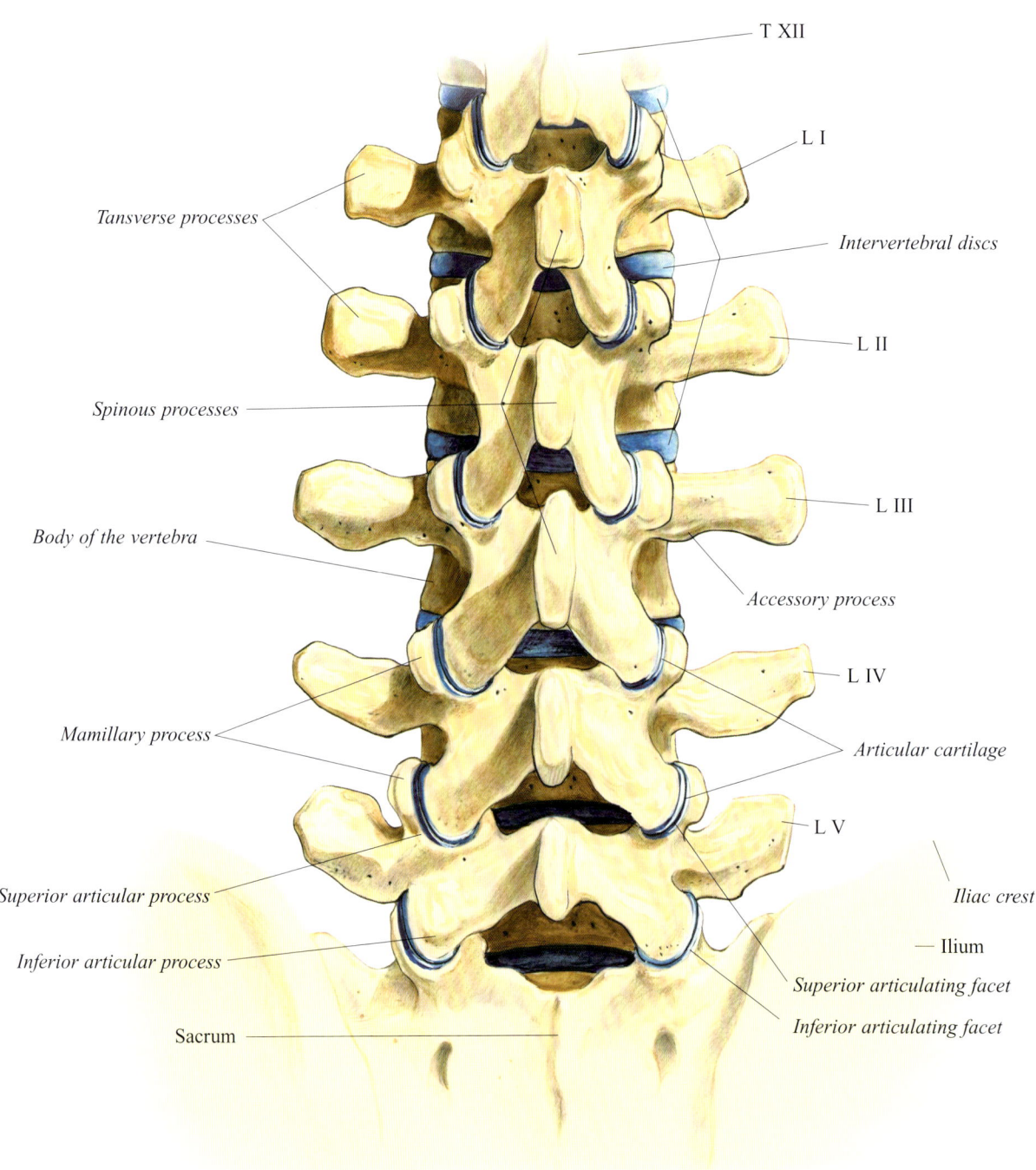

SKULL
LATERAL VIEW

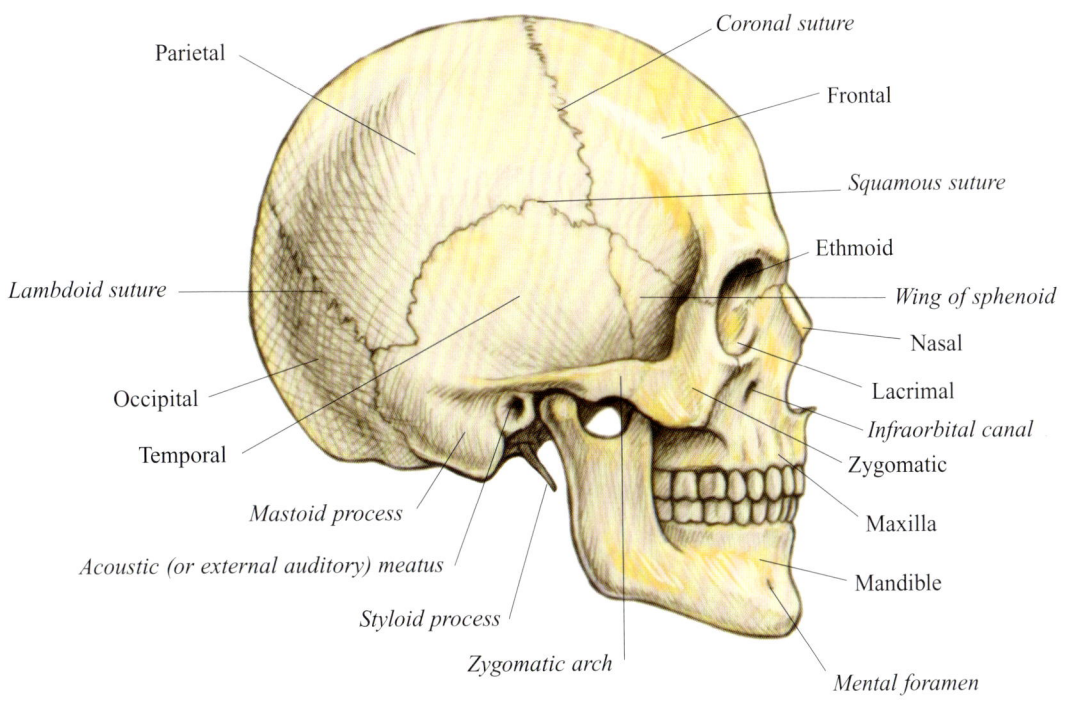

MEDIAN (SAGITTAL) SECTION

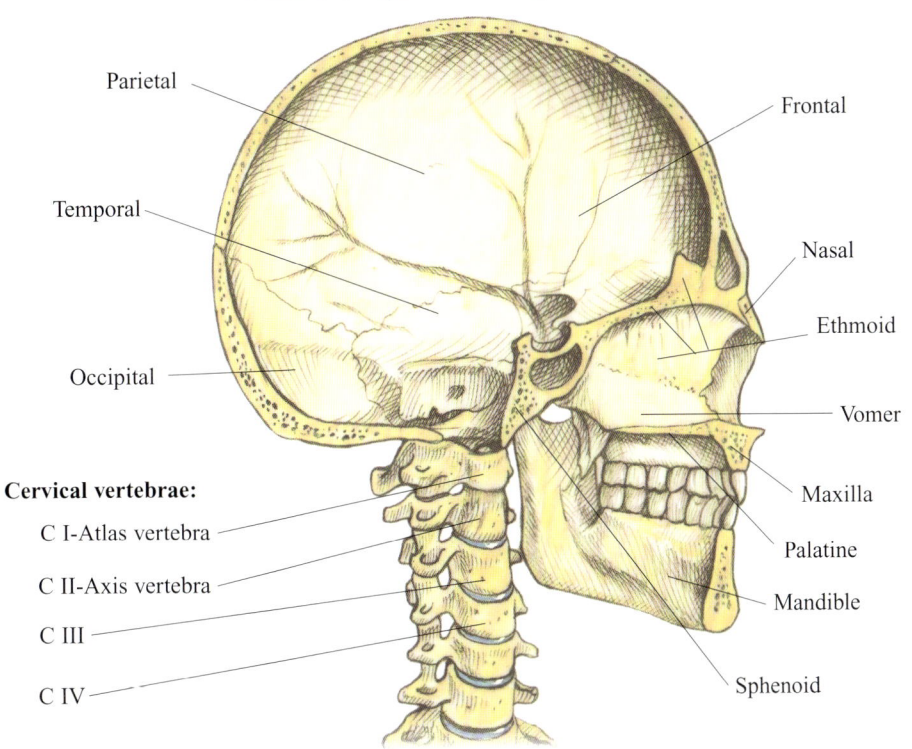

SKELETAL SYSTEM

SKULL
ANTERIOR VIEW

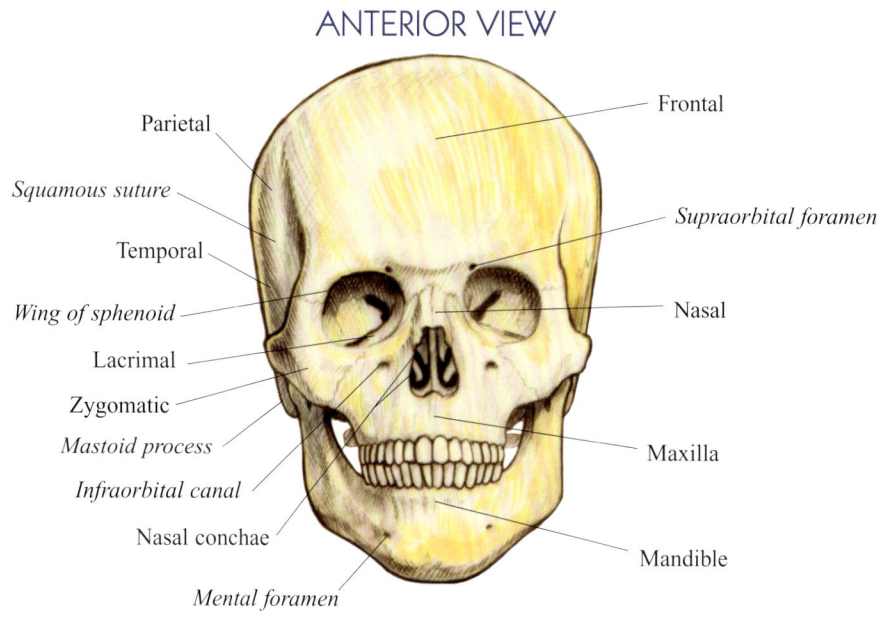

- Parietal
- *Squamous suture*
- Temporal
- *Wing of sphenoid*
- Lacrimal
- Zygomatic
- *Mastoid process*
- *Infraorbital canal*
- Nasal conchae
- *Mental foramen*
- Frontal
- *Supraorbital foramen*
- Nasal
- Maxilla
- Mandible

SKULL
INFERIOR VIEW

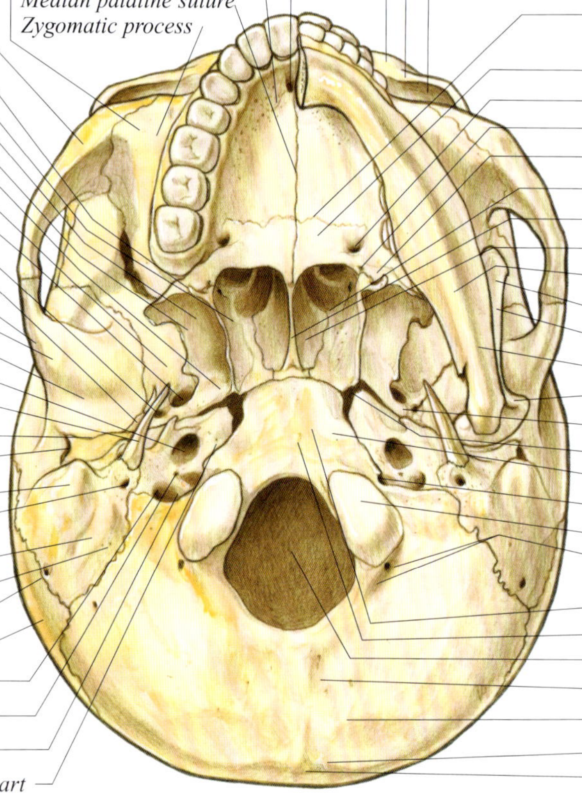

- Zygomatic bone
- **Sphenoid bone:**
  - Pterygoid process:
    - *Hamulus*
    - *Medial plate*
    - *Pterygoid fossa*
    - *Lateral plate*
    - *Scaphoid fossa*
  - *Greater wing*
  - *Foramen ovale*
  - *Foramen spinosum*
  - *Spine*
- **Temporal bone:**
  - *Zygomatic process*
  - *Articular tubercle*
  - *Mandibular fossa*
  - *Carotid canal*
  - *Styloid process*
  - *Petrotympanic fissure*
  - *External acoustic meatus*
  - *Mastoid process*
  - *Stylomastoid foramen*
  - *Mastoid groove for digastric m.*
  - *Groove for occipital a.*
  - *Mastoid foramen*
  - *Parietal bone*
  - *Mastoid canaliculus*
  - *Tympanic canaliculus*
  - *Jugular fossa*
  - *Petrous part*
- **Maxilla:**
  - *Incisive fossa*
  - *Palatine process*
  - *Median palatine suture*
  - *Zygomatic process*
- **Frontal bone:**
  - *Supraorbital ridge*
  - *Orbit*
  - *Transverse palatine suture*
- **Palatine bone:**
  - *Horizontal plate*
  - *Greater palatine foramen*
  - *Lesser palatine foramen*
  - *Posterior nasal spine*
  - *Choanae*
  - Vomer
  - *Ala*
- **Mandible****:
  - *Coronoid process*
  - *Ramus*
  - *Body*
  - *Head of condylar process*
  - *Groove for auditory tube*
  - *Foramen lacerum*
- **Occipital bone:**
  - *Hypoglossal canal*
  - *Occipital condyle*
  - *Condylar canal & fossa*
  - *Basilar part*
  - *Pharyngeal tubercle*
  - *Foramen magnum*
  - *External occipital crest*
  - *Inferior nuchal line*
  - *Superior nuchal line*
  - *External occipital protuberance*

ANTERIOR SKULL & CUT CERVICAL VERTEBRAE
POSTERIOR VIEW CORONAL SECTION

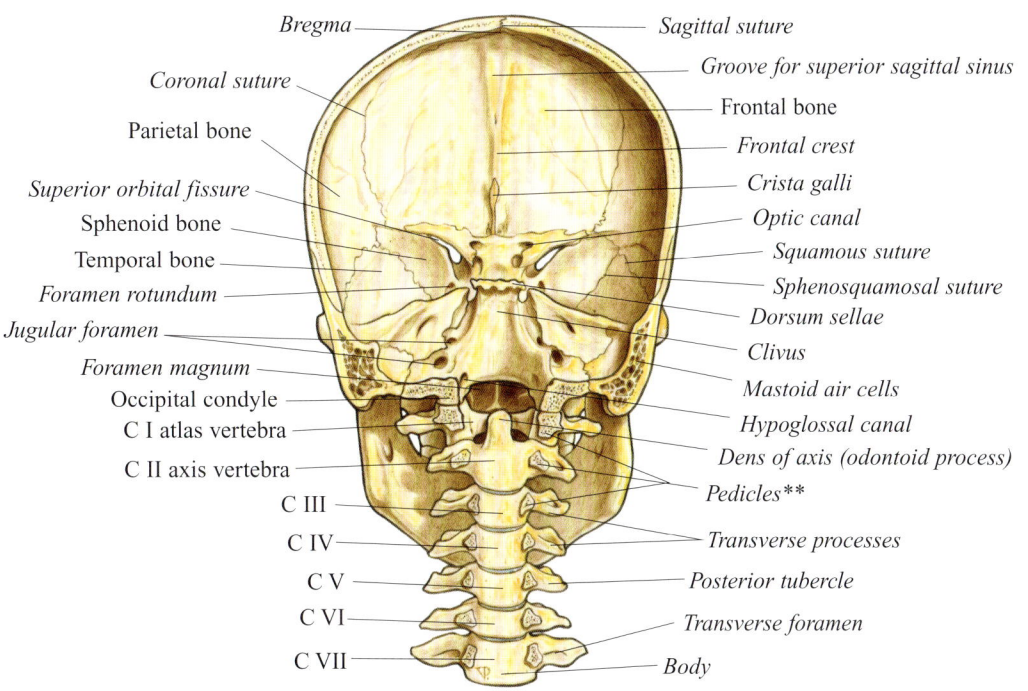

SKULL
POSTERIOR VIEW CORONAL SECTION

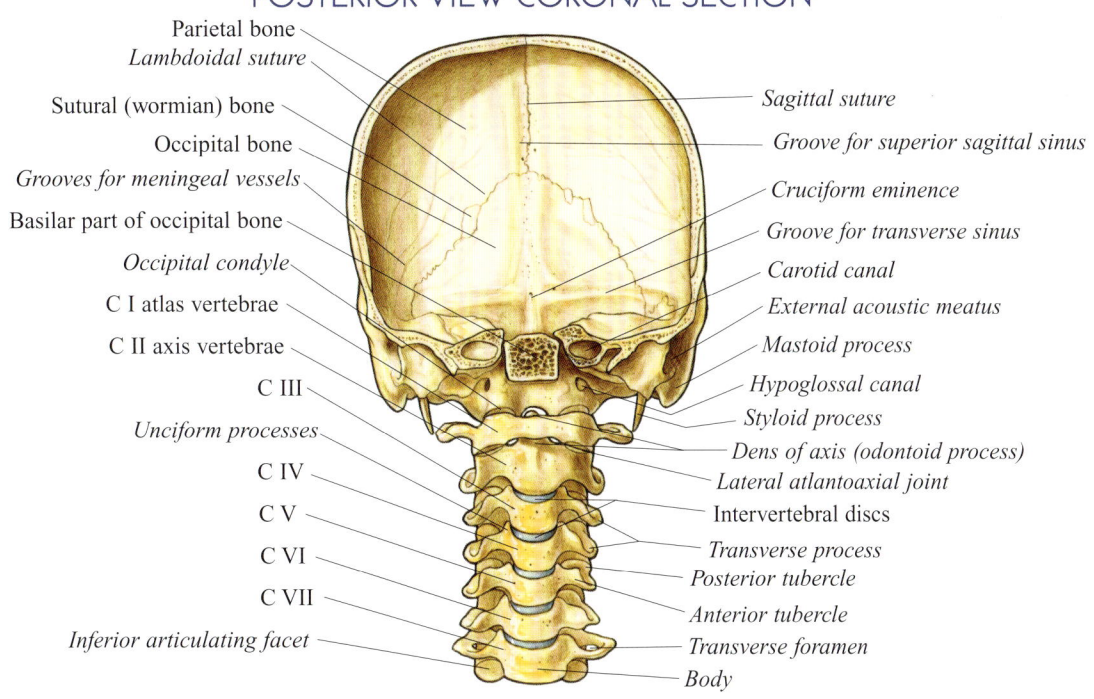

CERVICAL VERTEBRAE
ANTERIOR VIEW

SKELETAL SYSTEM

CLAVICLE
SUPERIOR VIEW

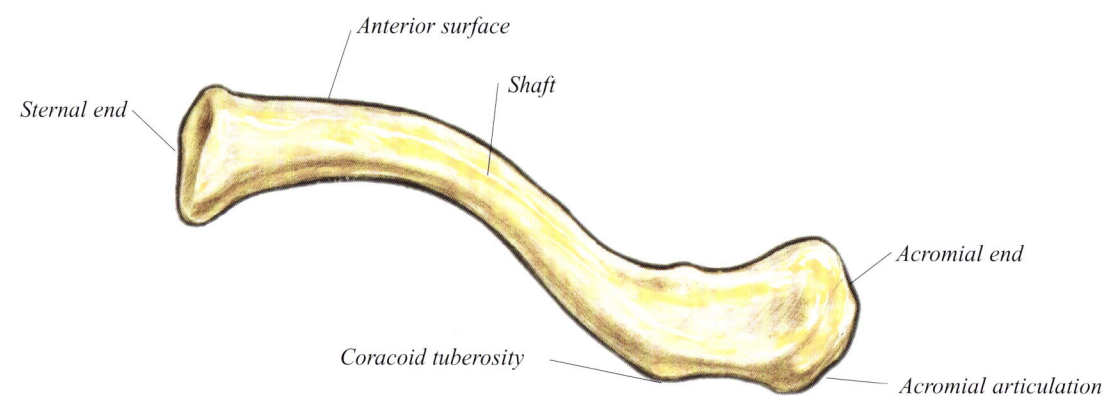

CLAVICLE
INFERIOR VIEW

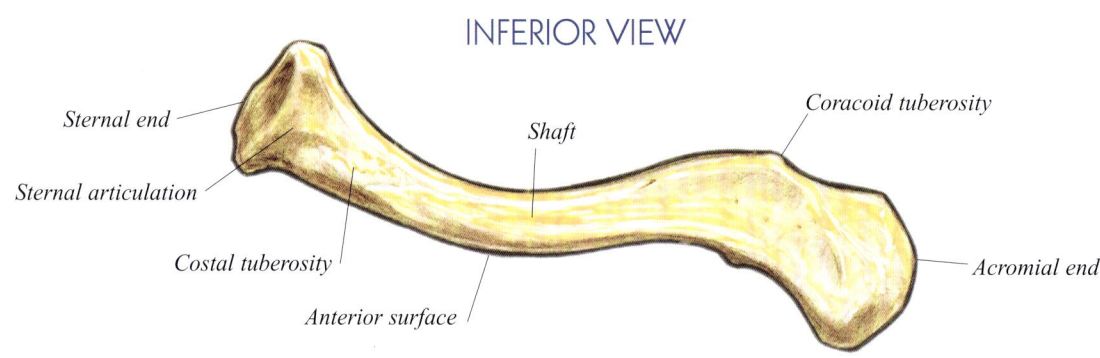

HYOID BONE

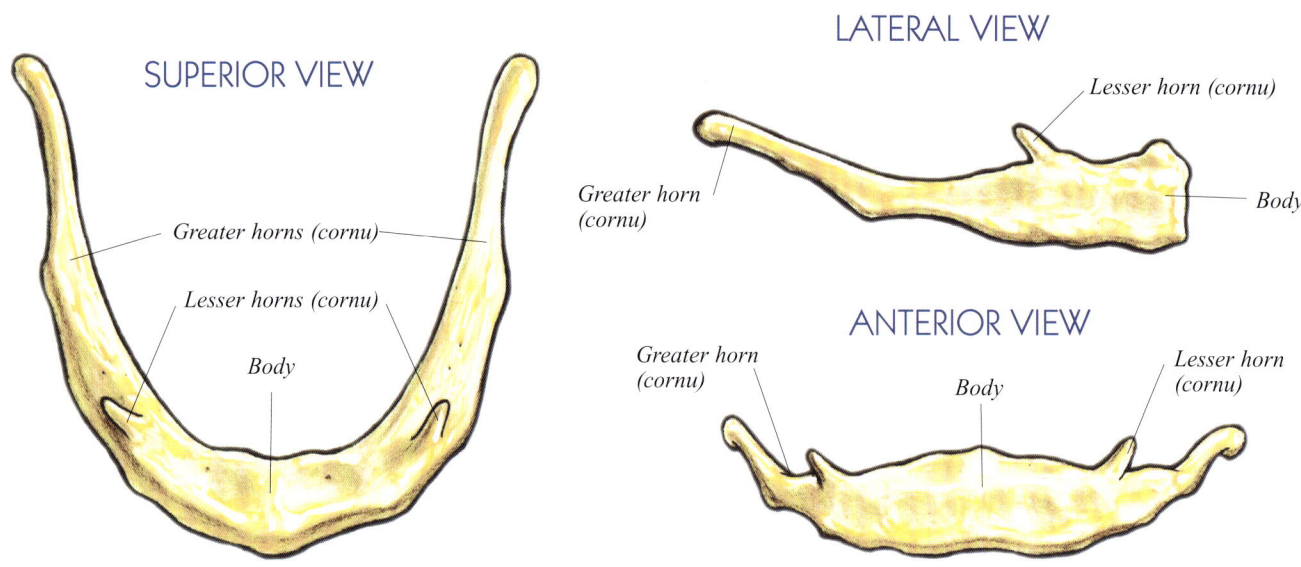

SCAPULA
ANTERIOR VIEW

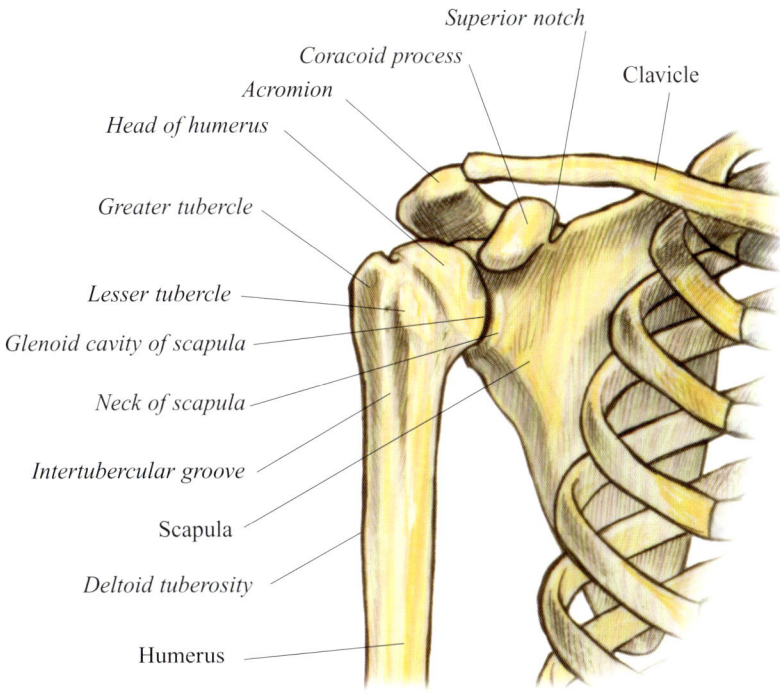

SCAPULA
POSTERIOR VIEW

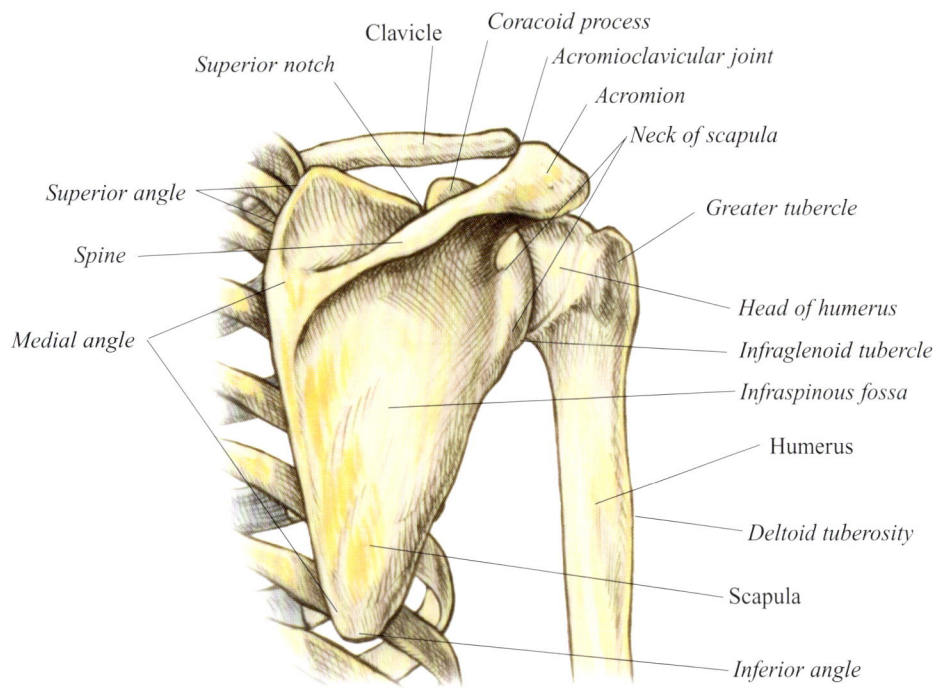

SKELETAL SYSTEM

THORACIC BONES

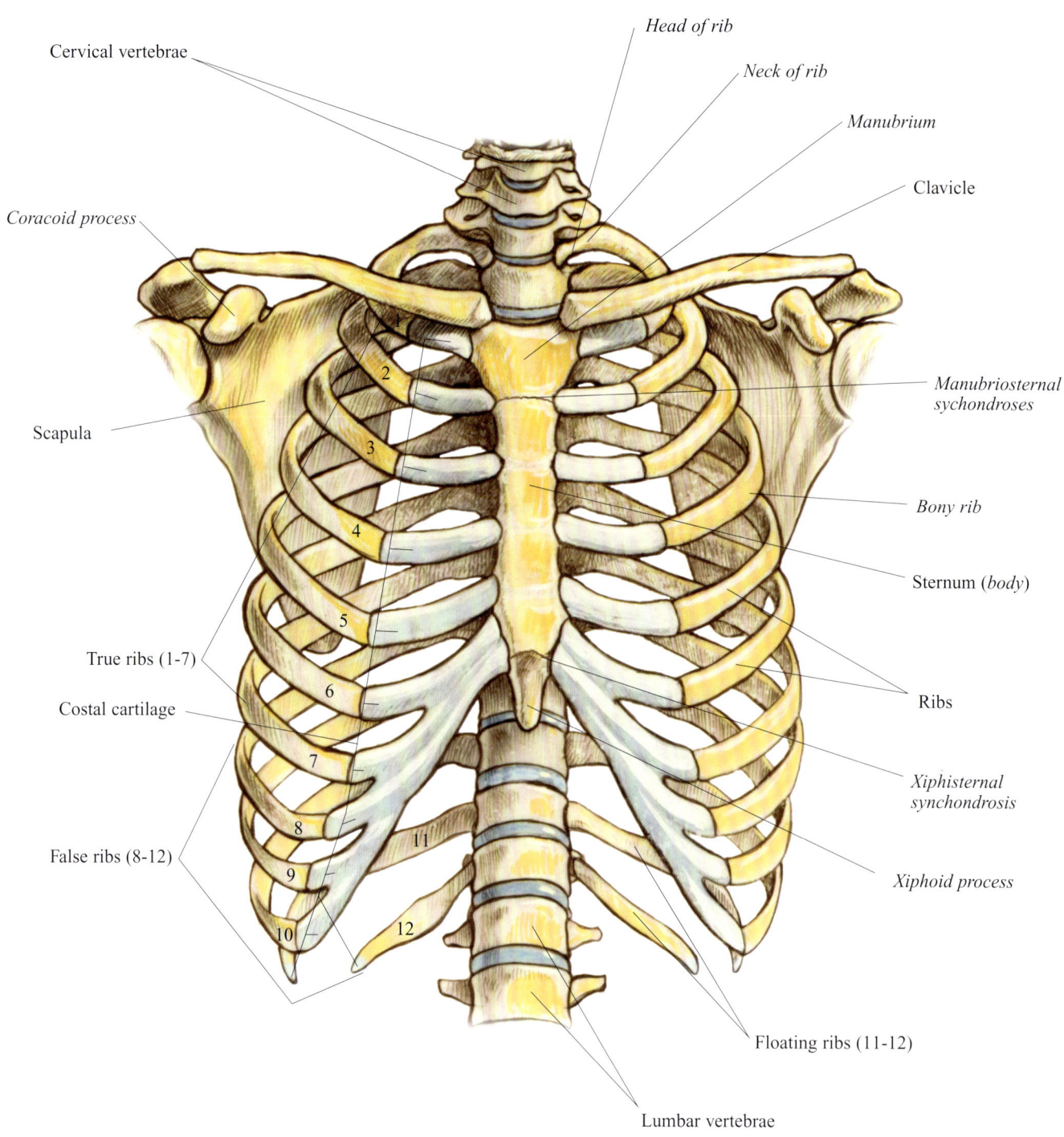

ANTERIOR VIEW

SKELETAL SYSTEM

LEFT HIP BONE
ANTERIOR VIEW

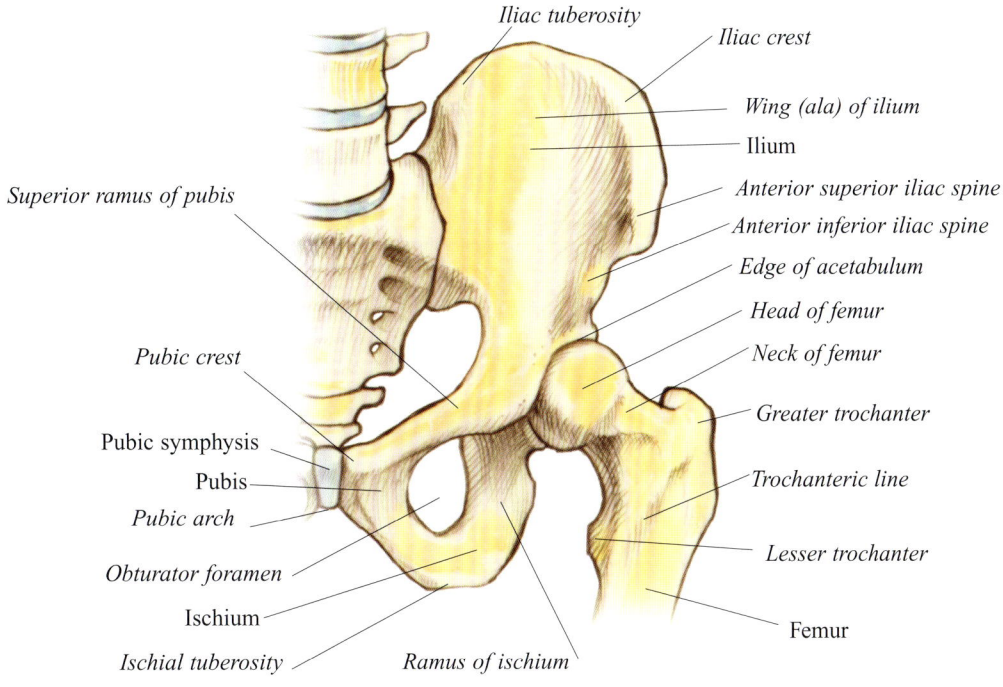

LEFT HIP BONE
POSTERIOR VIEW

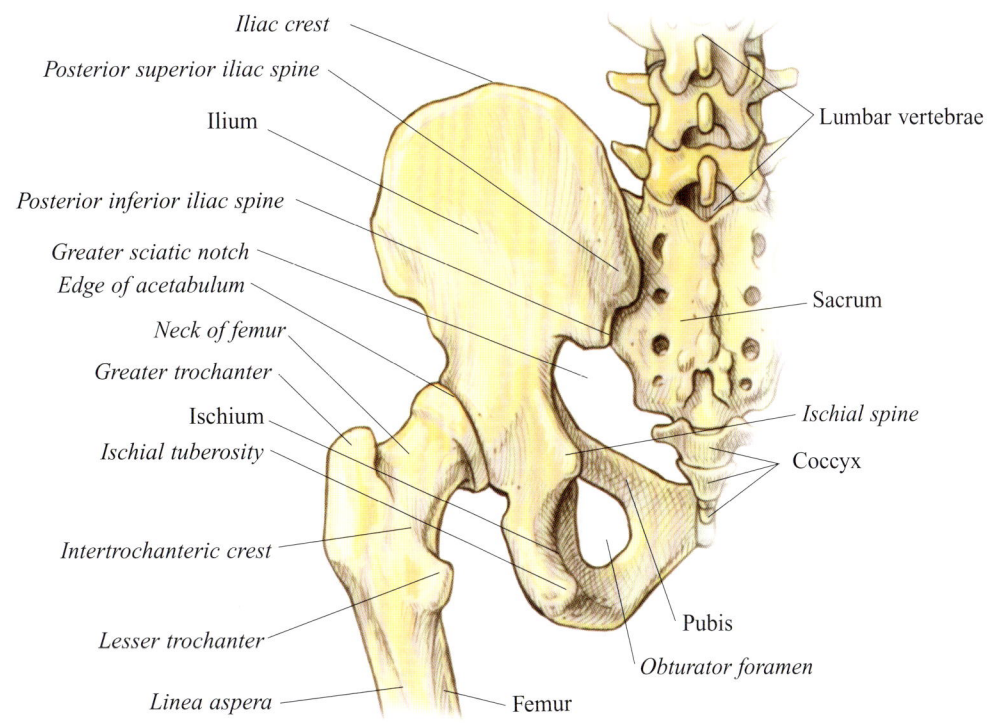

27

SKELETAL SYSTEM

ELBOWS

ANTERIOR VIEW   POSTERIOR VIEW

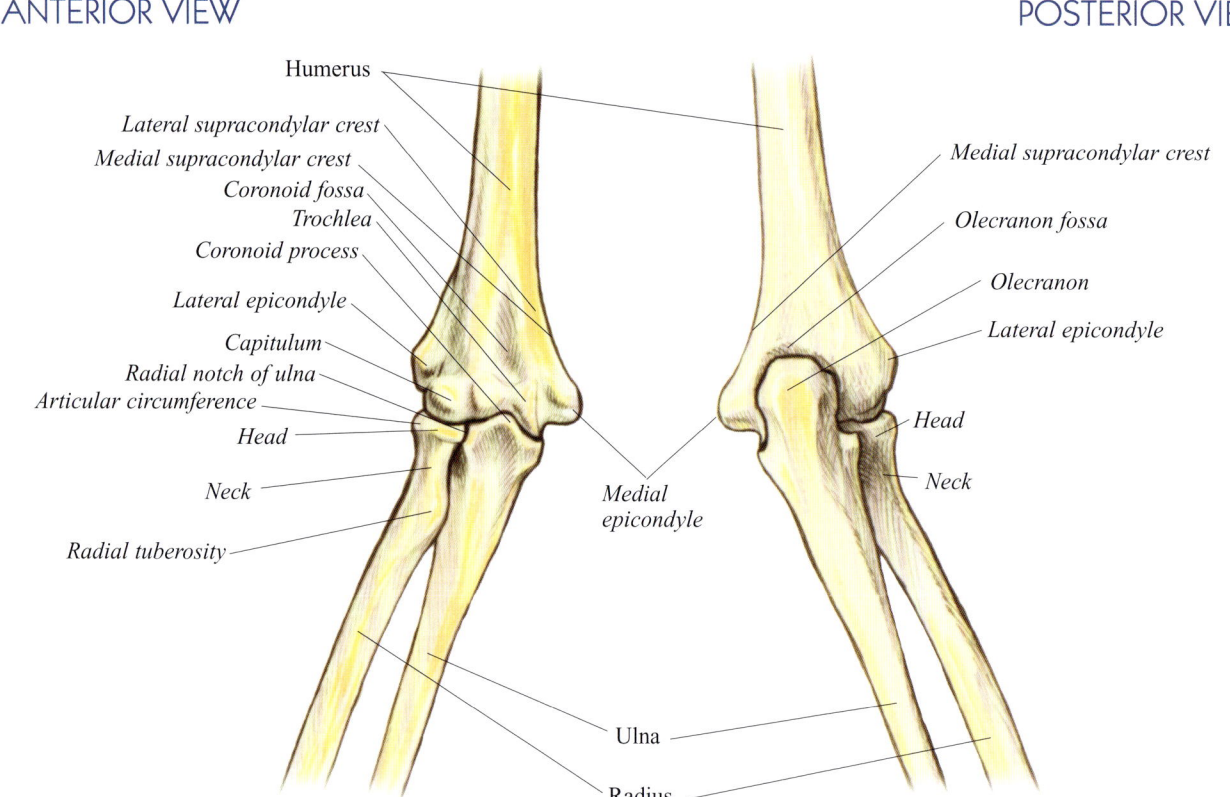

HANDS

PALMAR VIEW   DORSAL VIEW

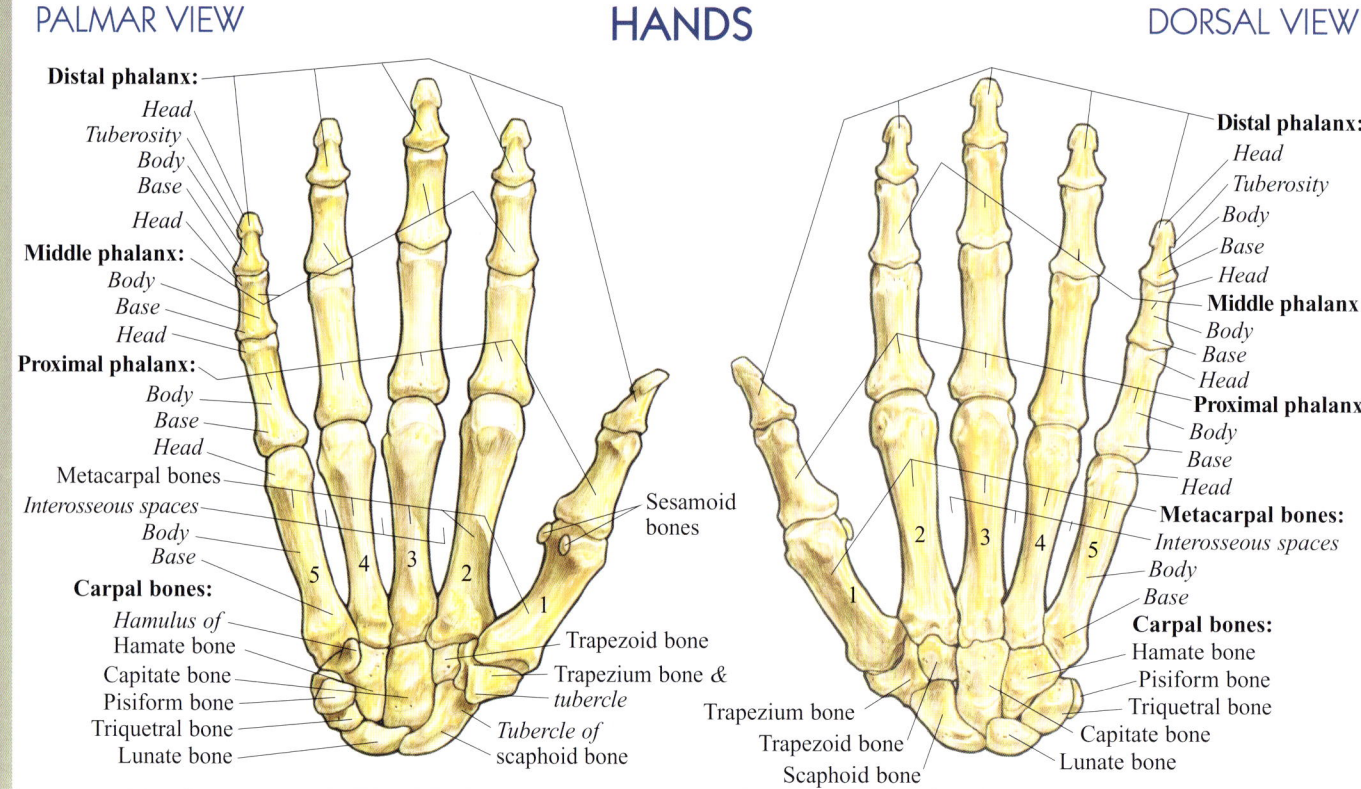

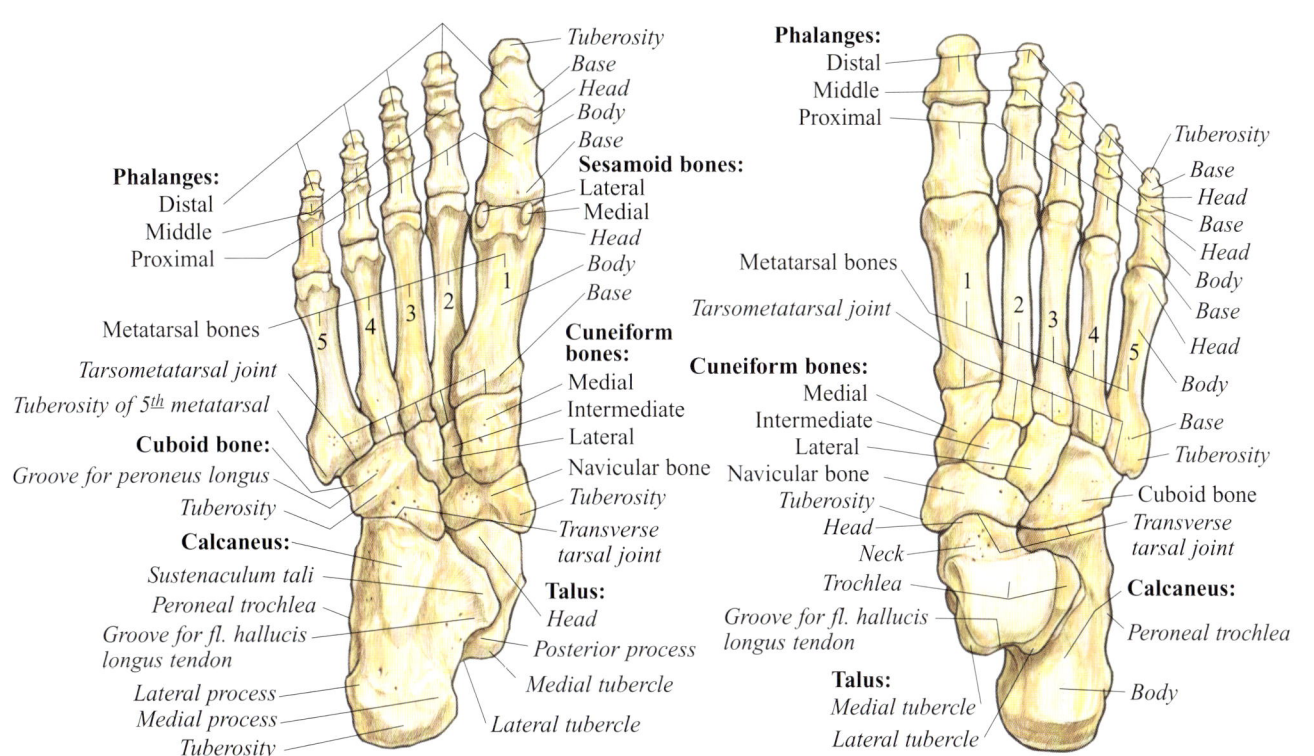

SKELETAL SYSTEM

BONE STRUCTURE

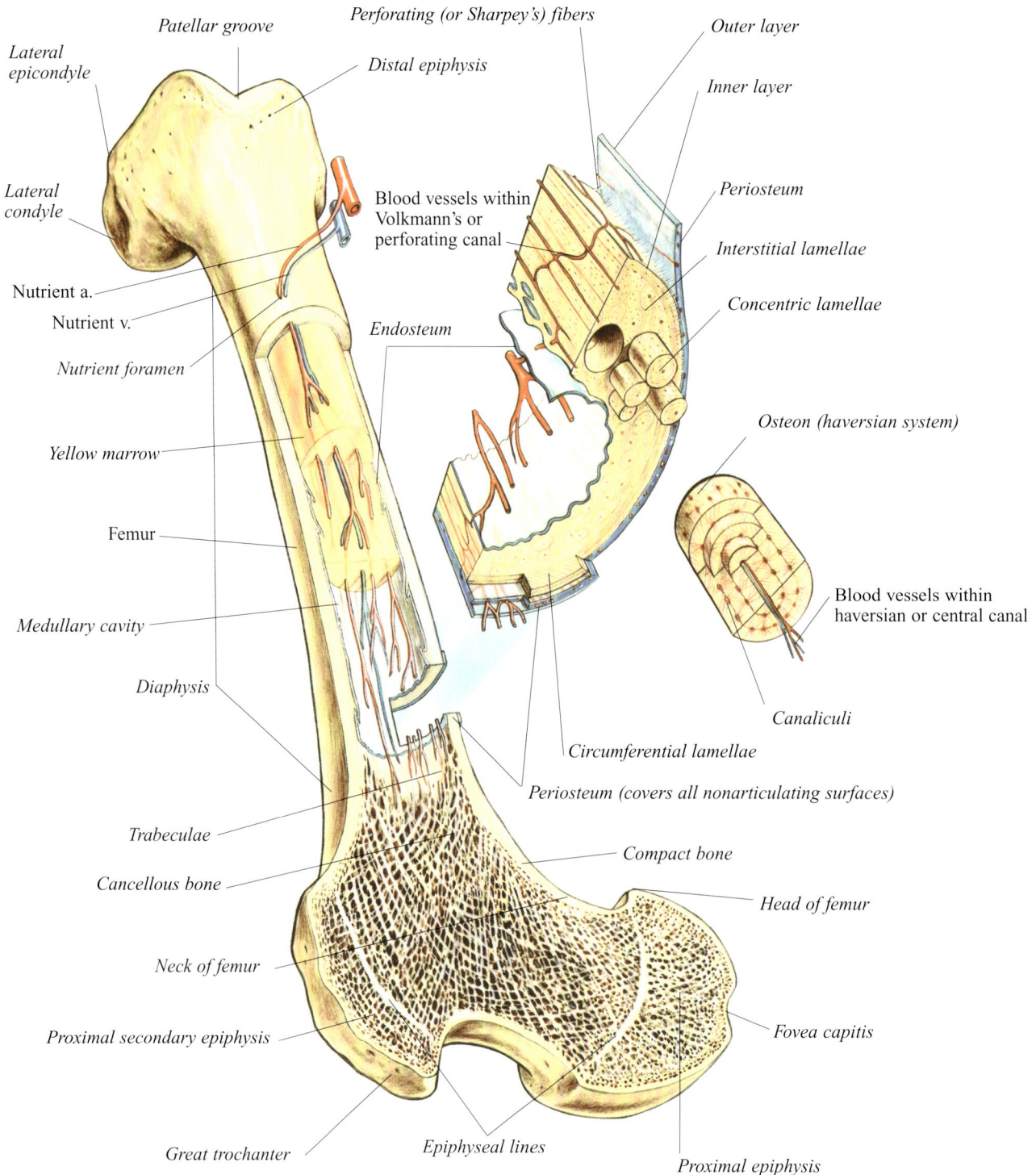

SKELETAL SYSTEM

NOTES

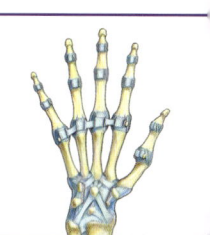

3
JOINTS & LIGAMENTS

JOINTS & LIGAMENTS

SPINE

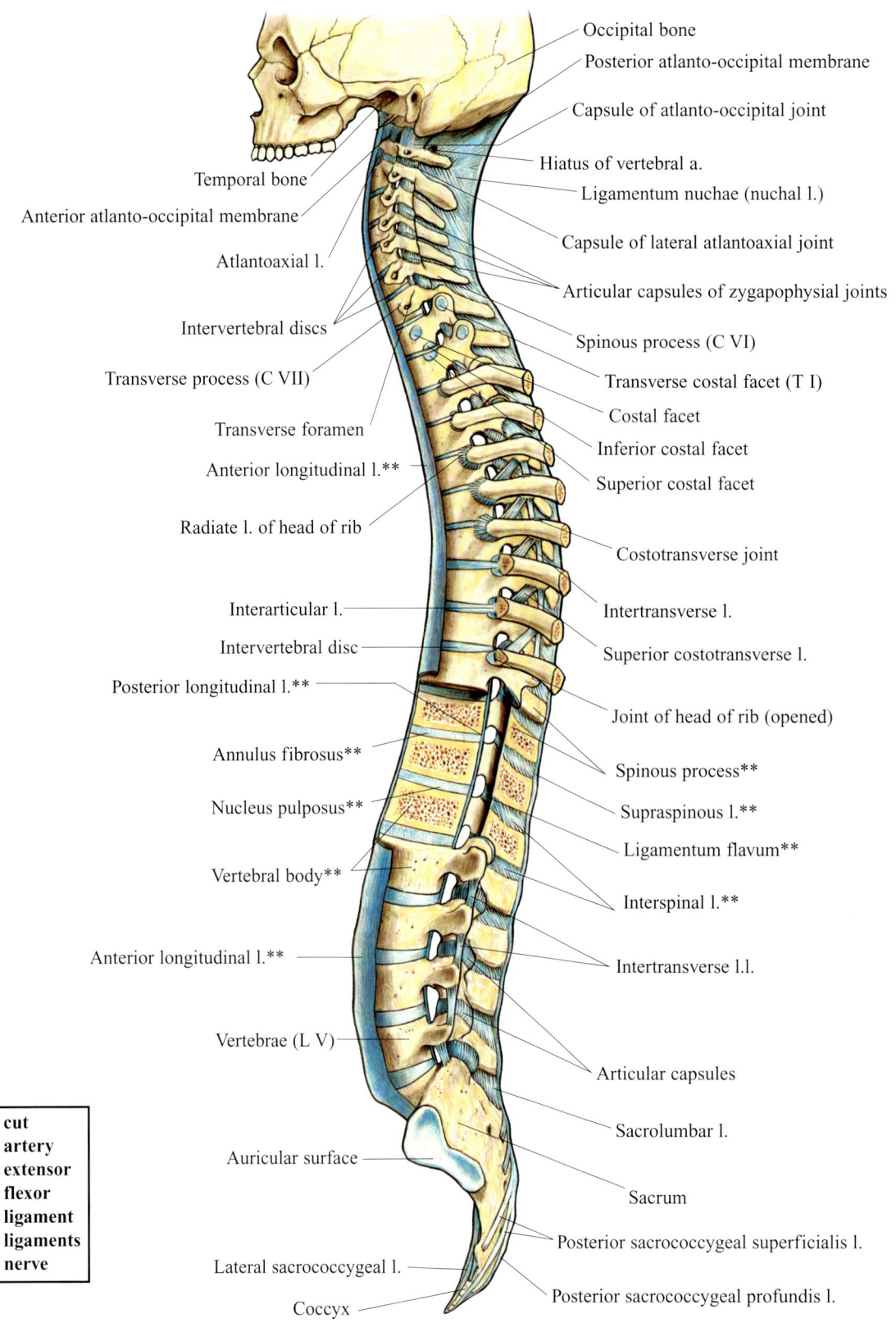

JOINTS & LIGAMENTS

TEMPOROMANDIBULAR & HYOID

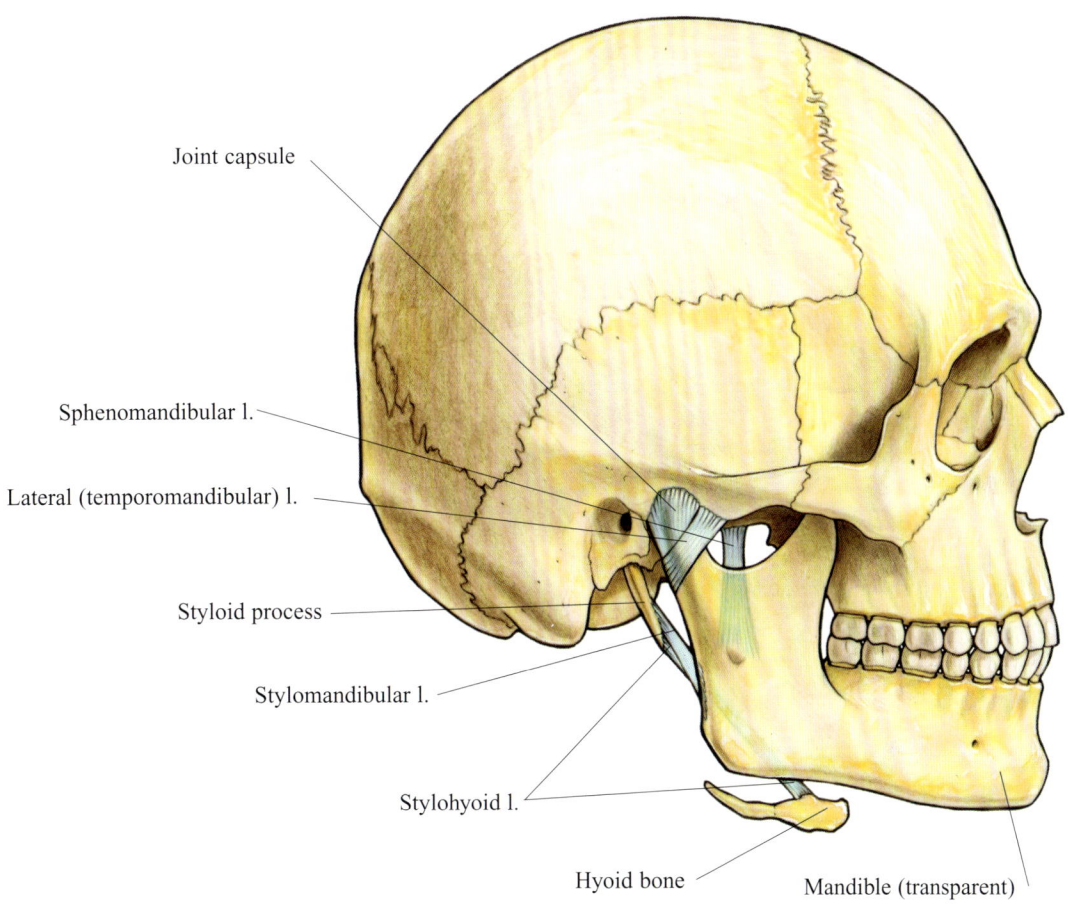

TEMPOROMANDIBULAR JOINT

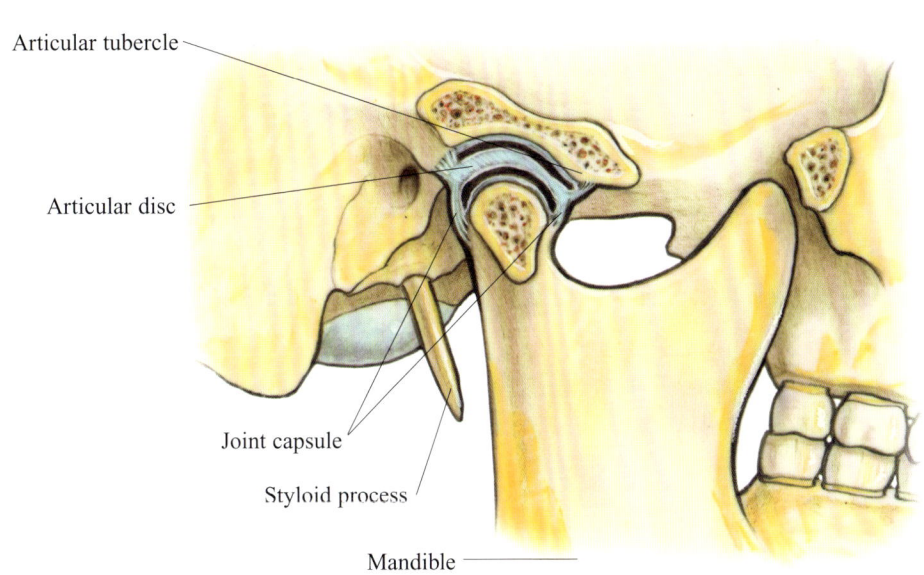

JOINTS & LIGAMENTS

CRANIOCERVICAL

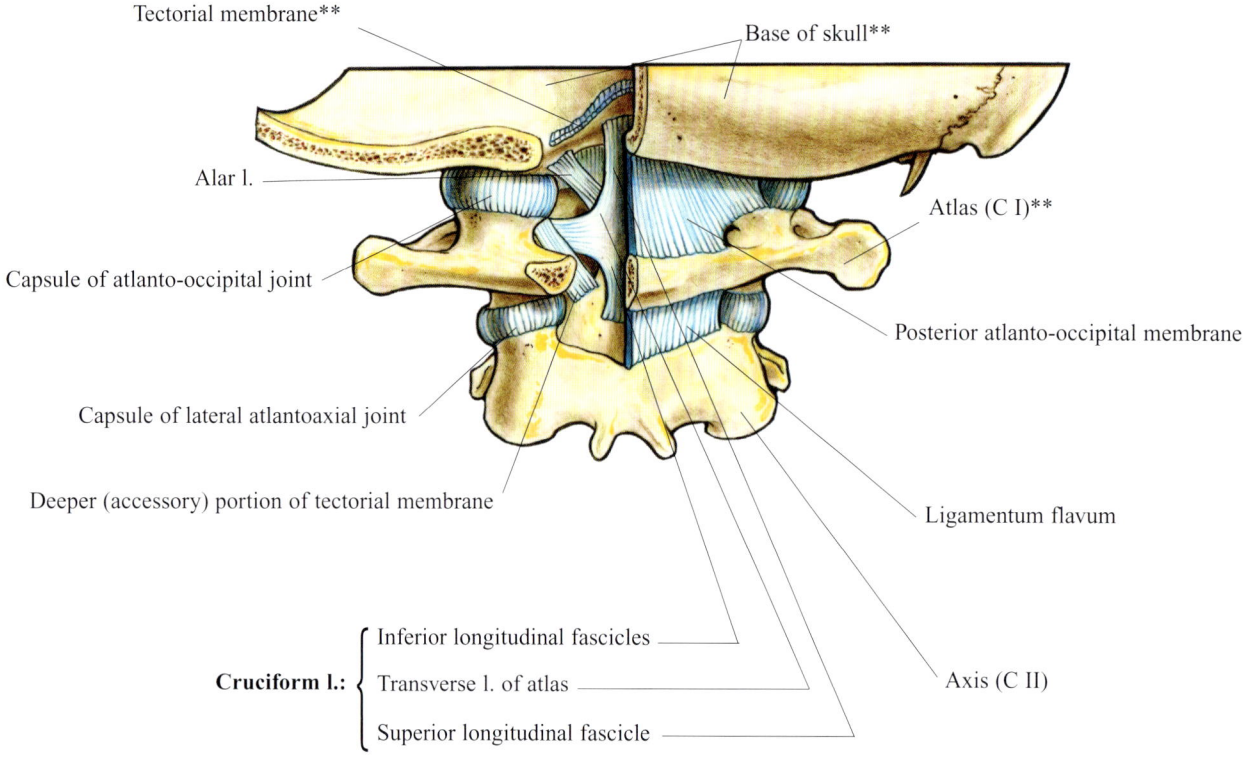

STERNOCLAVICULAR & SHOULDER

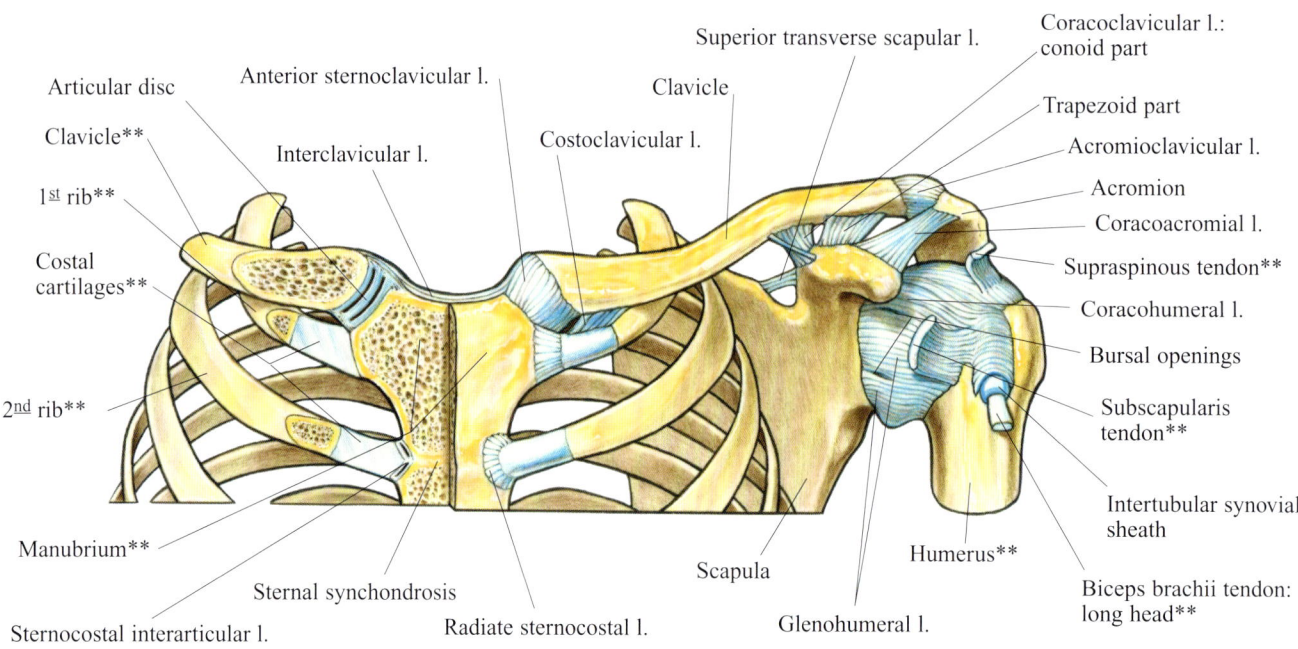

JOINTS & LIGAMENTS

ELBOW

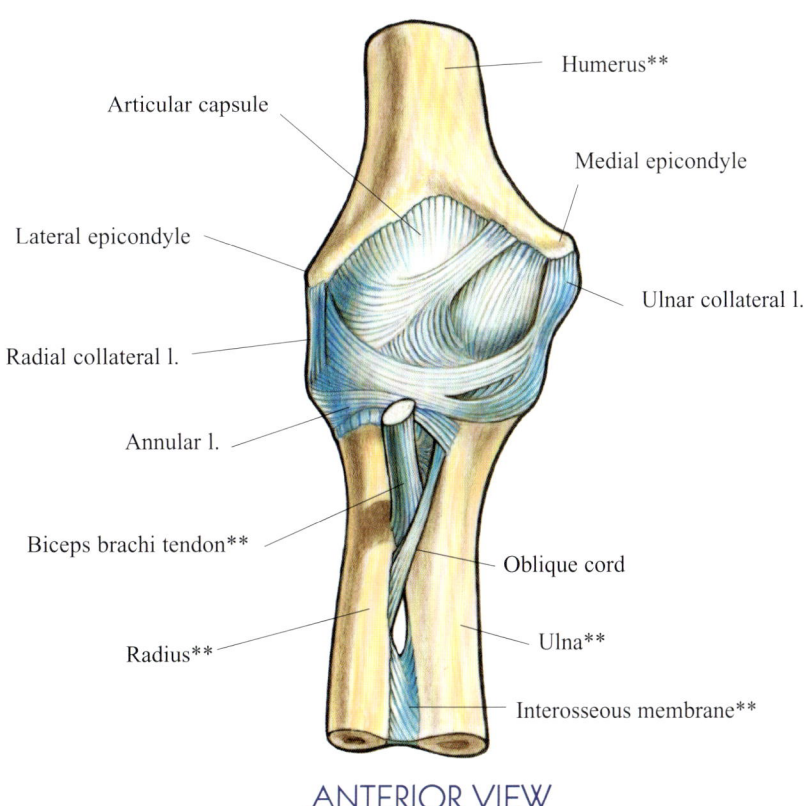

ANTERIOR VIEW

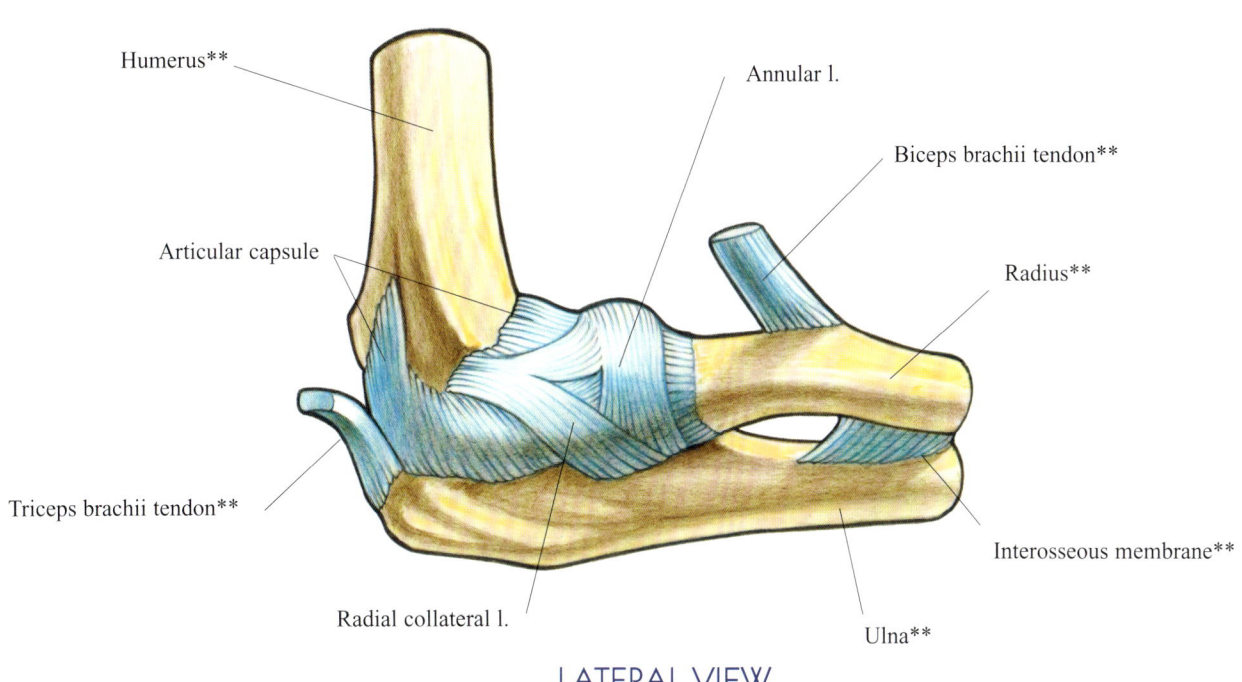

LATERAL VIEW

JOINTS & LIGAMENTS

WRIST & HAND

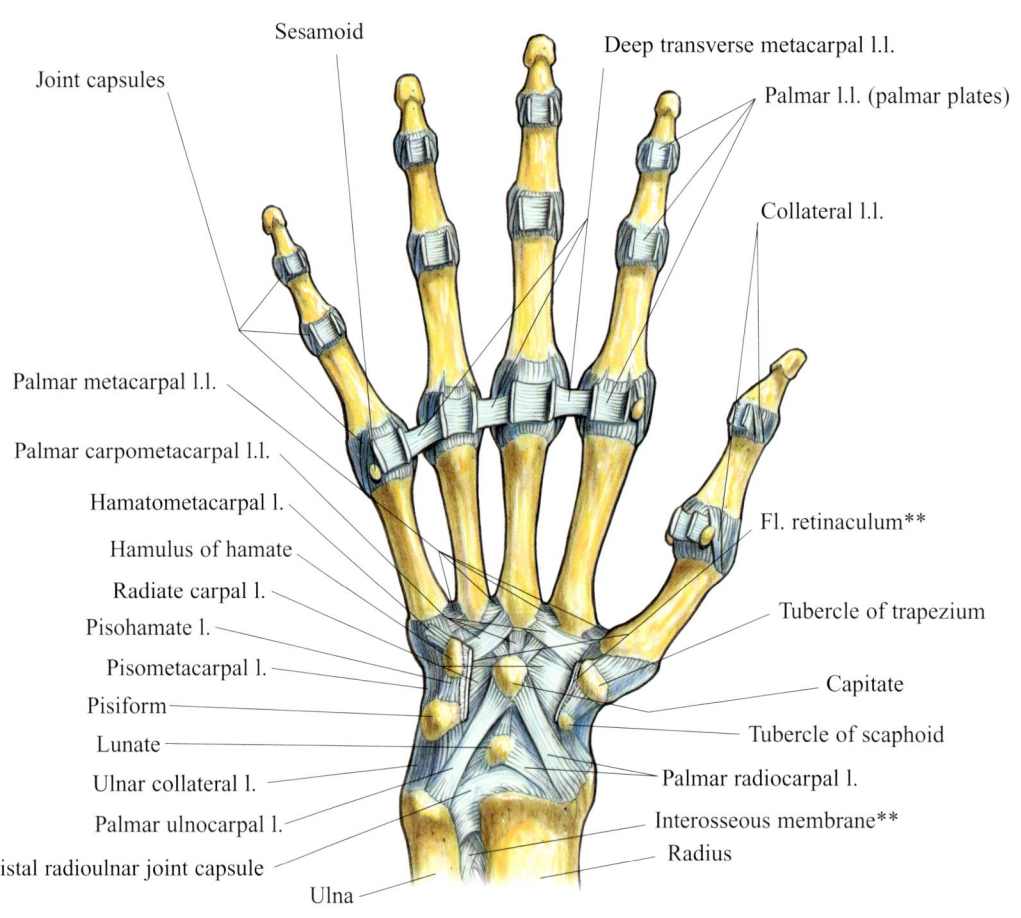

PALMAR VIEW

WRIST

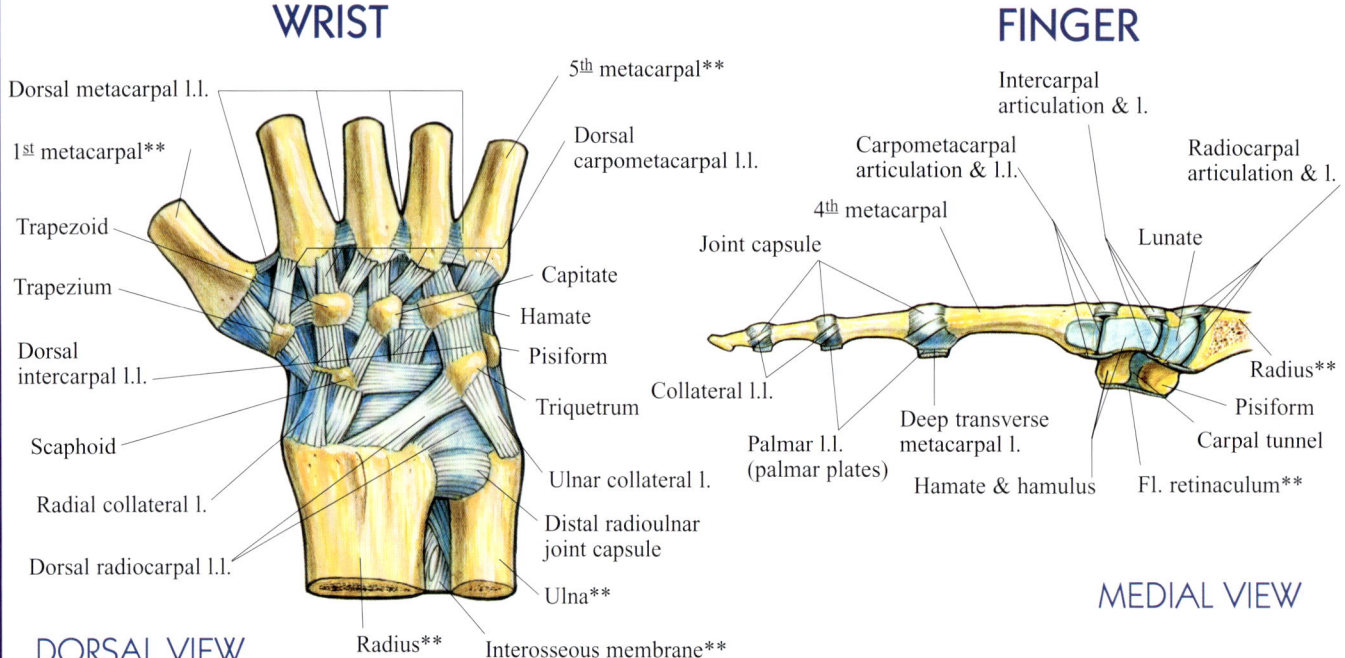

DORSAL VIEW

FINGER

MEDIAL VIEW

JOINTS & LIGAMENTS

LUMBAR SPINE

CONNECTIVE COMPONENTS OF THE PELVIS

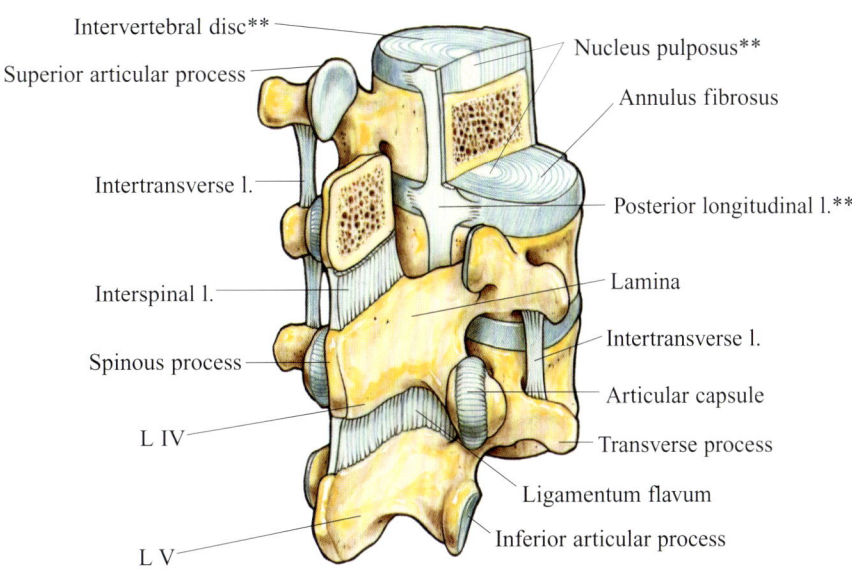

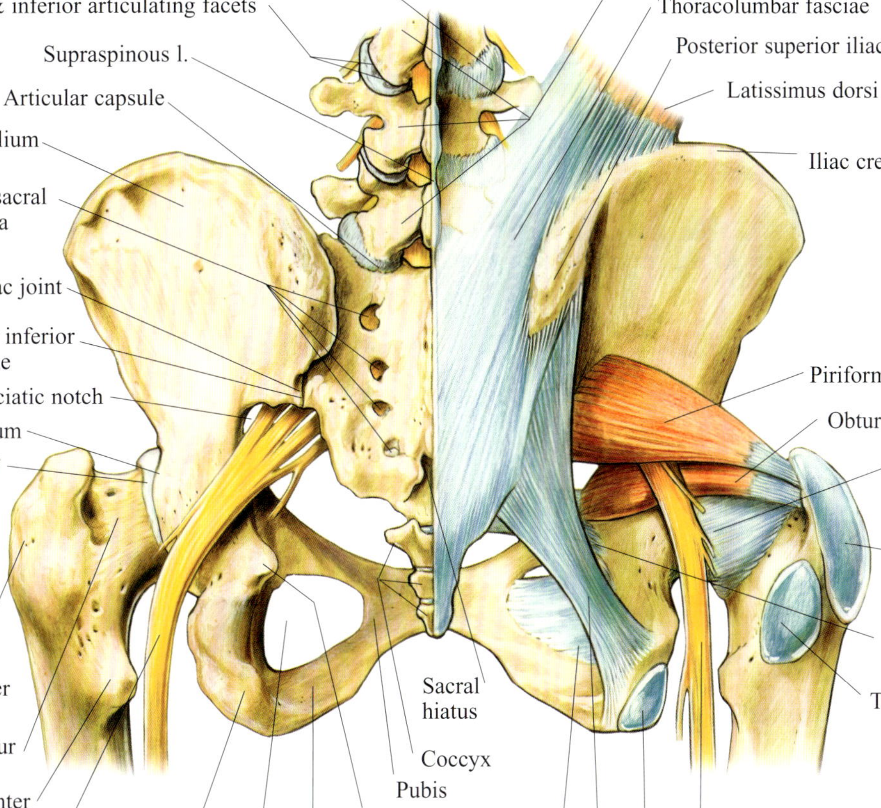

JOINTS & LIGAMENTS

HIP LIGAMENTS

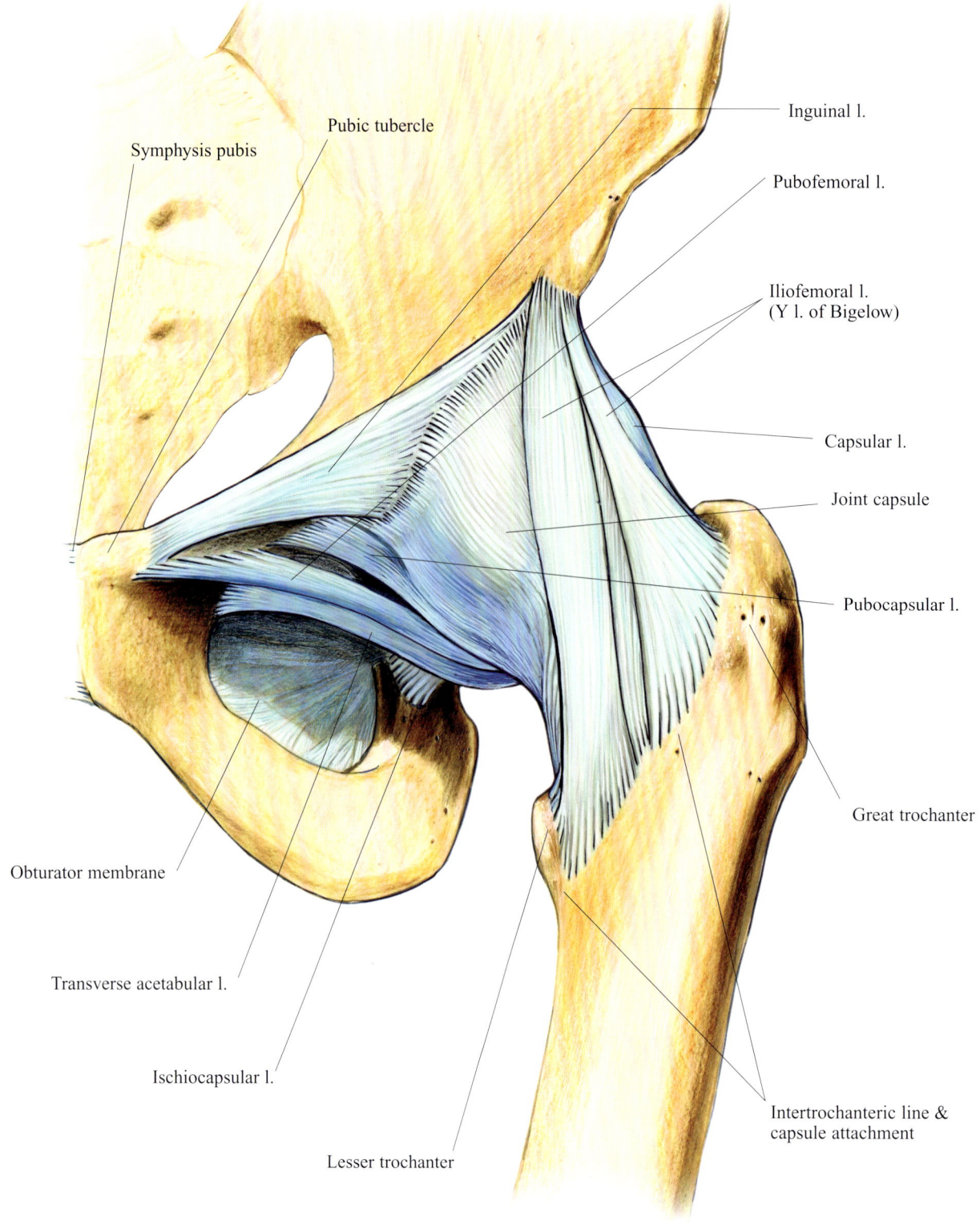

JOINTS & LIGAMENTS

HIP LIGAMENTS (OPENED)

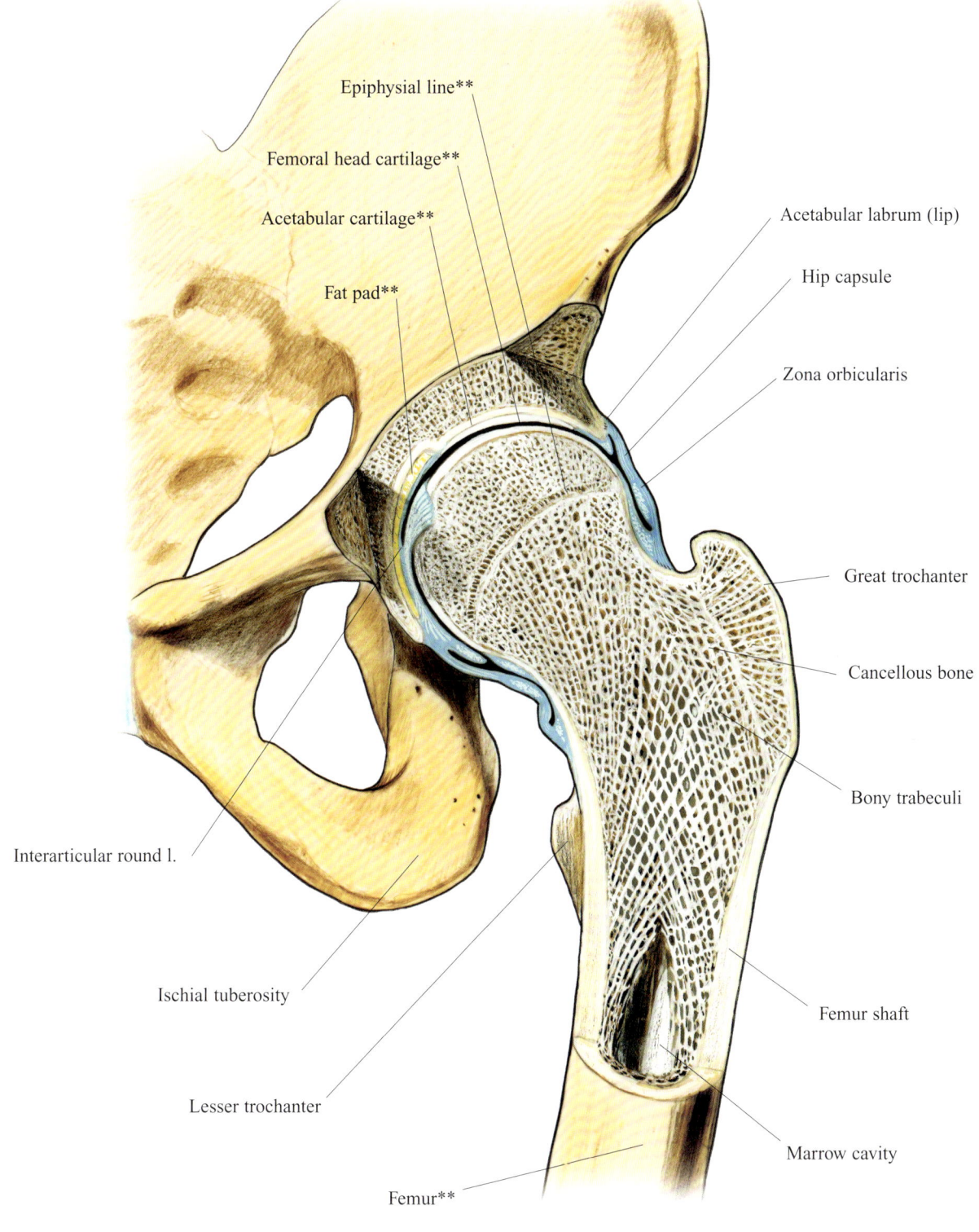

JOINTS & LIGAMENTS

PELVIS
SUPERIOR VIEW

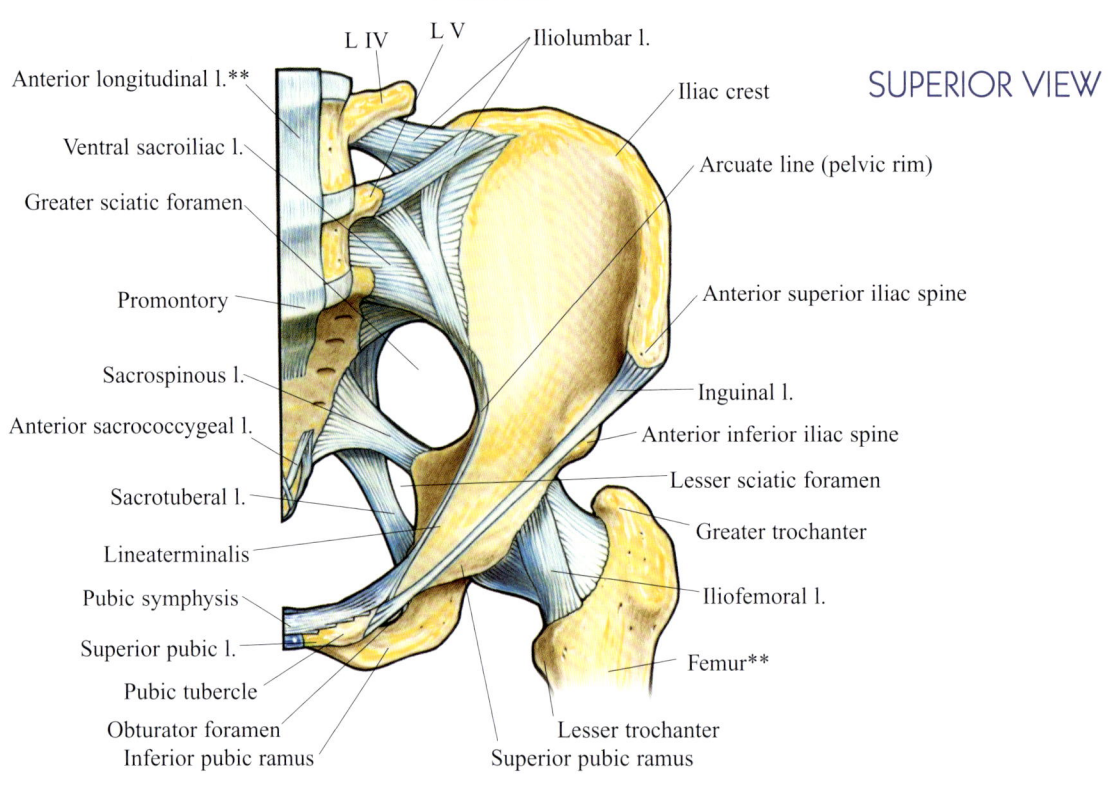

PELVIS
POSTERIOR VIEW

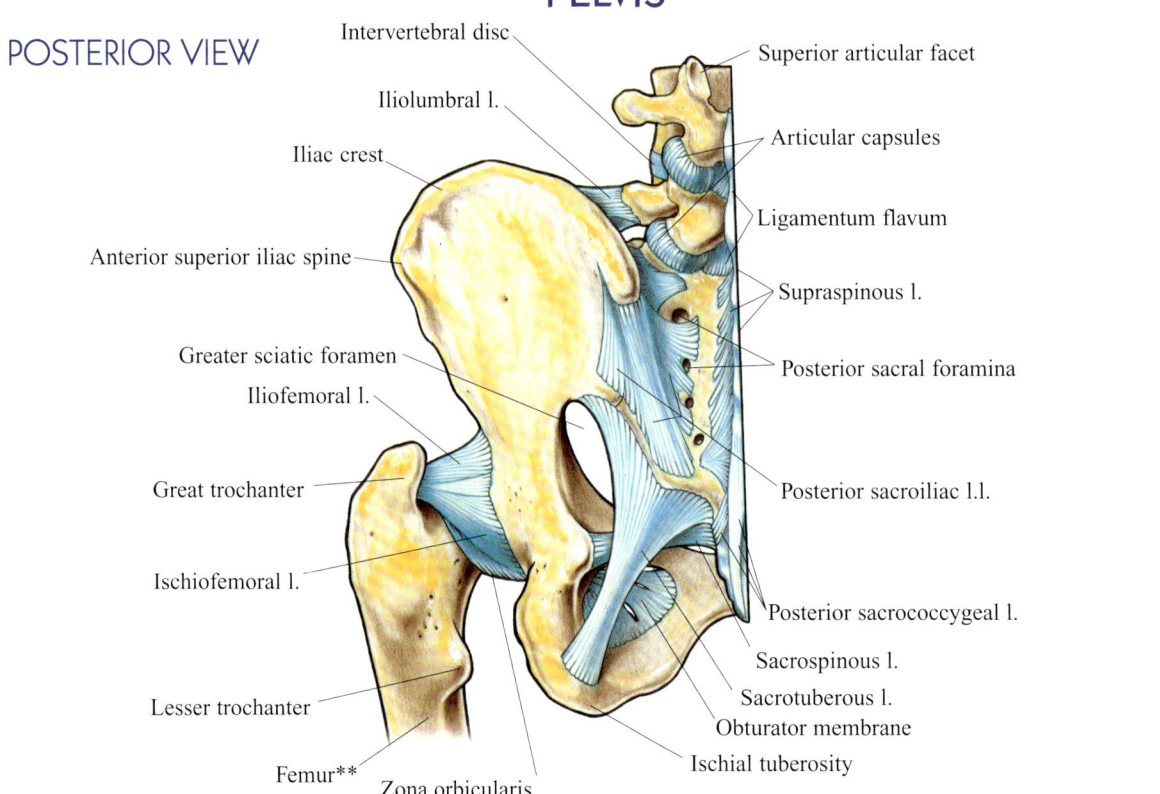

JOINTS & LIGAMENTS

KNEE LIGAMENTS

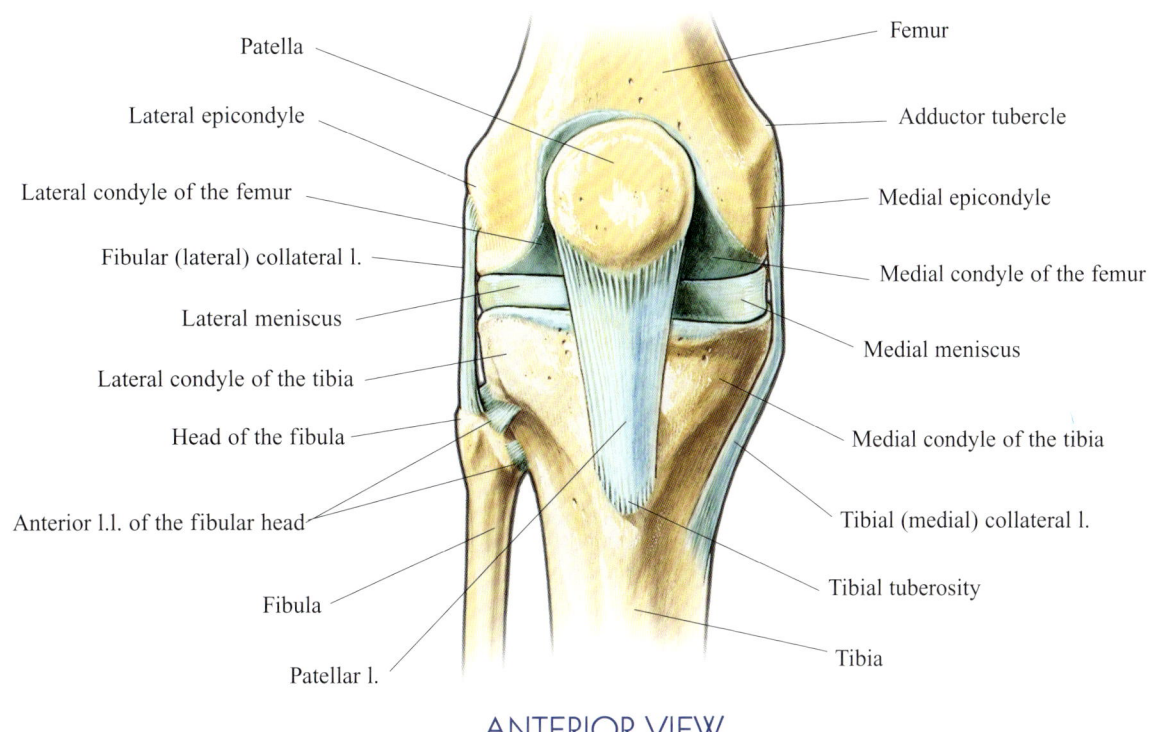

ANTERIOR VIEW

KNEE LIGAMENTS

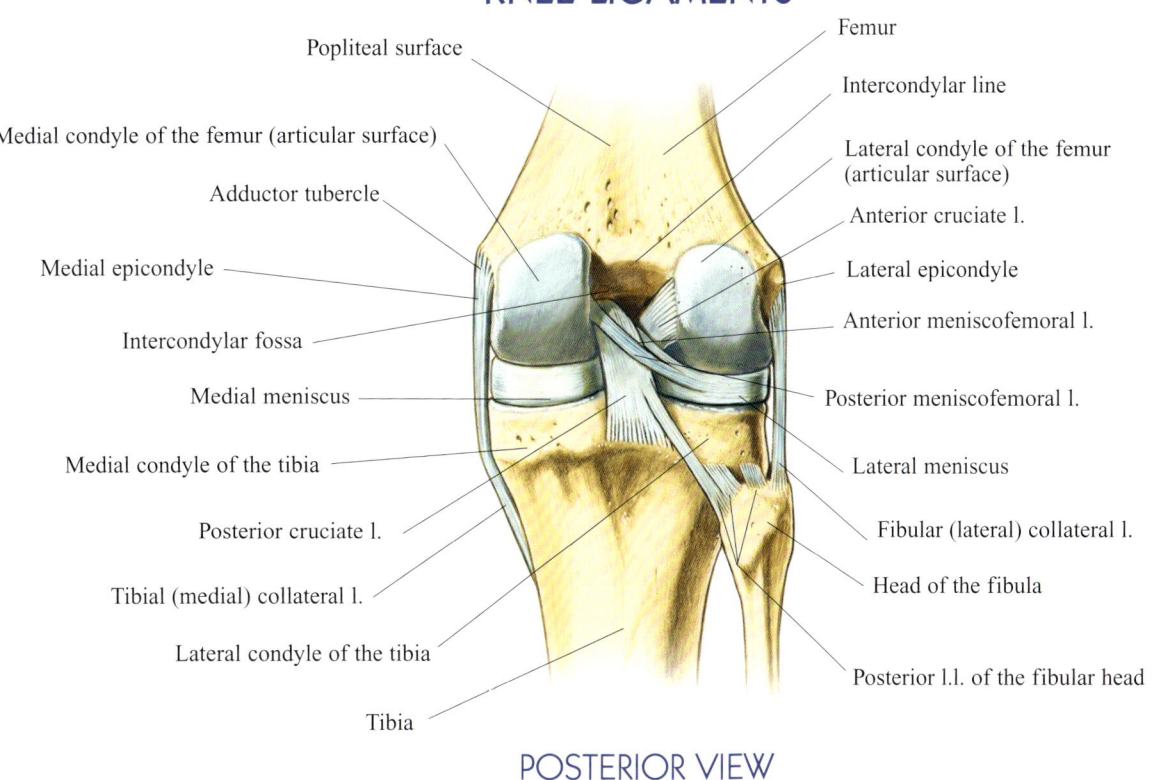

POSTERIOR VIEW

JOINTS & LIGAMENTS

RIGHT FOOT

LATERAL VIEW

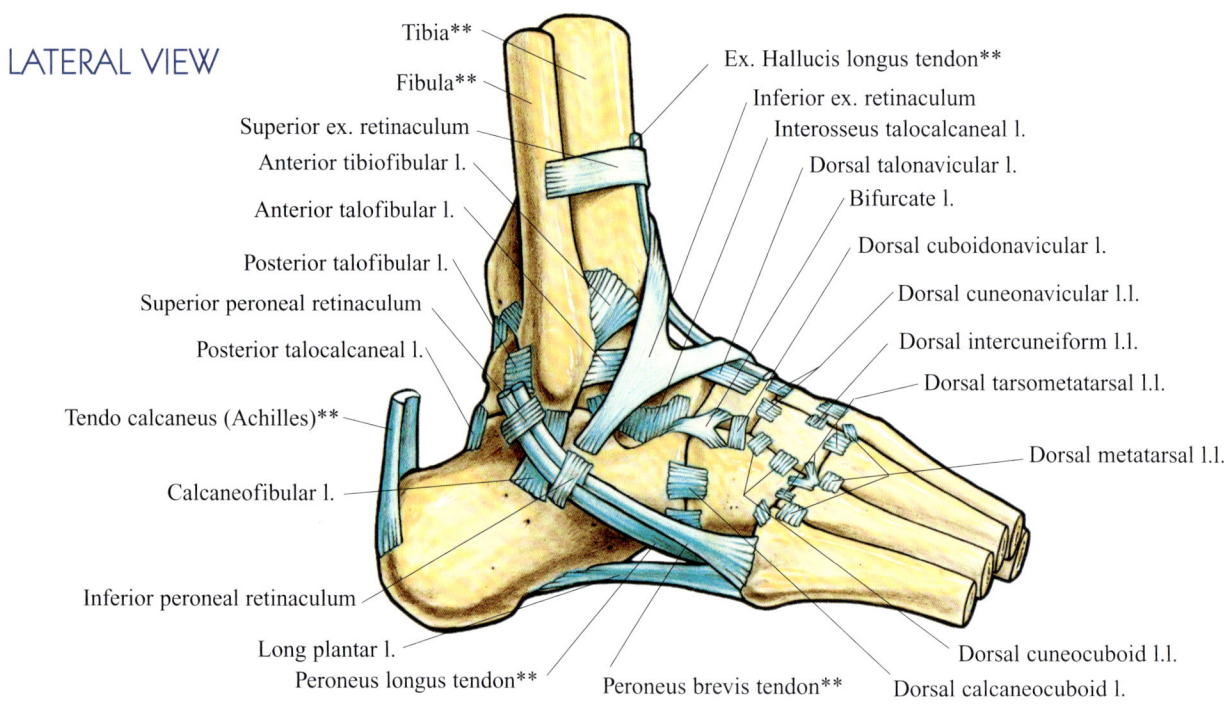

RIGHT FOOT

MEDIAL VIEW

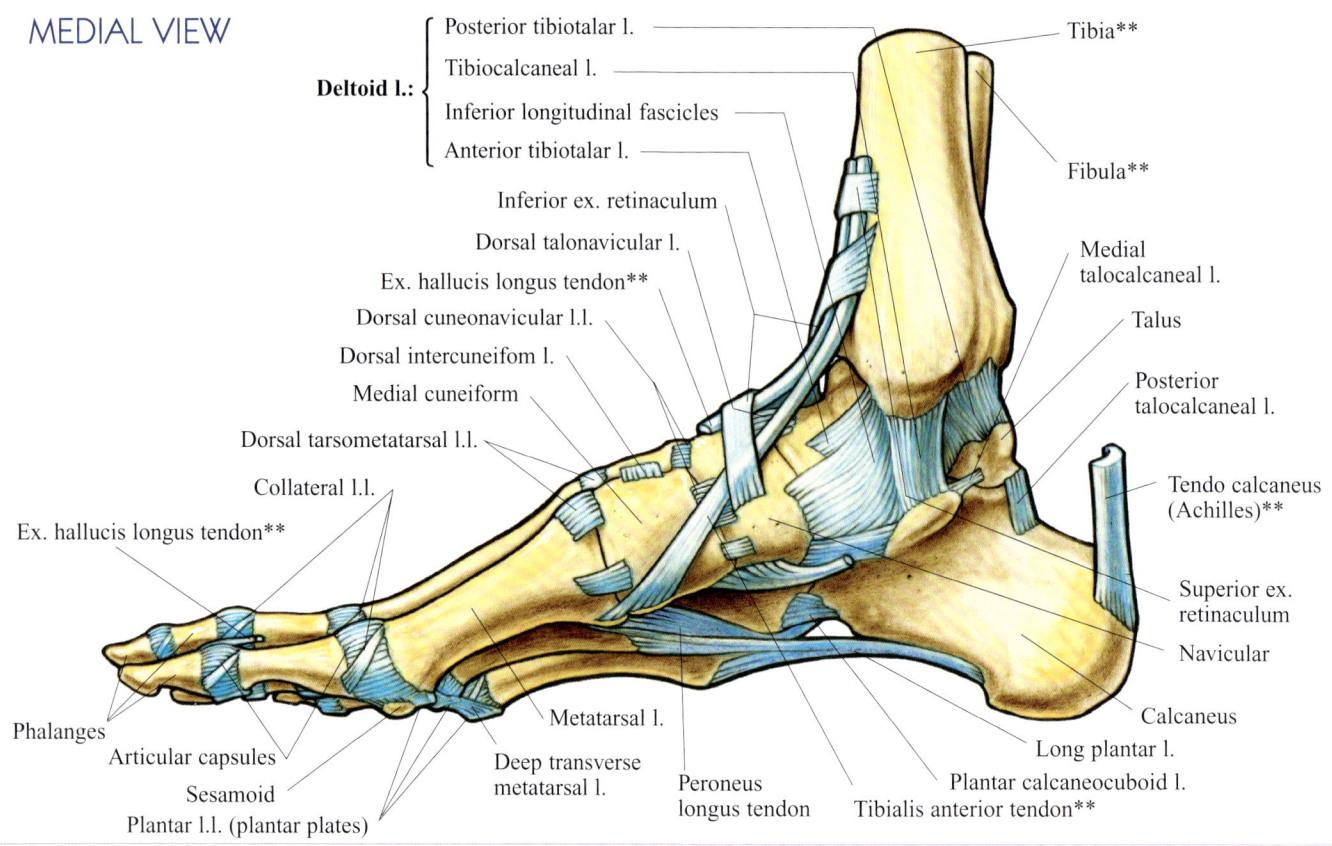

JOINTS & LIGAMENTS

RIGHT FOOT

INFERIOR VIEW

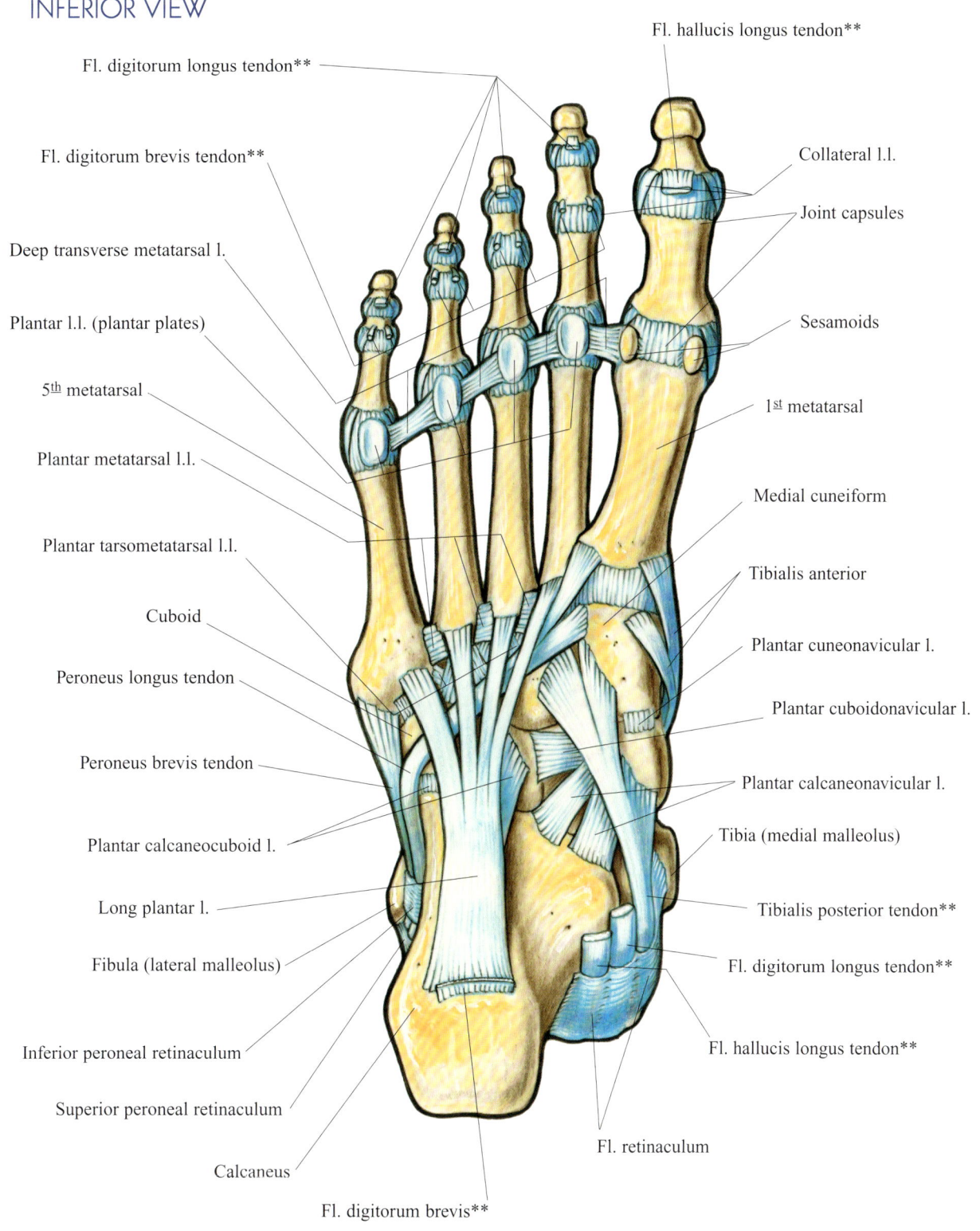

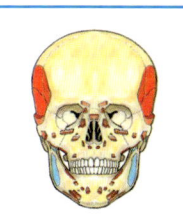

4
ORIGINS & INSERTIONS

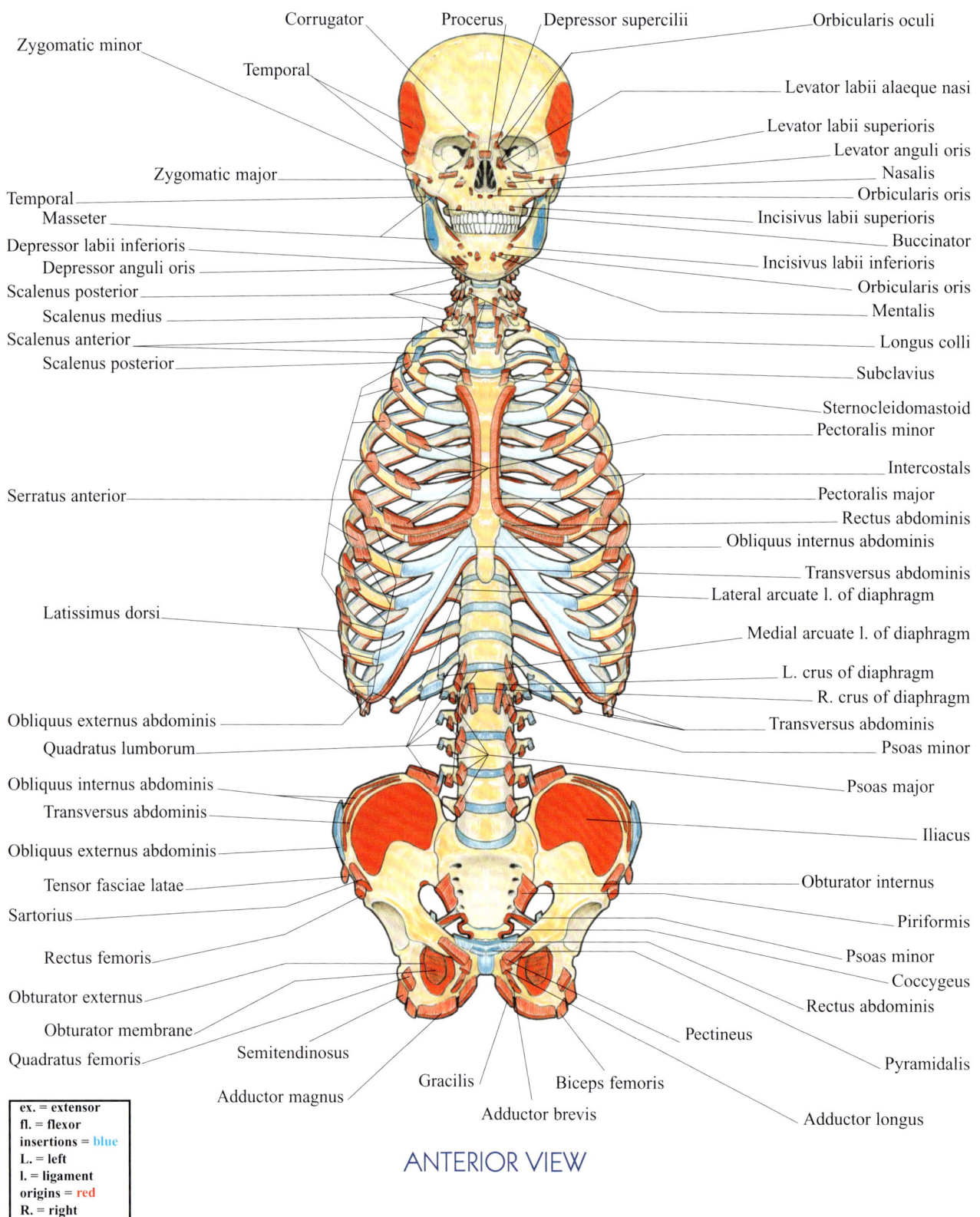

HEAD & TRUNK

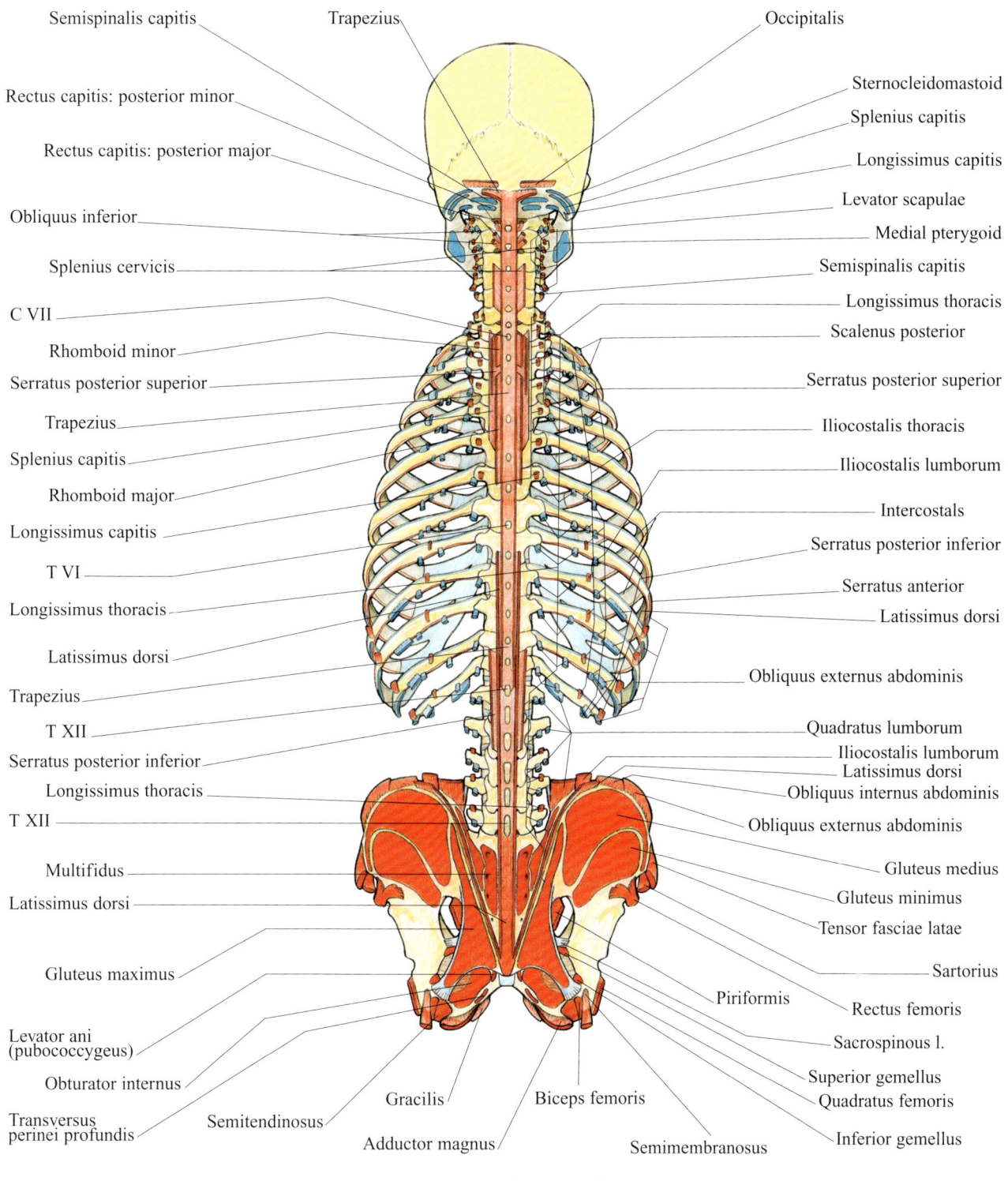

POSTERIOR VIEW

ARM

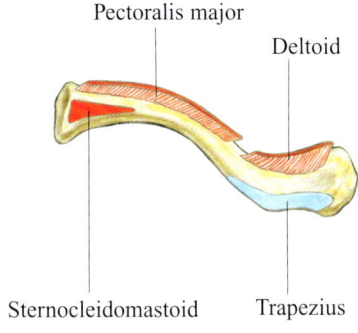

CLAVICLE
SUPERIOR VIEW
ANTERIOR

POSTERIOR

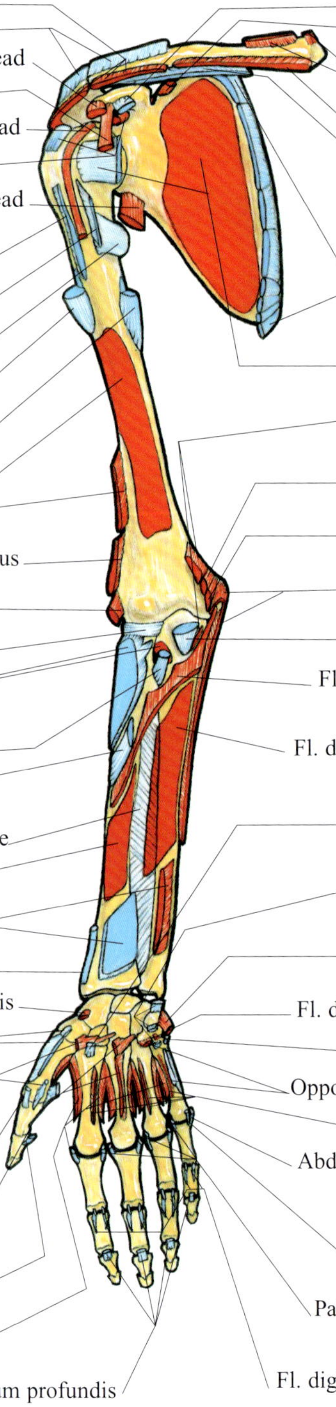

ANTERIOR VIEW

ARM

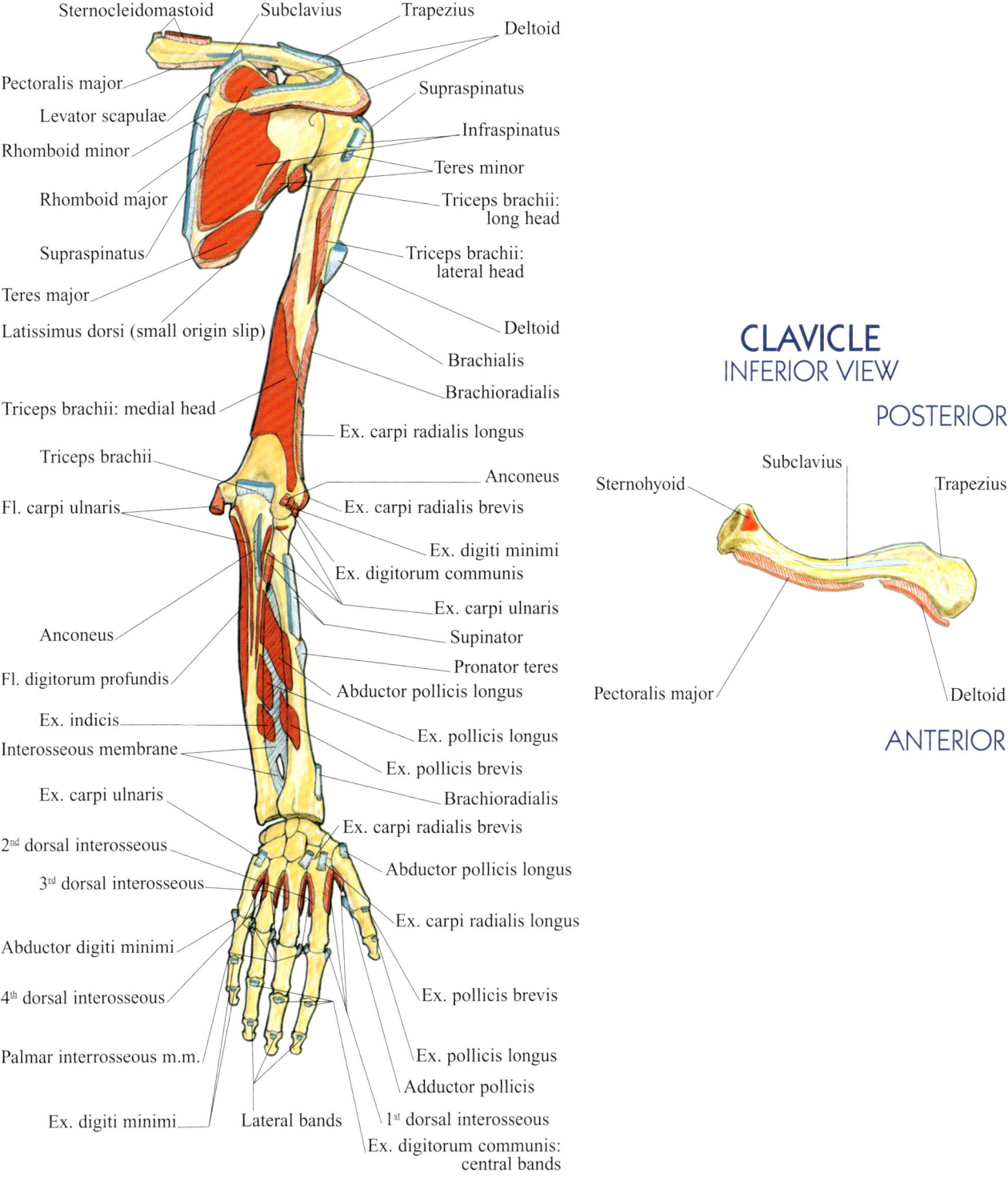

POSTERIOR VIEW

CLAVICLE
INFERIOR VIEW

HAND

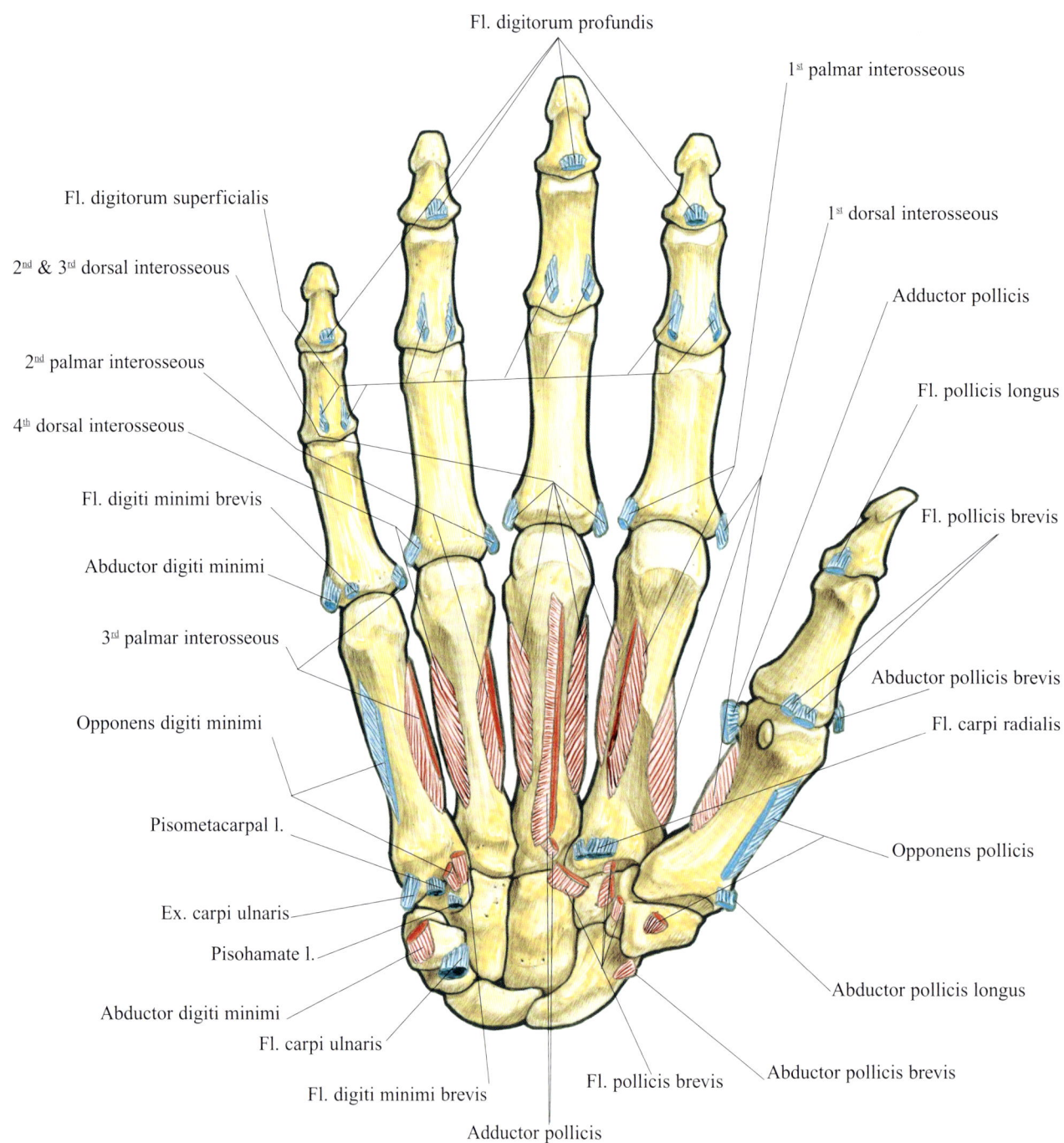

PALMAR VIEW

HAND

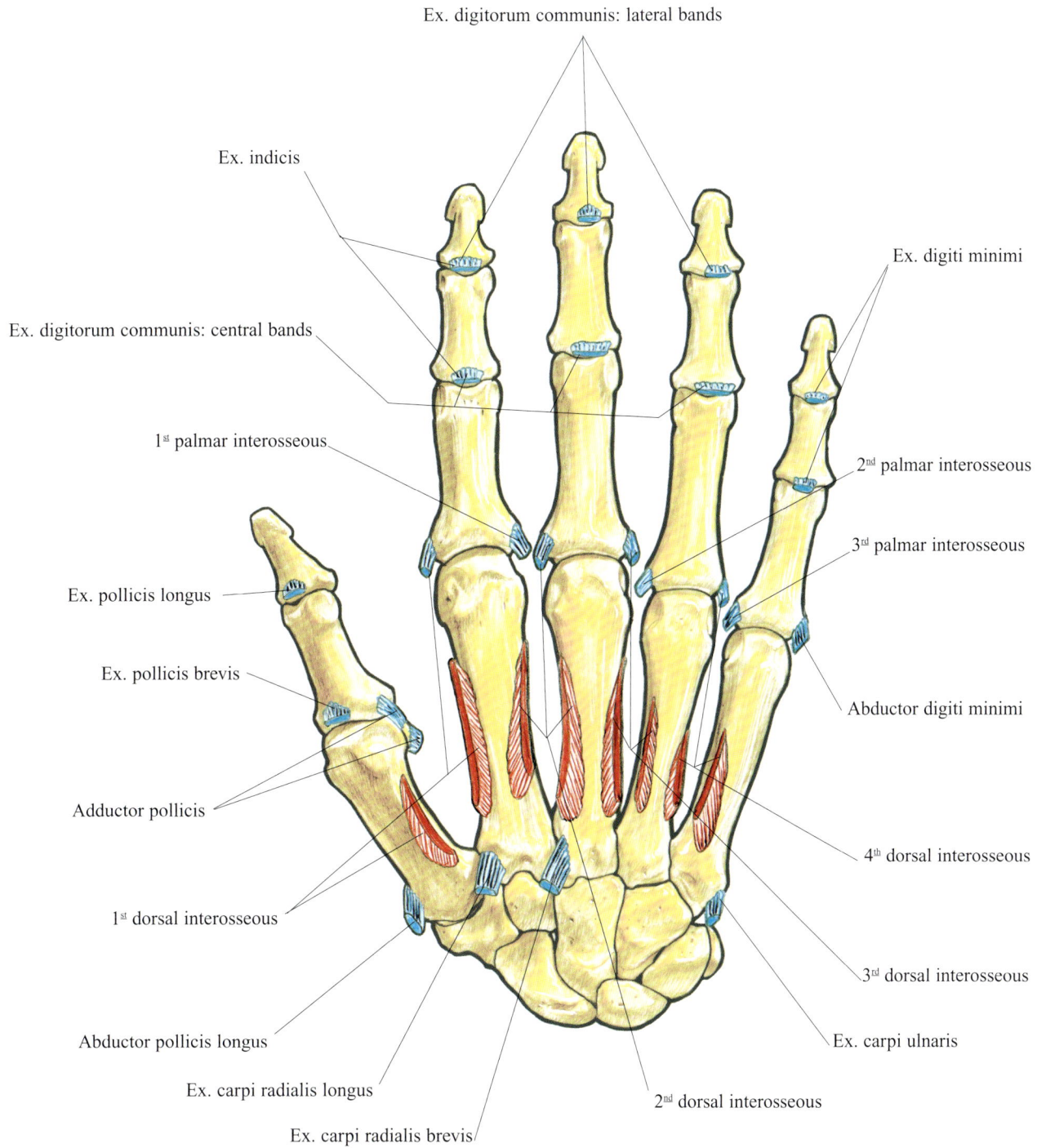

DORSAL VIEW

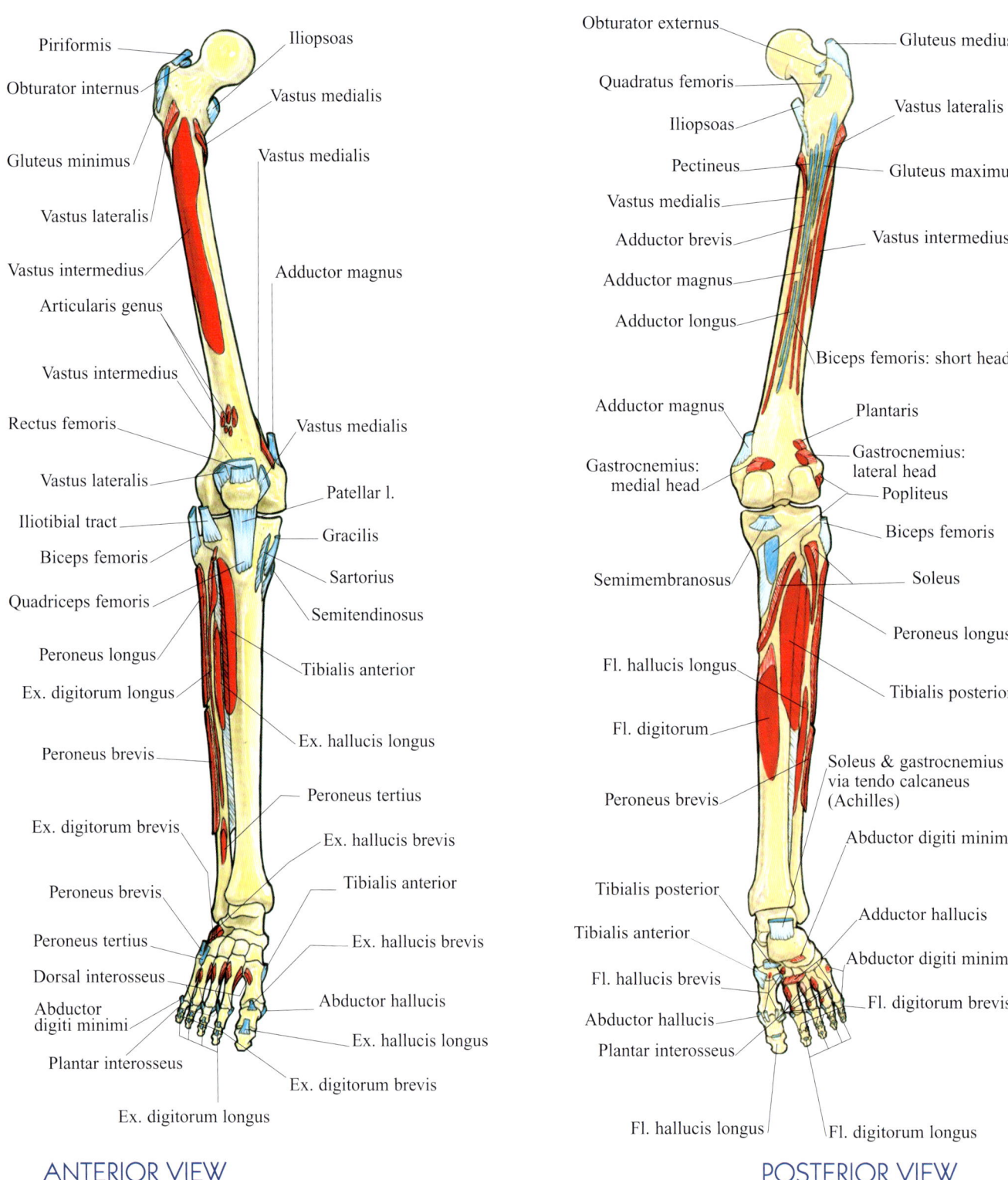

FOOT

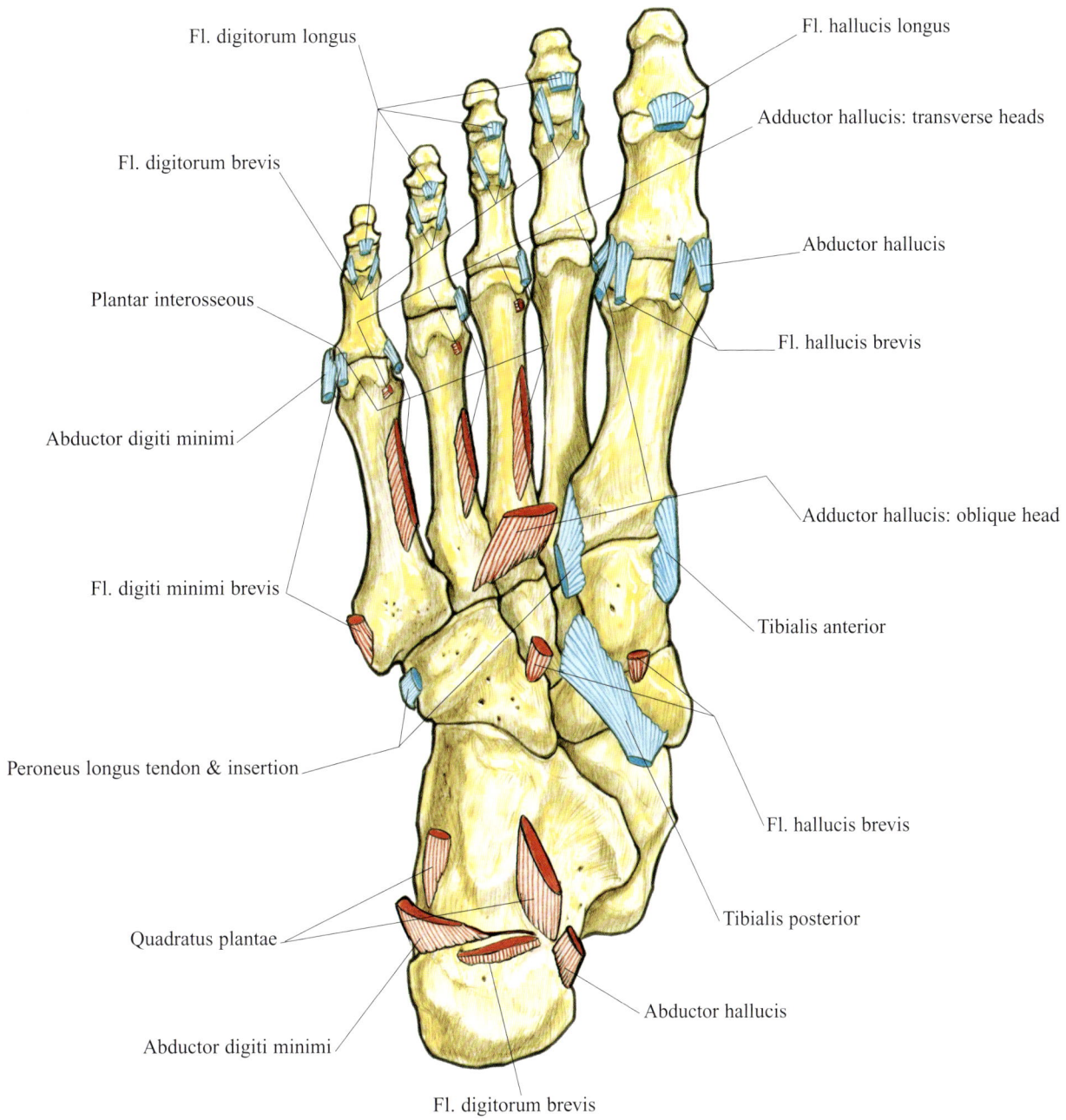

PLANTAR VIEW

FOOT

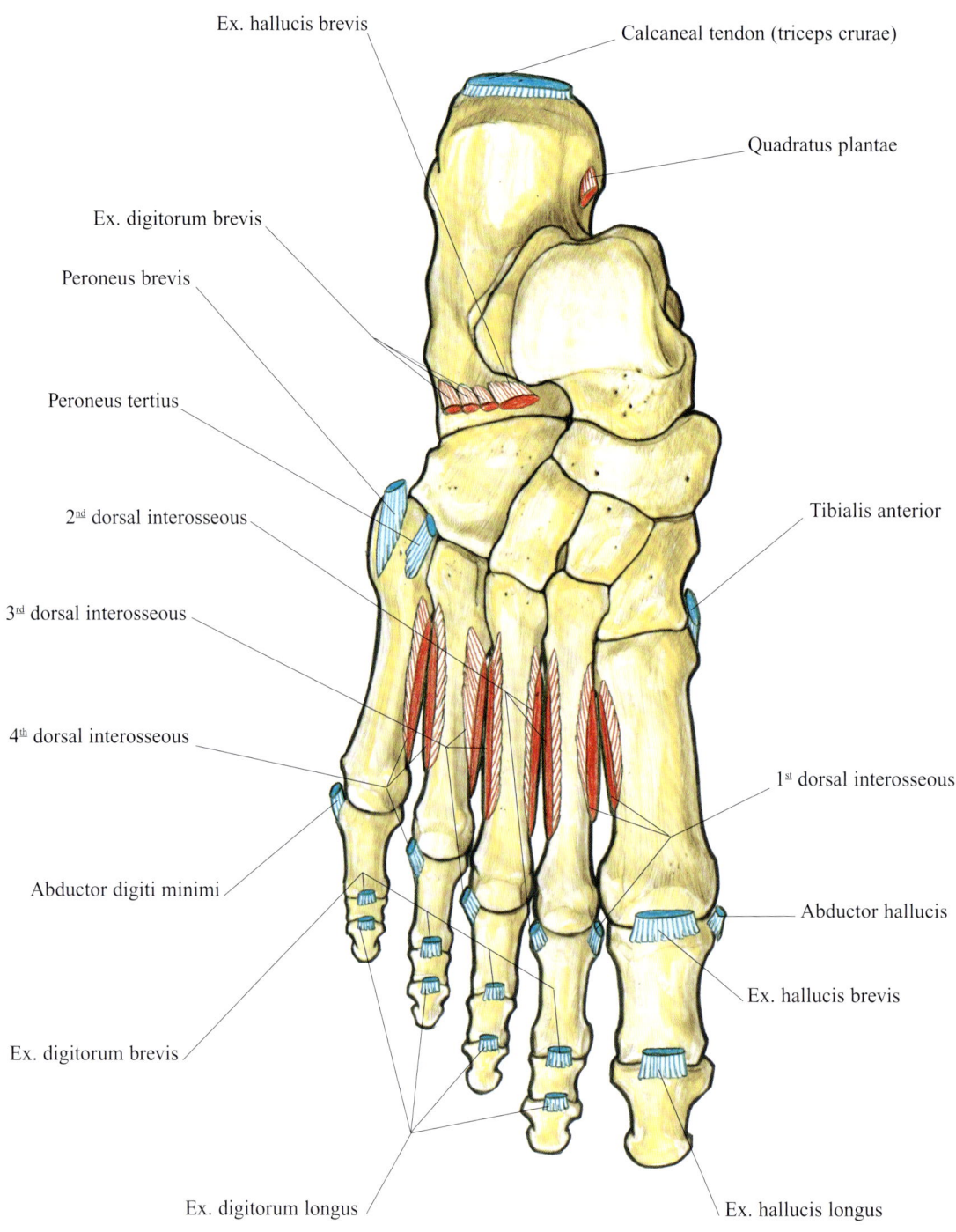

DORSAL VIEW

BASE OF SKULL

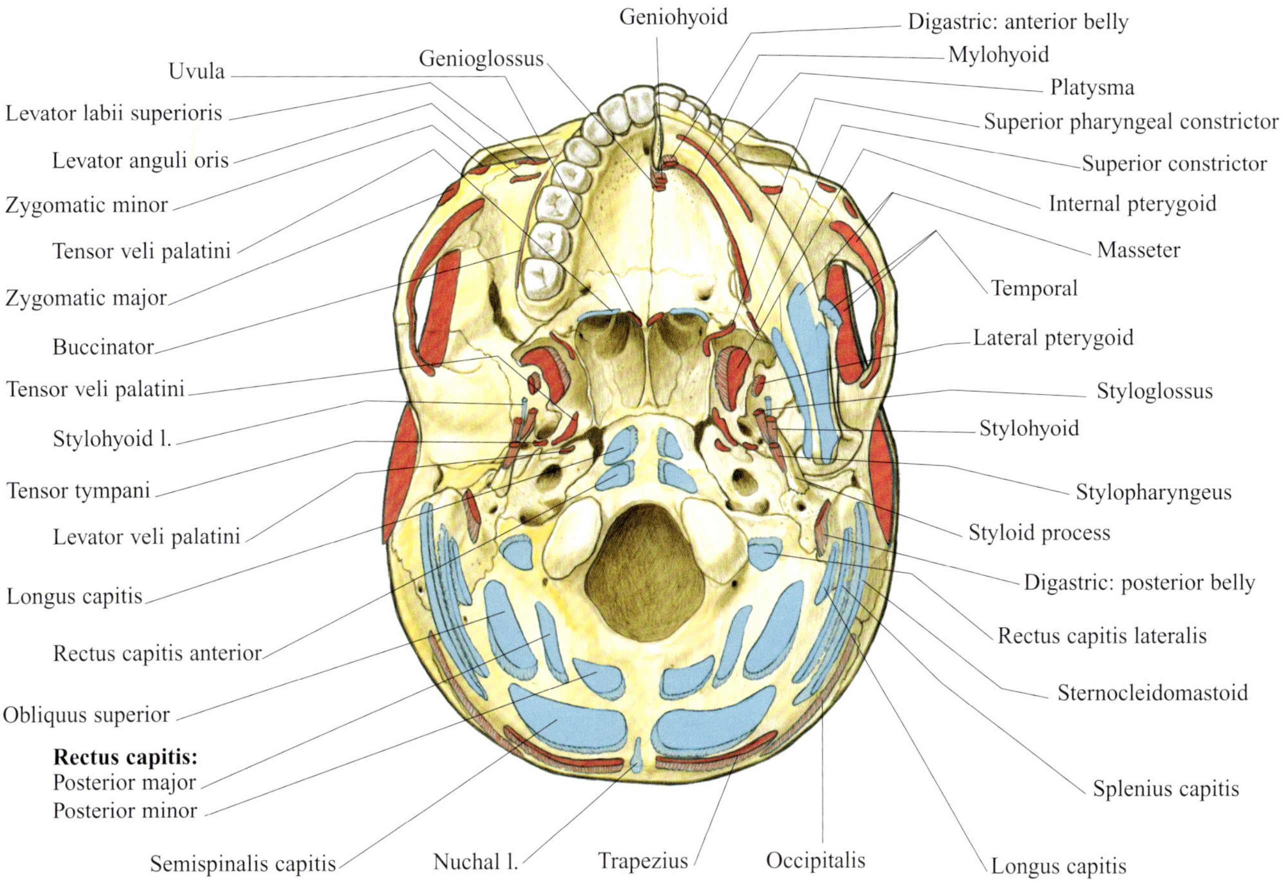

- Geniohyoid
- Genioglossus
- Uvula
- Levator labii superioris
- Levator anguli oris
- Zygomatic minor
- Tensor veli palatini
- Zygomatic major
- Buccinator
- Tensor veli palatini
- Stylohyoid l.
- Tensor tympani
- Levator veli palatini
- Longus capitis
- Rectus capitis anterior
- Obliquus superior
- **Rectus capitis:** Posterior major / Posterior minor
- Semispinalis capitis
- Nuchal l.
- Trapezius
- Occipitalis
- Digastric: anterior belly
- Mylohyoid
- Platysma
- Superior pharyngeal constrictor
- Superior constrictor
- Internal pterygoid
- Masseter
- Temporal
- Lateral pterygoid
- Styloglossus
- Stylohyoid
- Stylopharyngeus
- Styloid process
- Digastric: posterior belly
- Rectus capitis lateralis
- Sternocleidomastoid
- Splenius capitis
- Longus capitis

HYOID BONE

- Middle pharyngeal constrictor
- Stylohyoid l.
- Thyrohyoid
- Omohyoid
- Geniohyoid
- Sternohyoid
- Hyoglossus
- Hyoepiglottic l.
- Mylohyoid
- Stylohyoid
- Intermediate digastric tendon

SUPERIOR VIEW

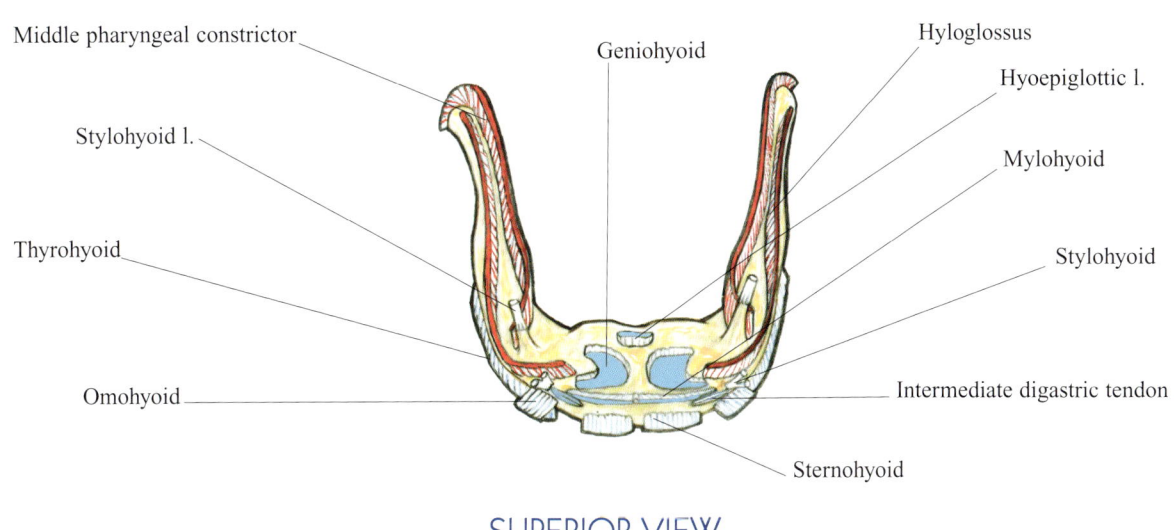

57

5

MUSCULAR SYSTEM

MUSCULAR SYSTEM

SURFACE MUSCLES

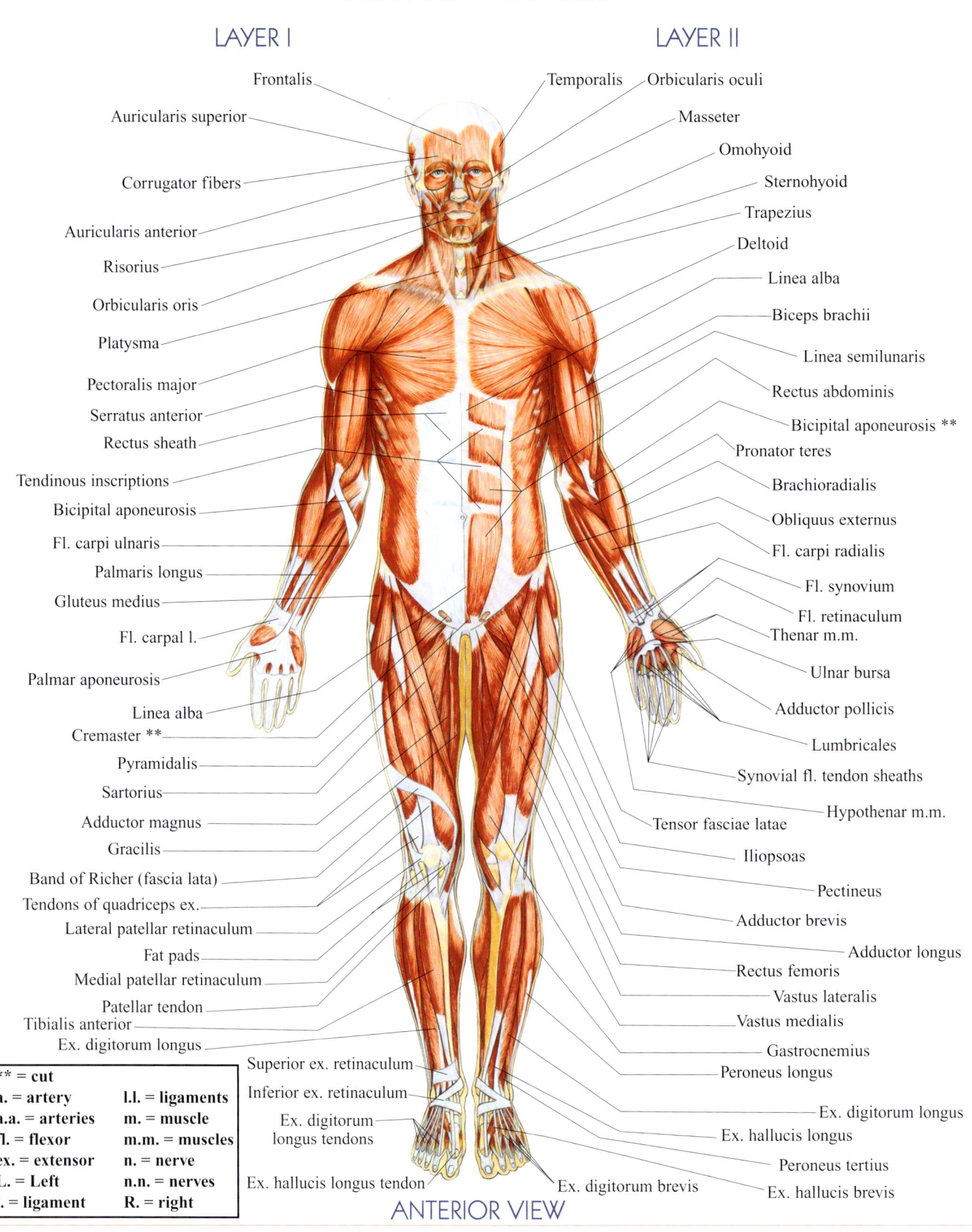

LAYER I — LAYER II

ANTERIOR VIEW

** = cut
a. = artery
a.a. = arteries
fl. = flexor
ex. = extensor
L. = Left
l. = ligament
l.l. = ligaments
m. = muscle
m.m. = muscles
n. = nerve
n.n. = nerves
R. = right

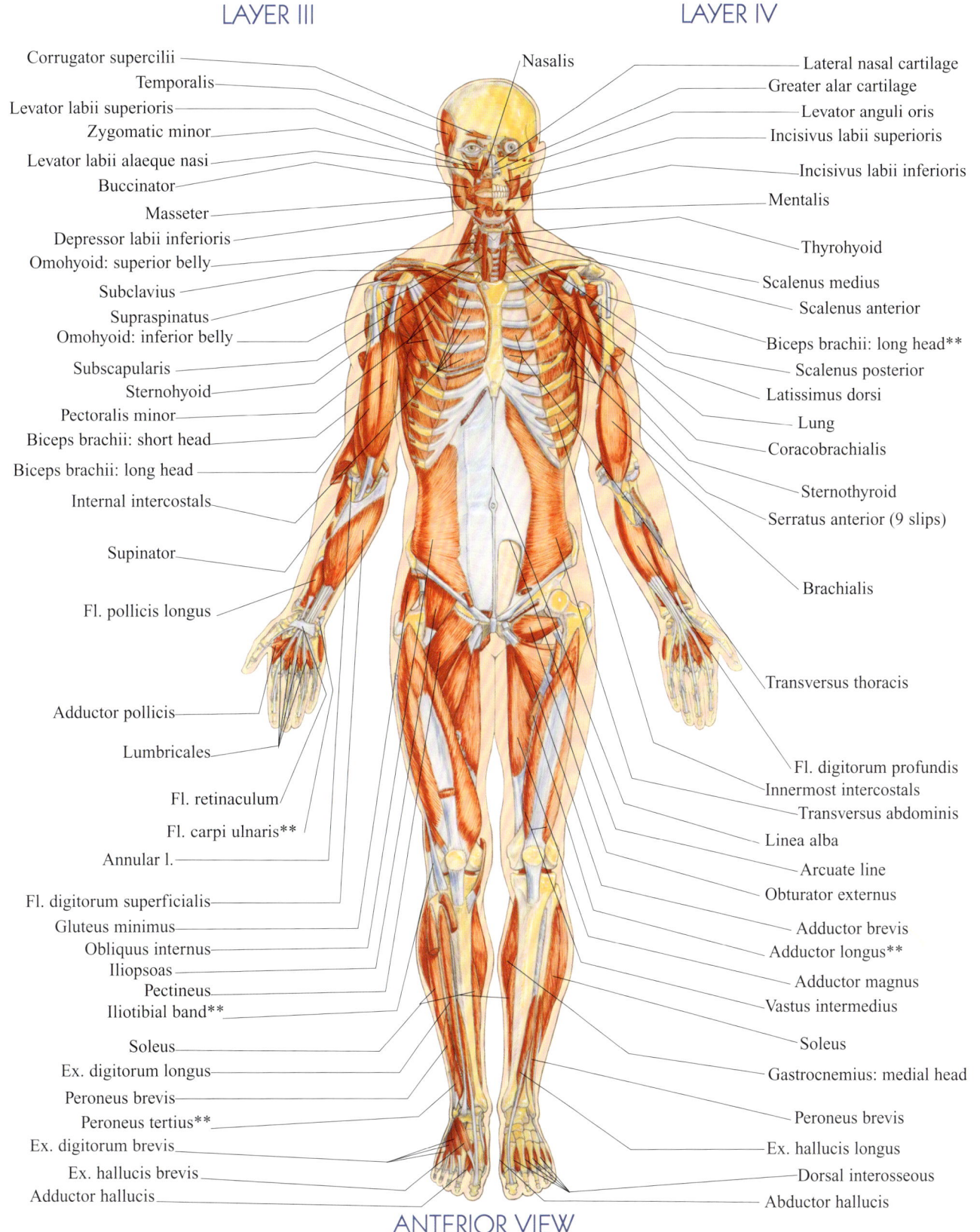

MUSCULAR SYSTEM

DEEP MUSCLES

LAYER V　　　　　　　　　　　　　LAYER VI

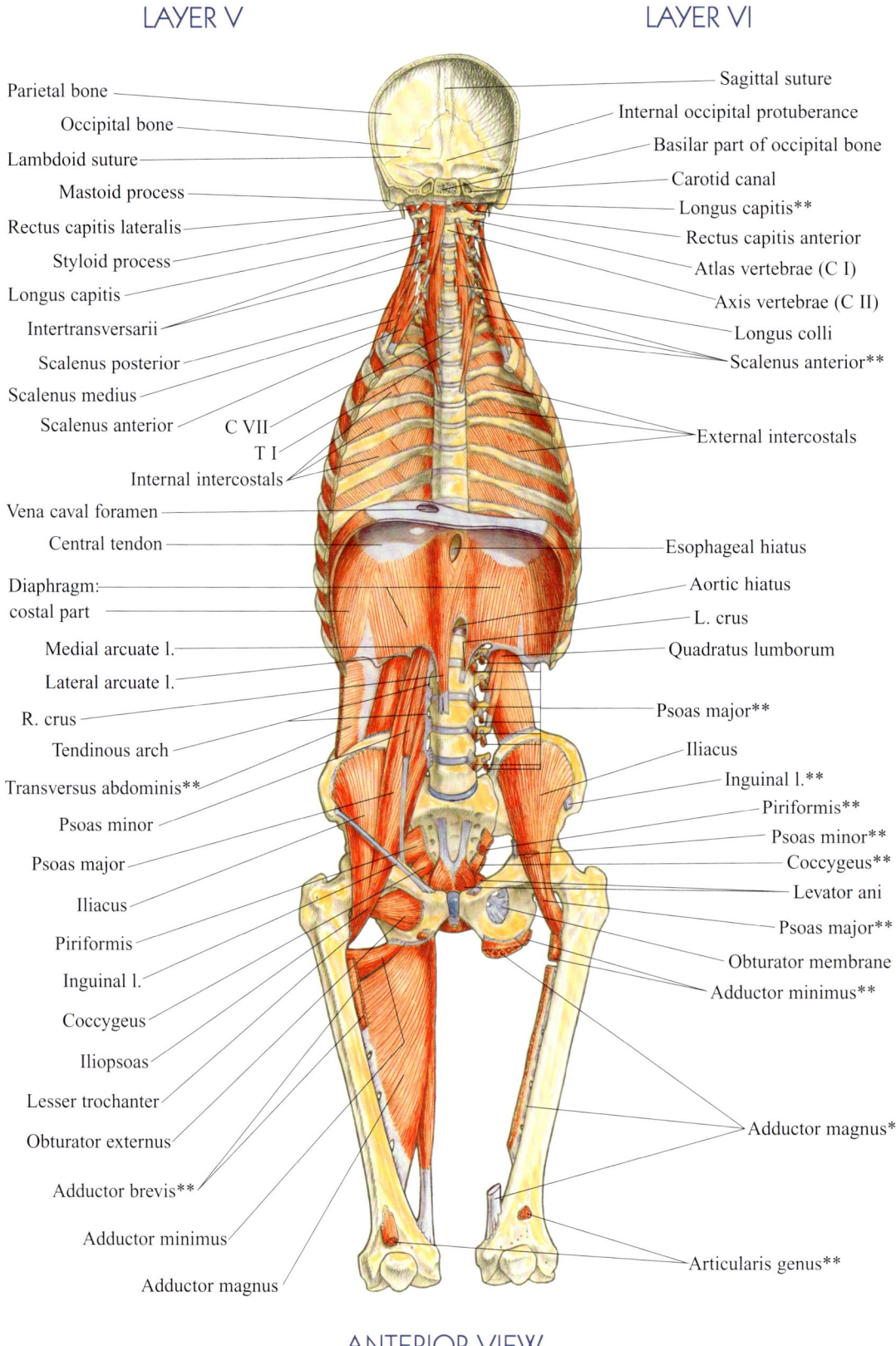

ANTERIOR VIEW

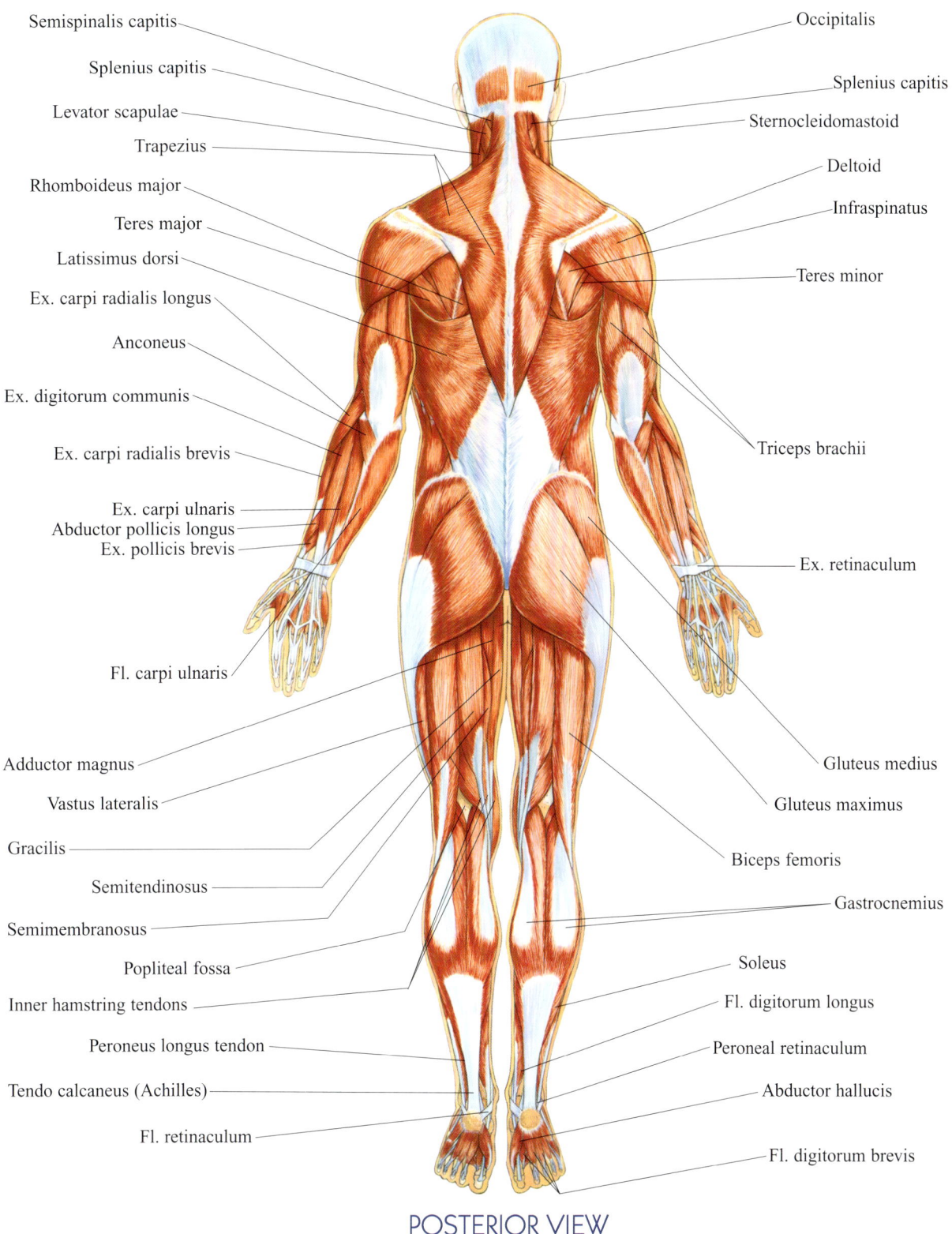

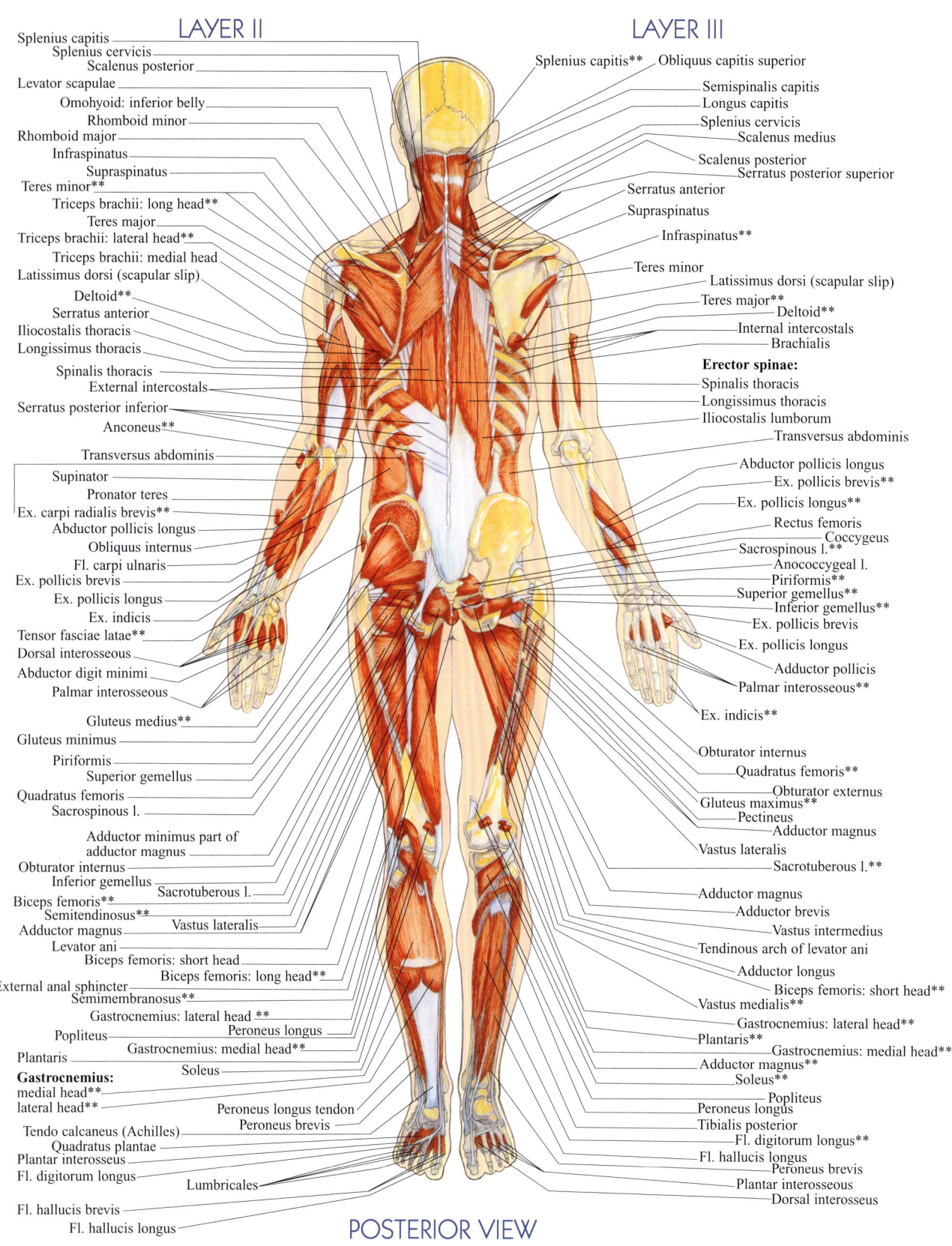

MUSCULAR SYSTEM

DEEP MUSCLES

LAYER IV   LAYER V

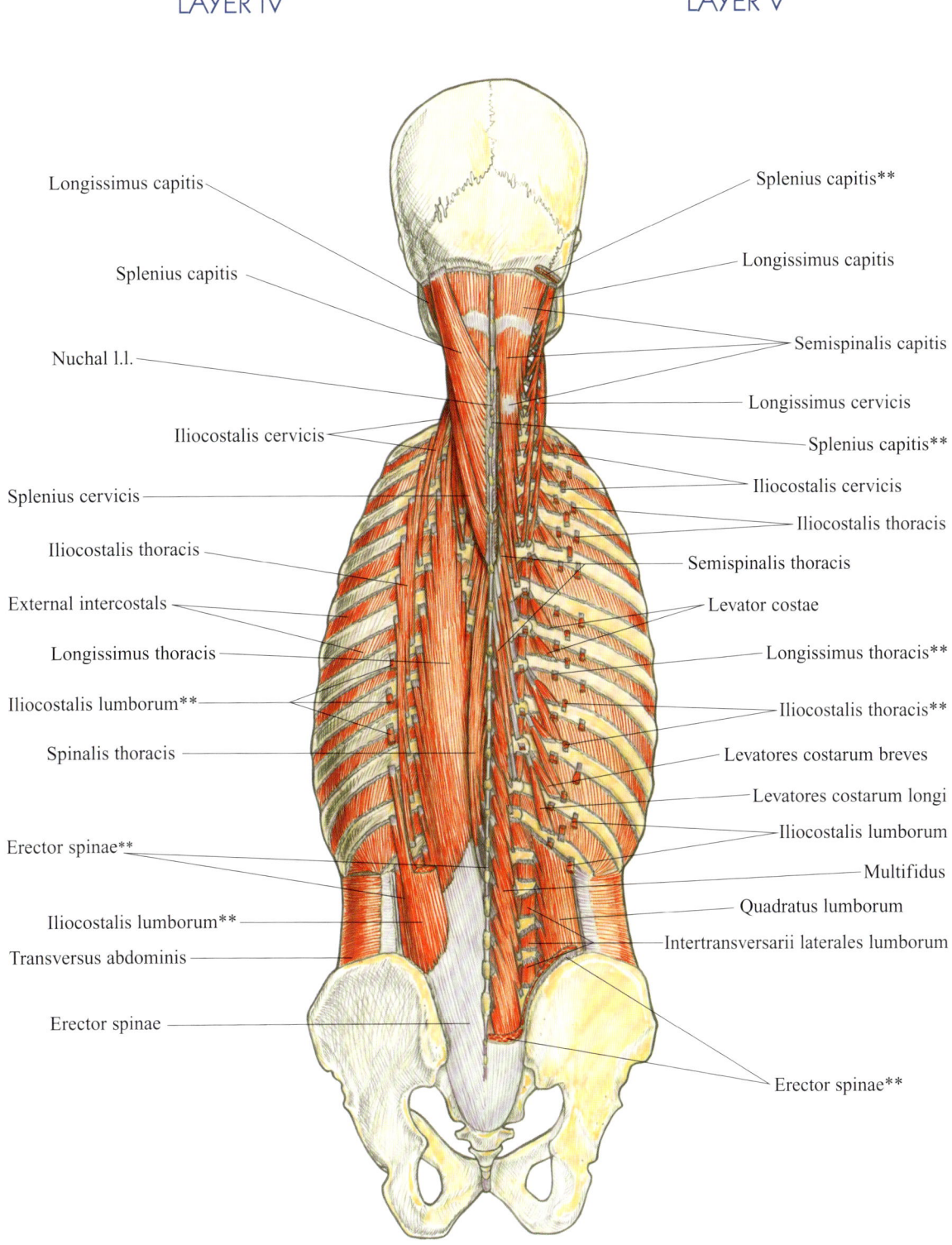

POSTERIOR VIEW

MUSCULAR SYSTEM

DEEP MUSCLES

LAYER VI　　　　　　　　　　　　　LAYER VII

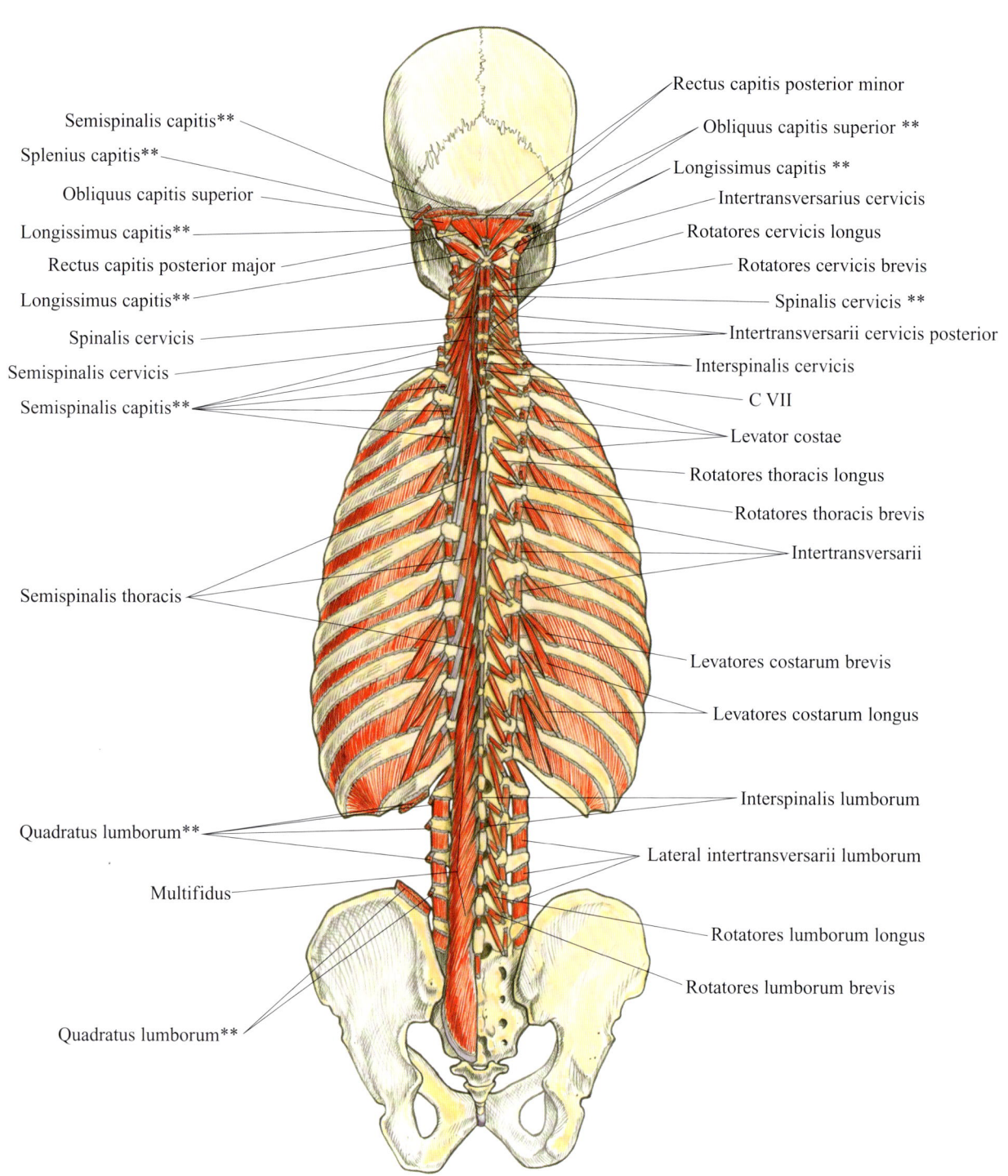

POSTERIOR VIEW

MUSCULAR SYSTEM

SURFACE MUSCLES
LAYER I

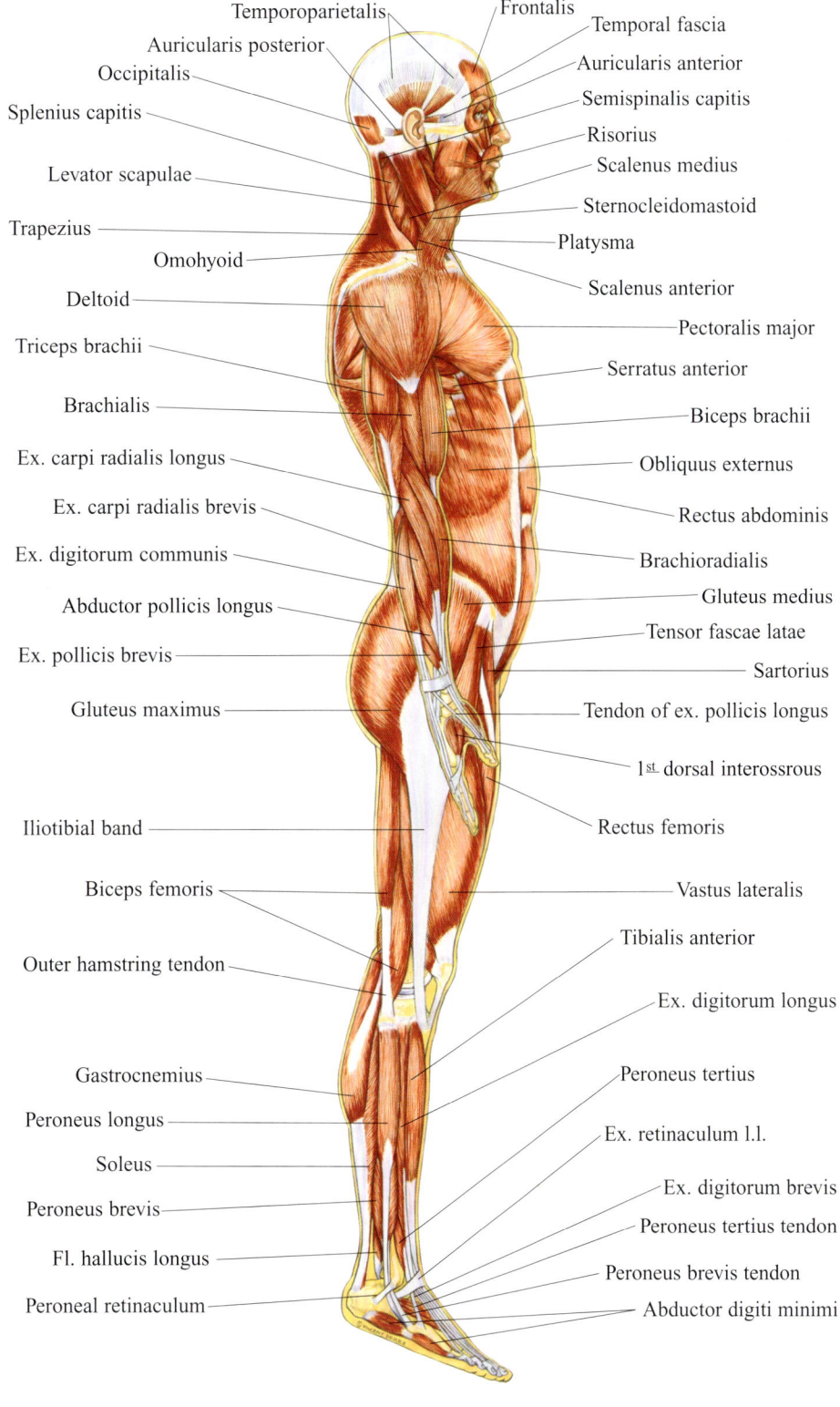

LATERAL VIEW

MUSCULAR SYSTEM

SURFACE MUSCLES
LAYER IA

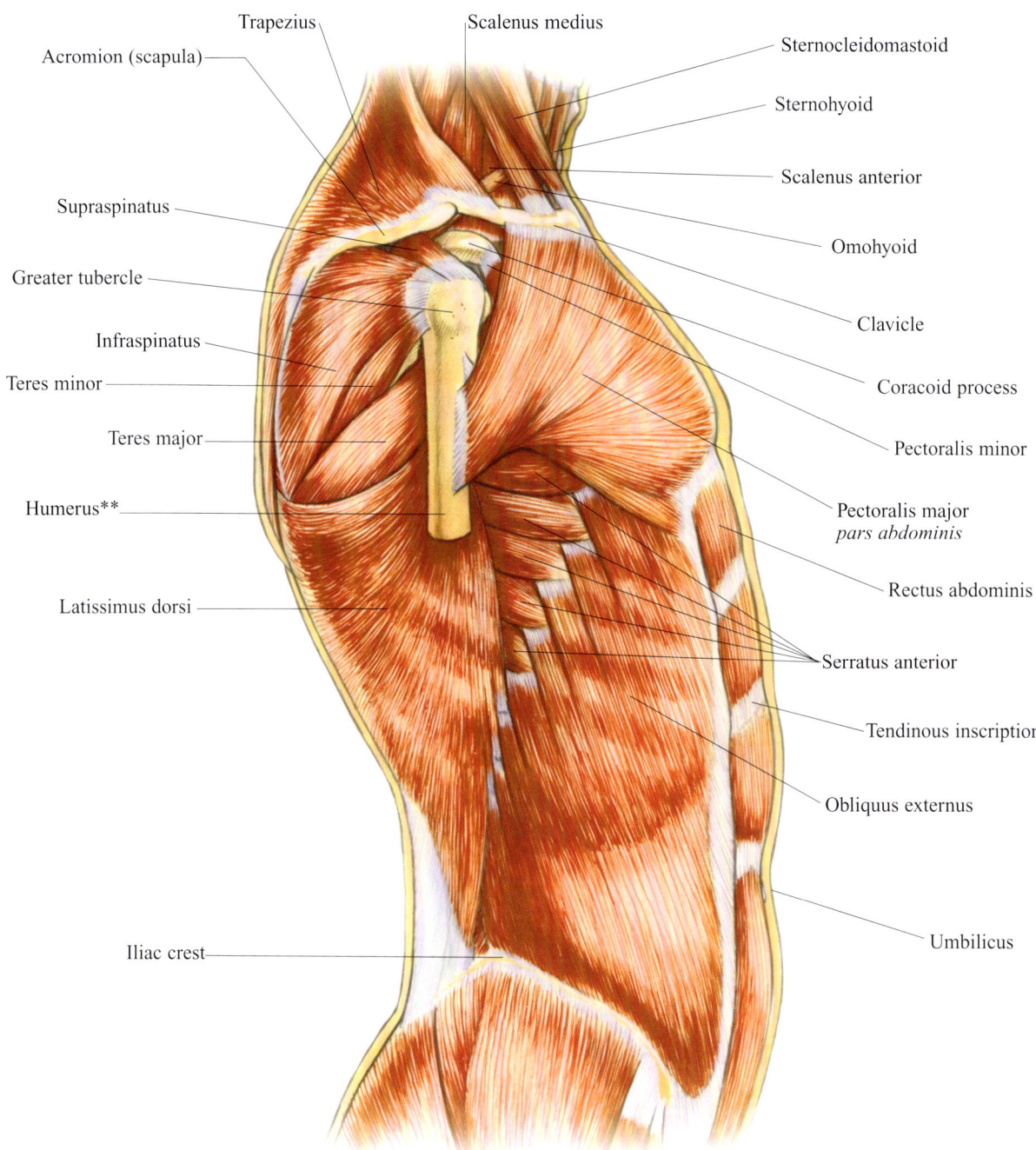

LATERAL VIEW

68

MUSCULAR SYSTEM

DEEP MUSCLES
LAYER II

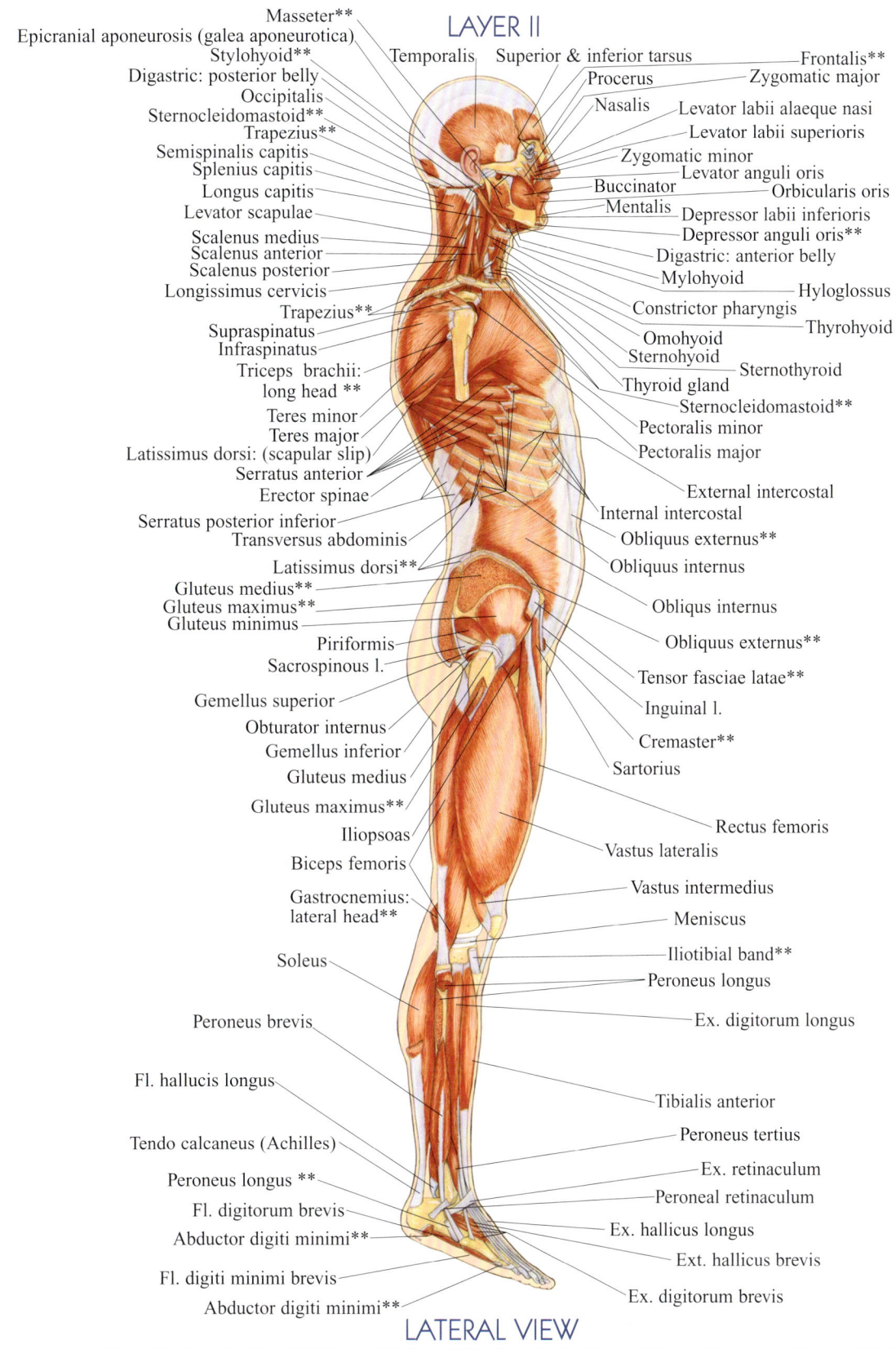

LATERAL VIEW

MUSCULAR SYSTEM

DEEP MUSCLES
LAYER III

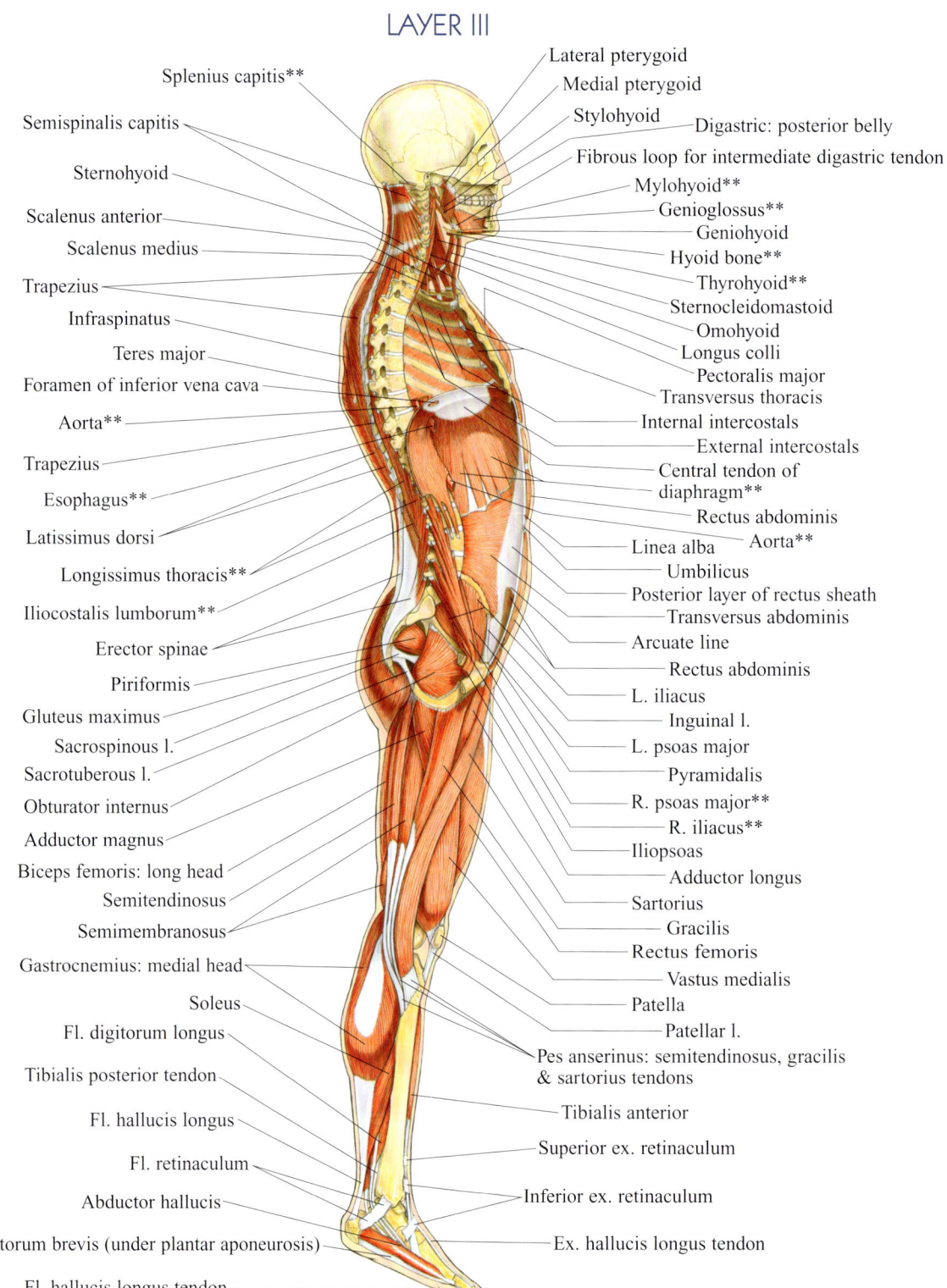

LATERAL VIEW (MEDIAL LEG)

MUSCULAR SYSTEM

HEAD MUSCLES

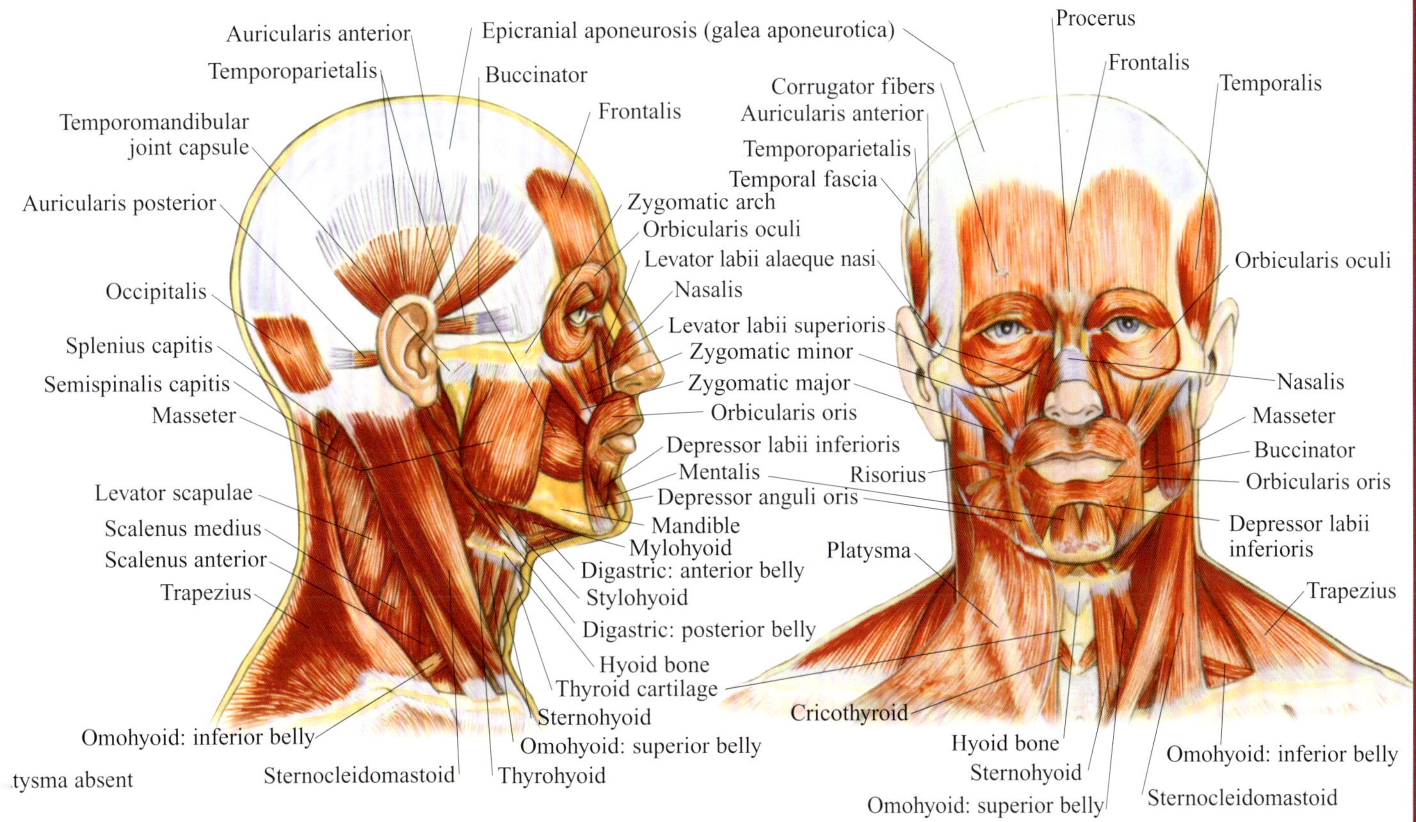

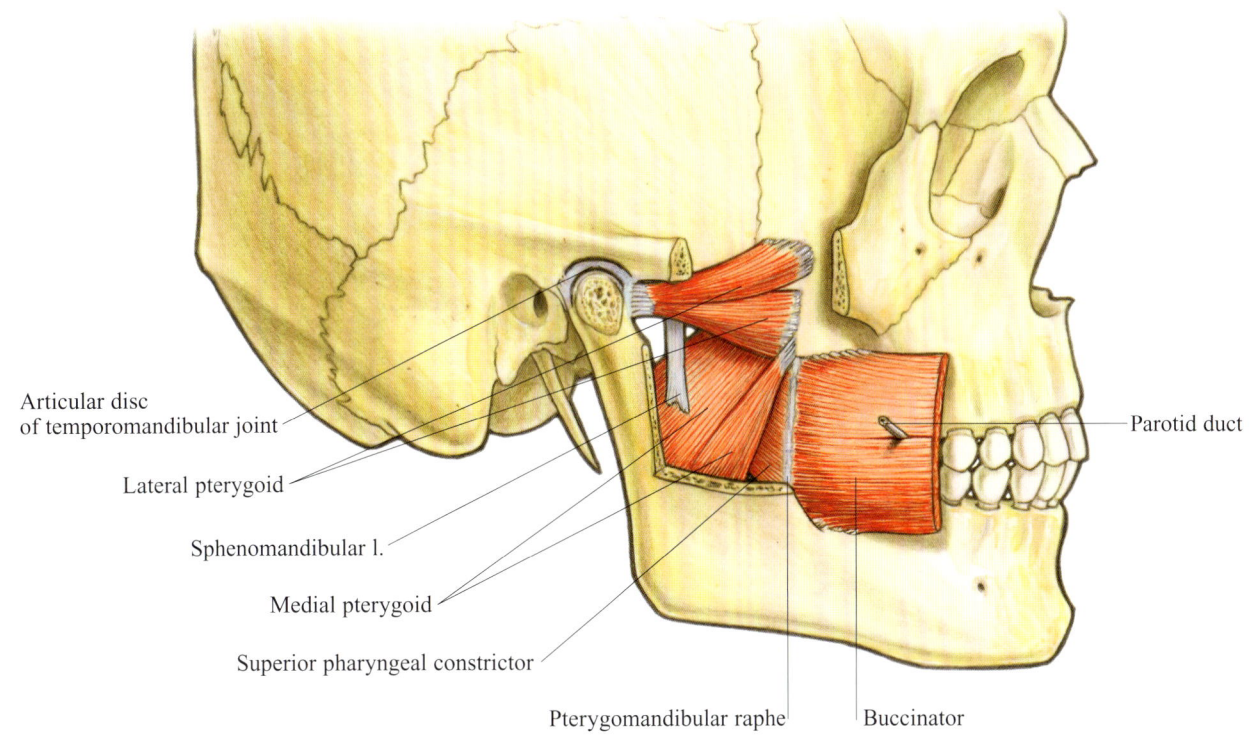

MUSCULAR SYSTEM

MUSCLES OF THE EYE
EXTRINSIC EYE MUSCLES

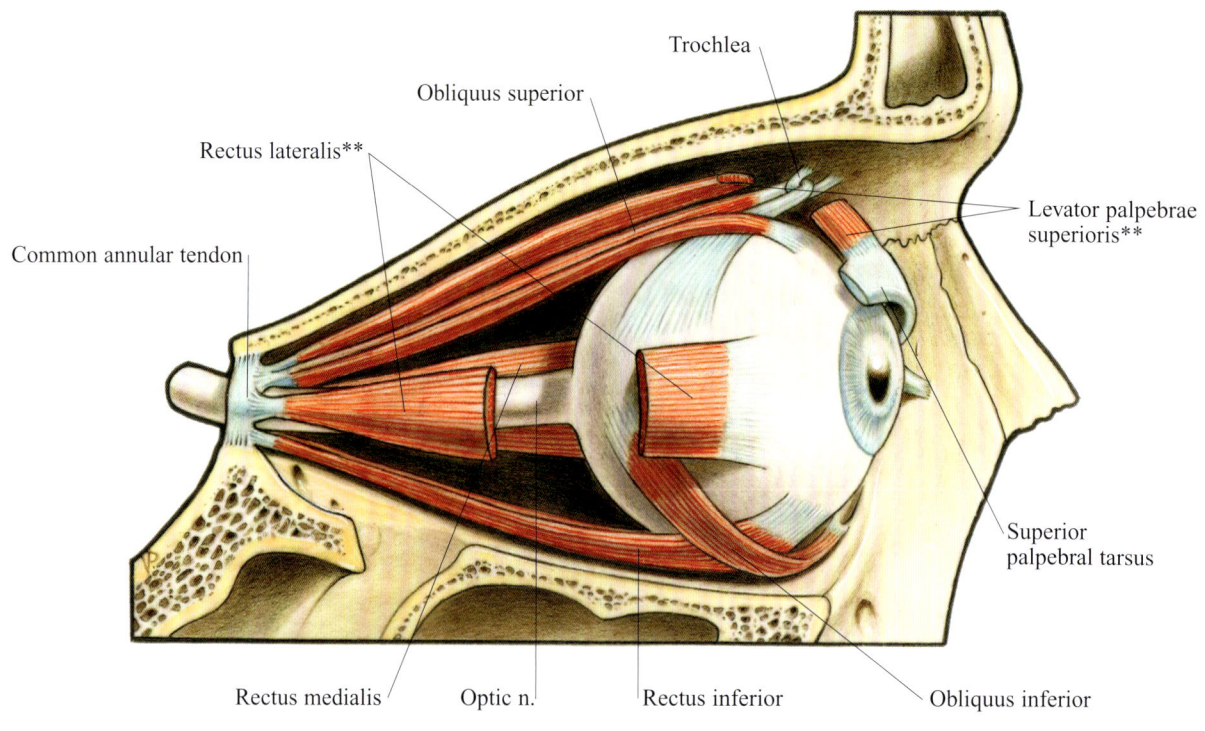

RIGHT LATERAL VIEW

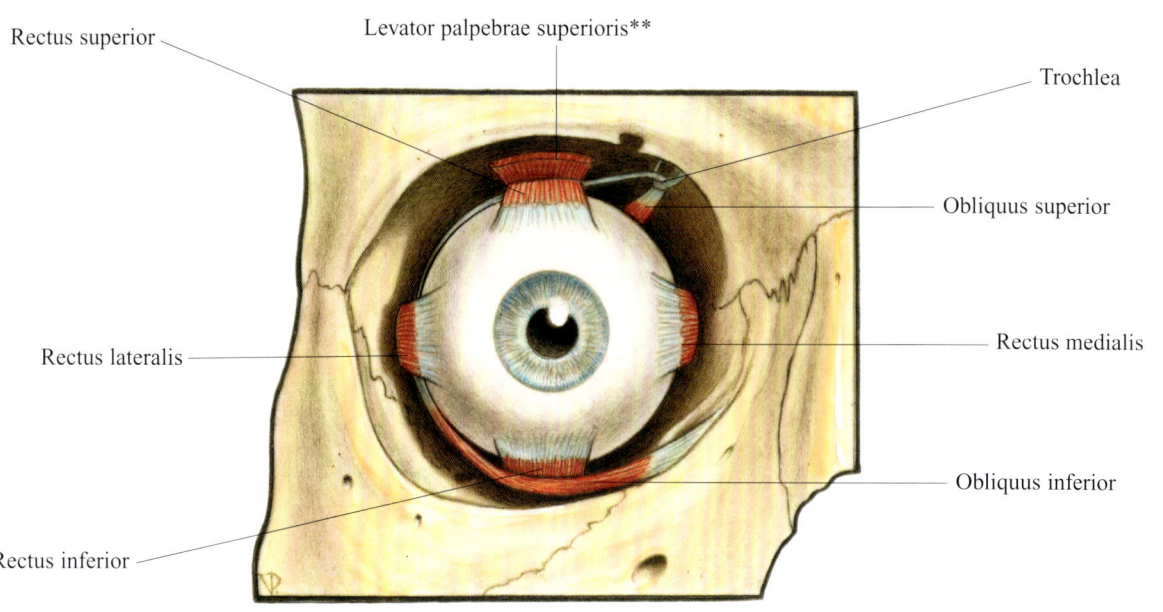

ANTERIOR VIEW

MUSCULAR SYSTEM

DEEP NECK MUSCLE

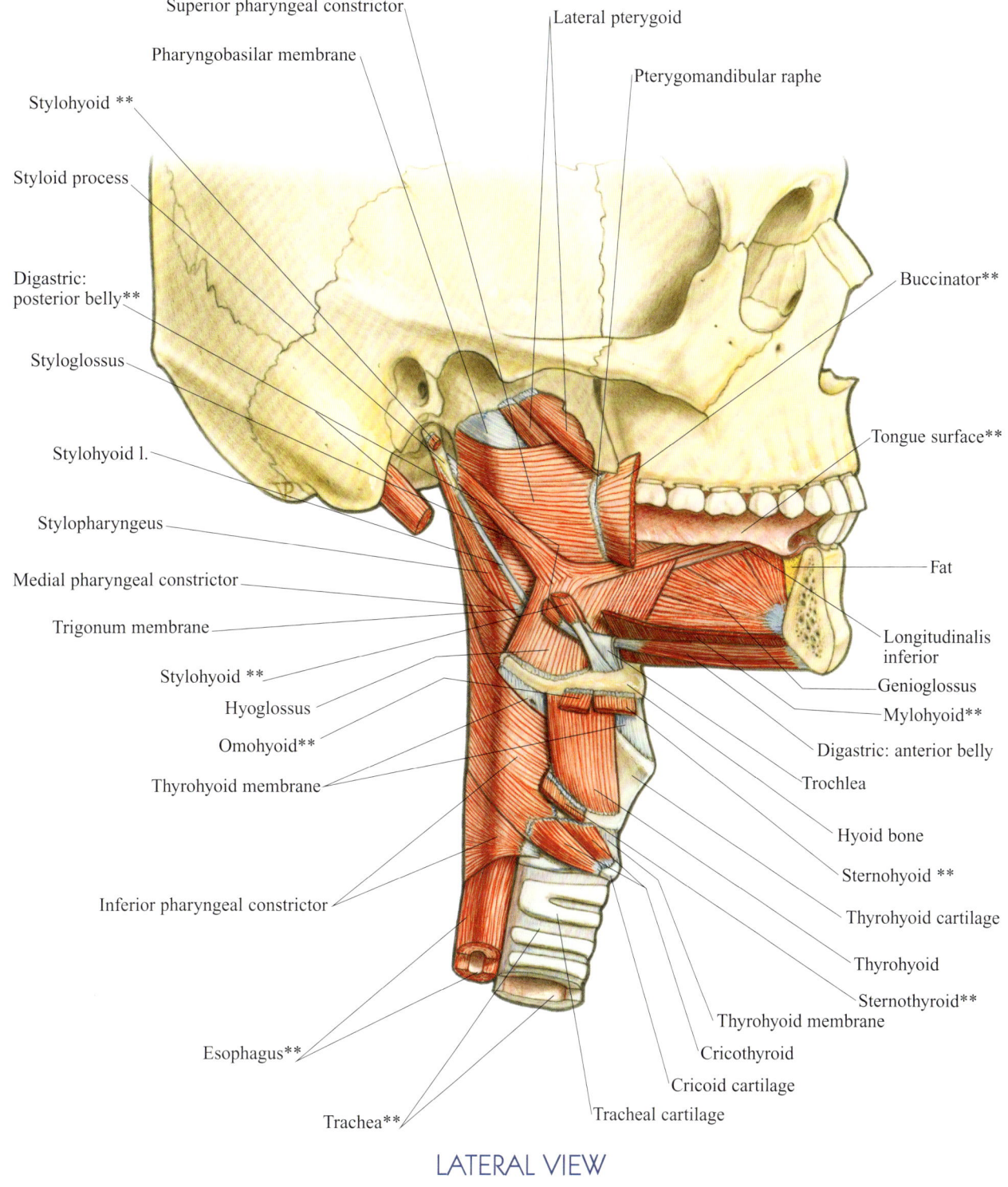

LATERAL VIEW

MUSCULAR SYSTEM

MUSCLES OF RESPIRATION

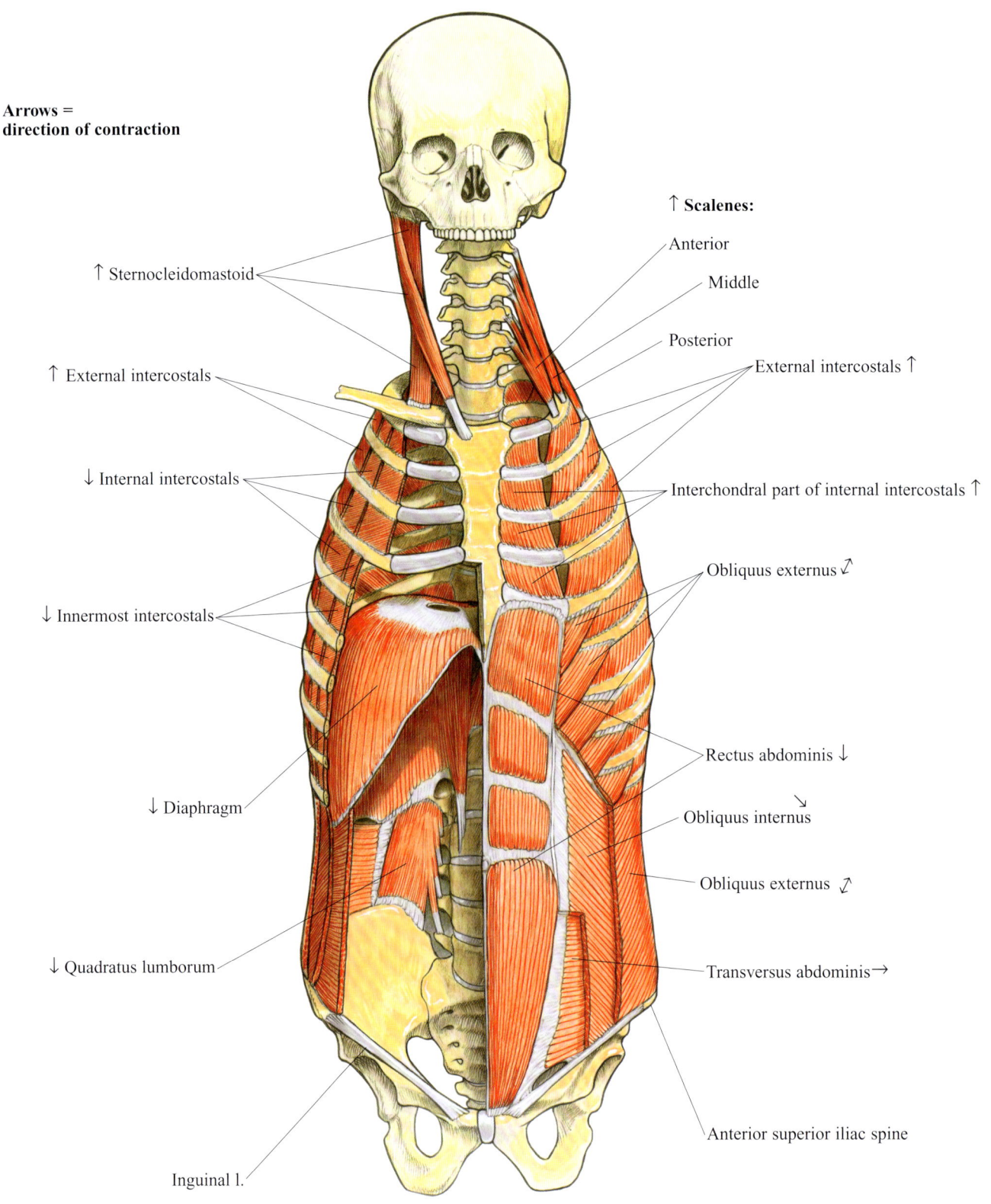

MUSCULAR SYSTEM

COMPONENTS OF THE HAND
DORSAL VIEW

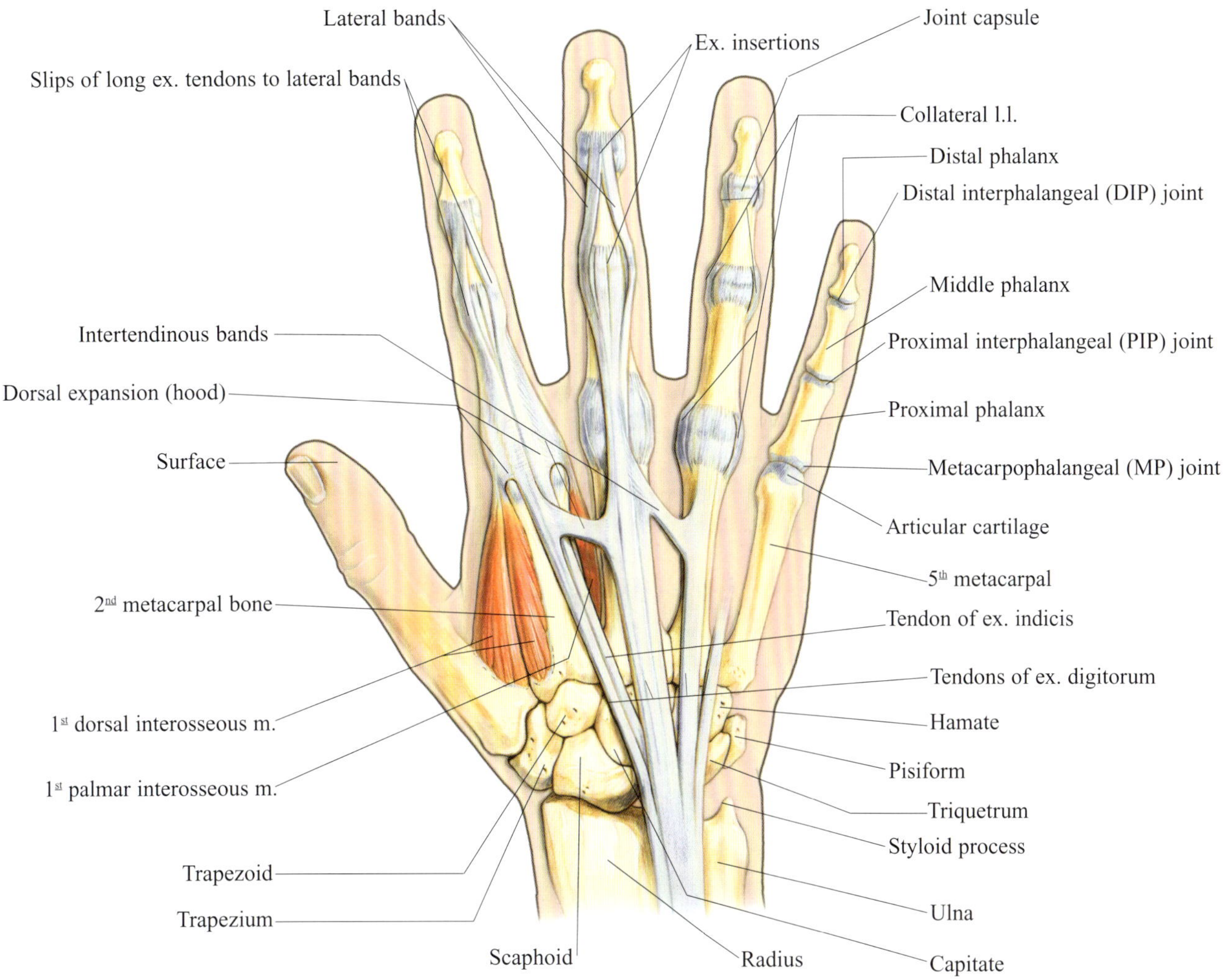

COMPONENTS OF THE FINGER
CROSS SECTION

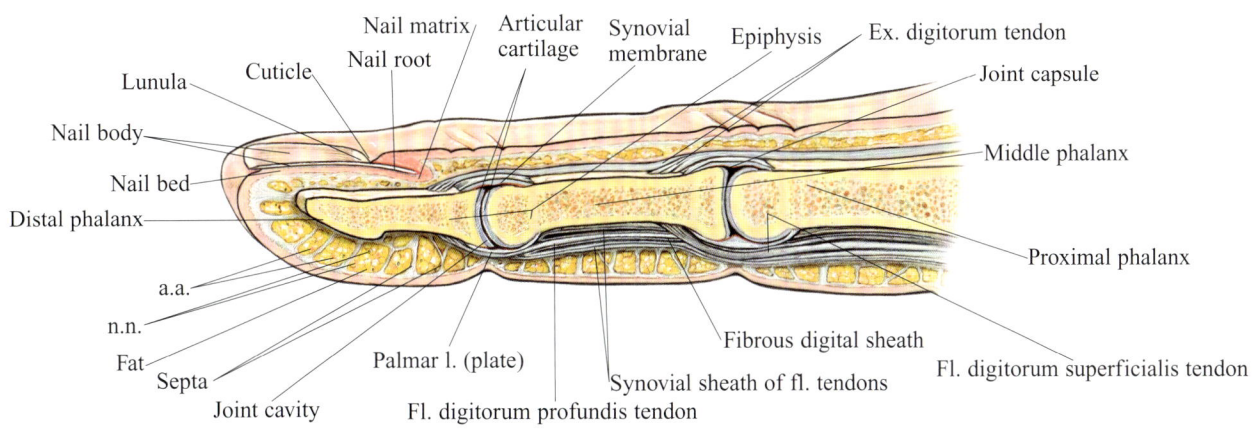

75

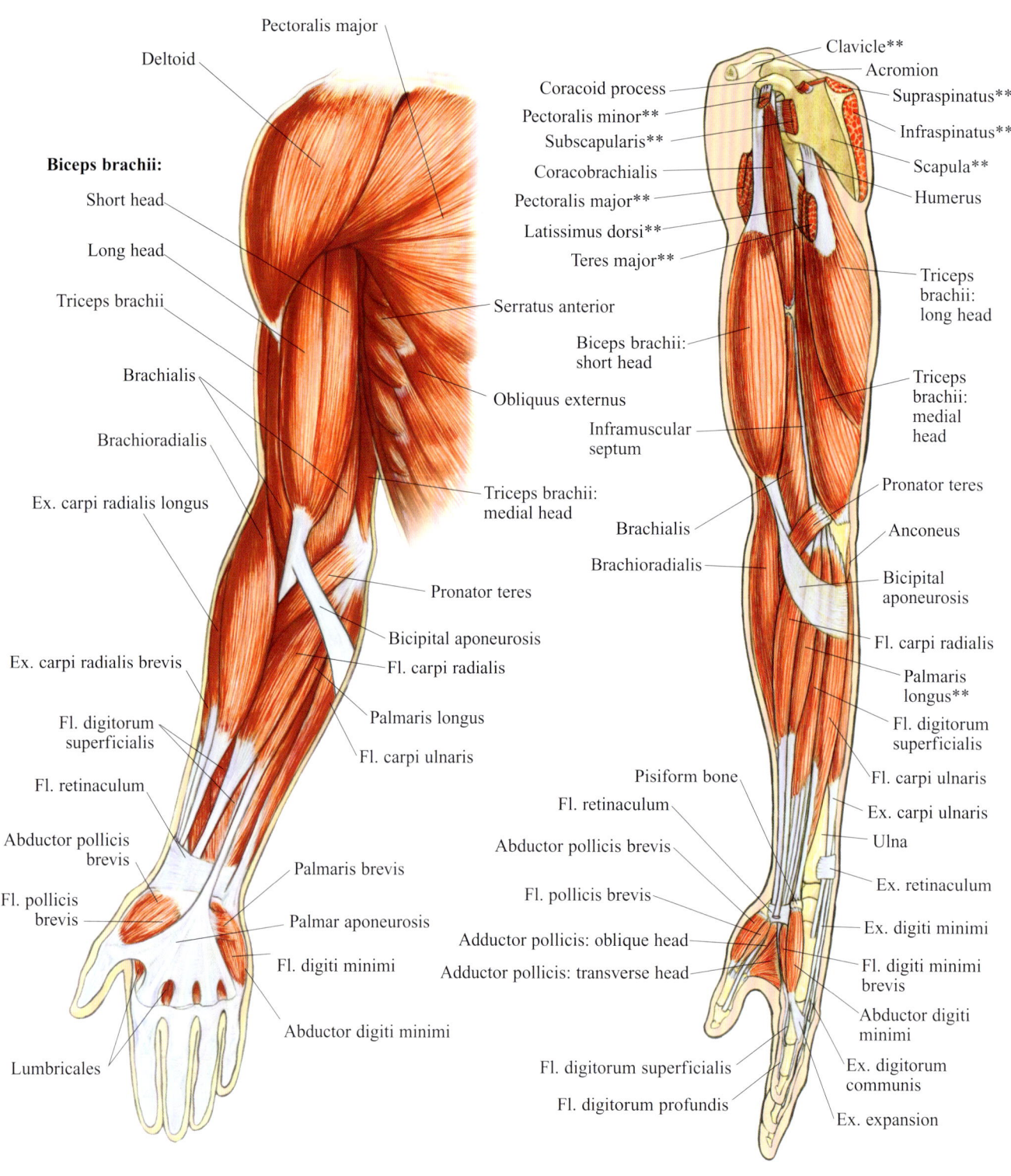

MUSCULAR SYSTEM

ARM & HAND MUSCLES

ANTERIOR VIEW

MEDIAL VIEW

MUSCULAR SYSTEM

ARM & HAND MUSCLES

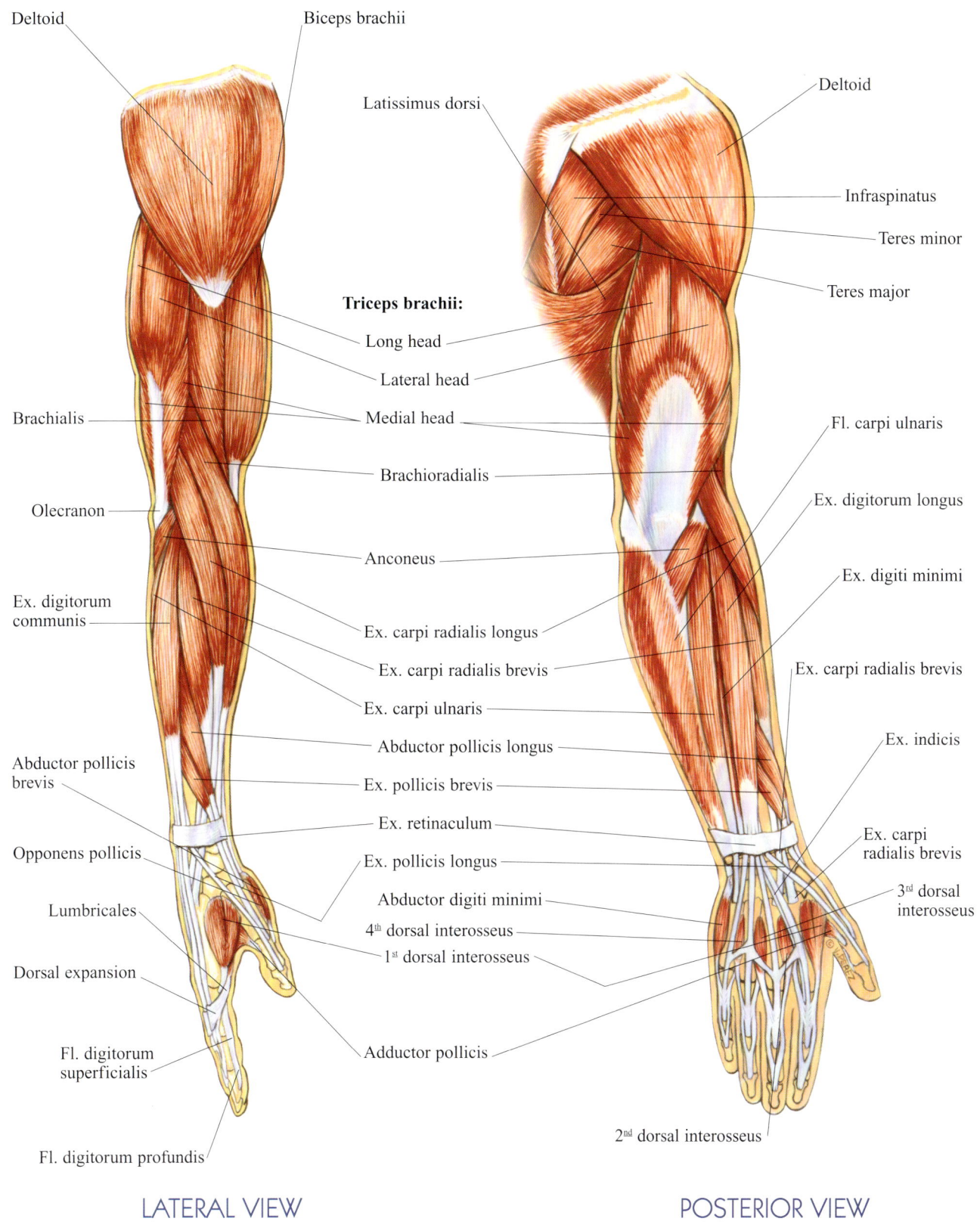

LATERAL VIEW   POSTERIOR VIEW

77

MUSCULAR SYSTEM

PALMAR HAND

LAYER I

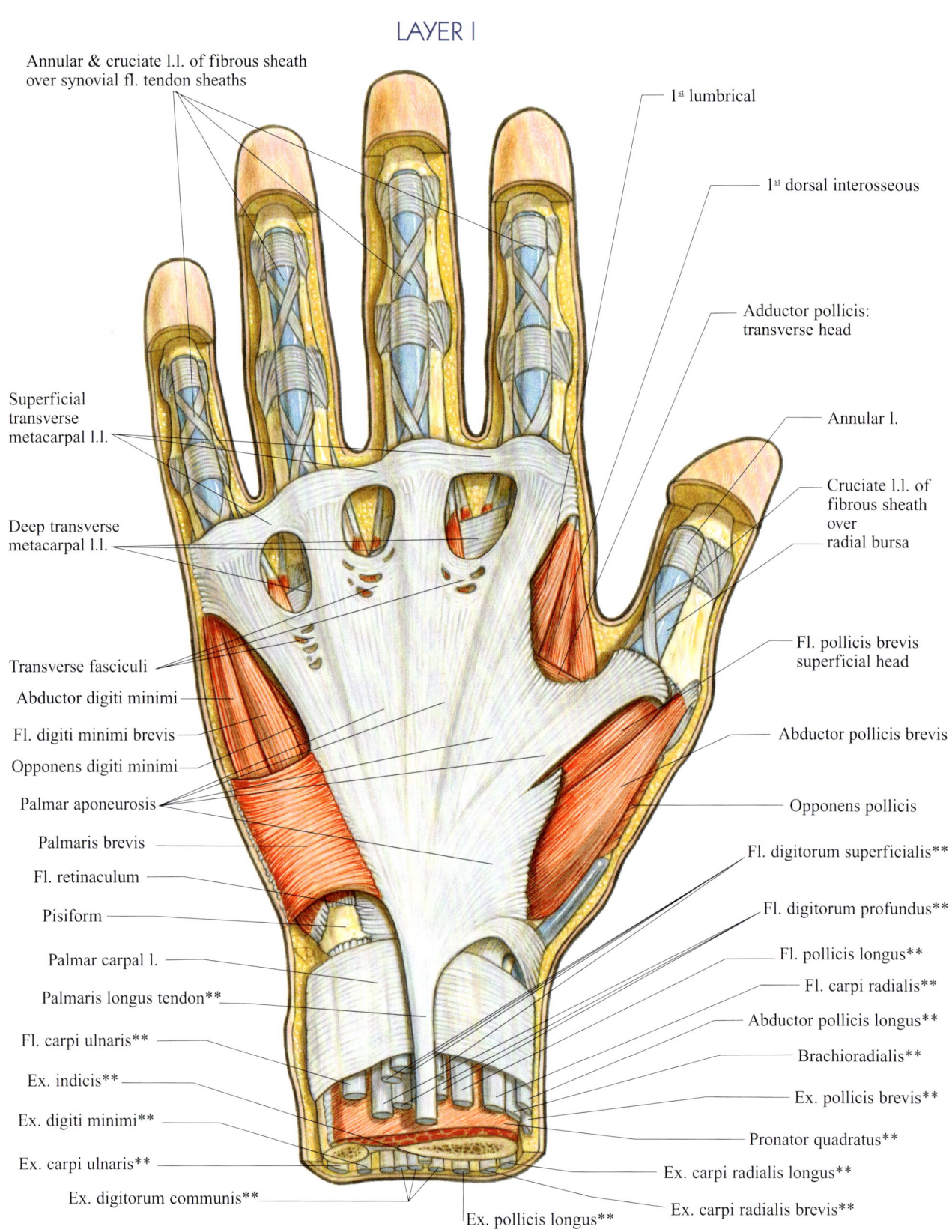

MUSCULAR SYSTEM

PALMAR HAND
LAYER II

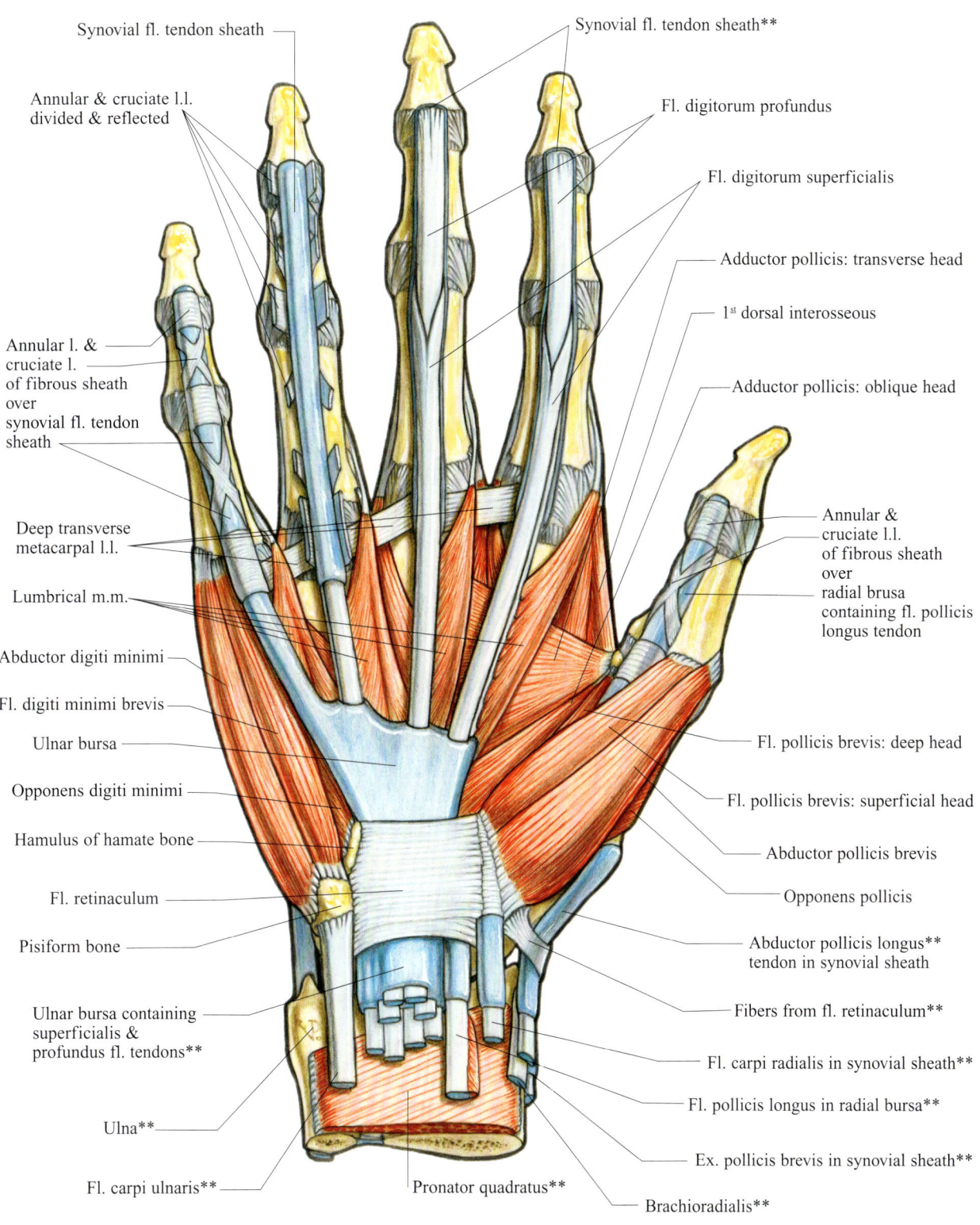

79

MUSCULAR SYSTEM

PALMAR HAND
LAYER III

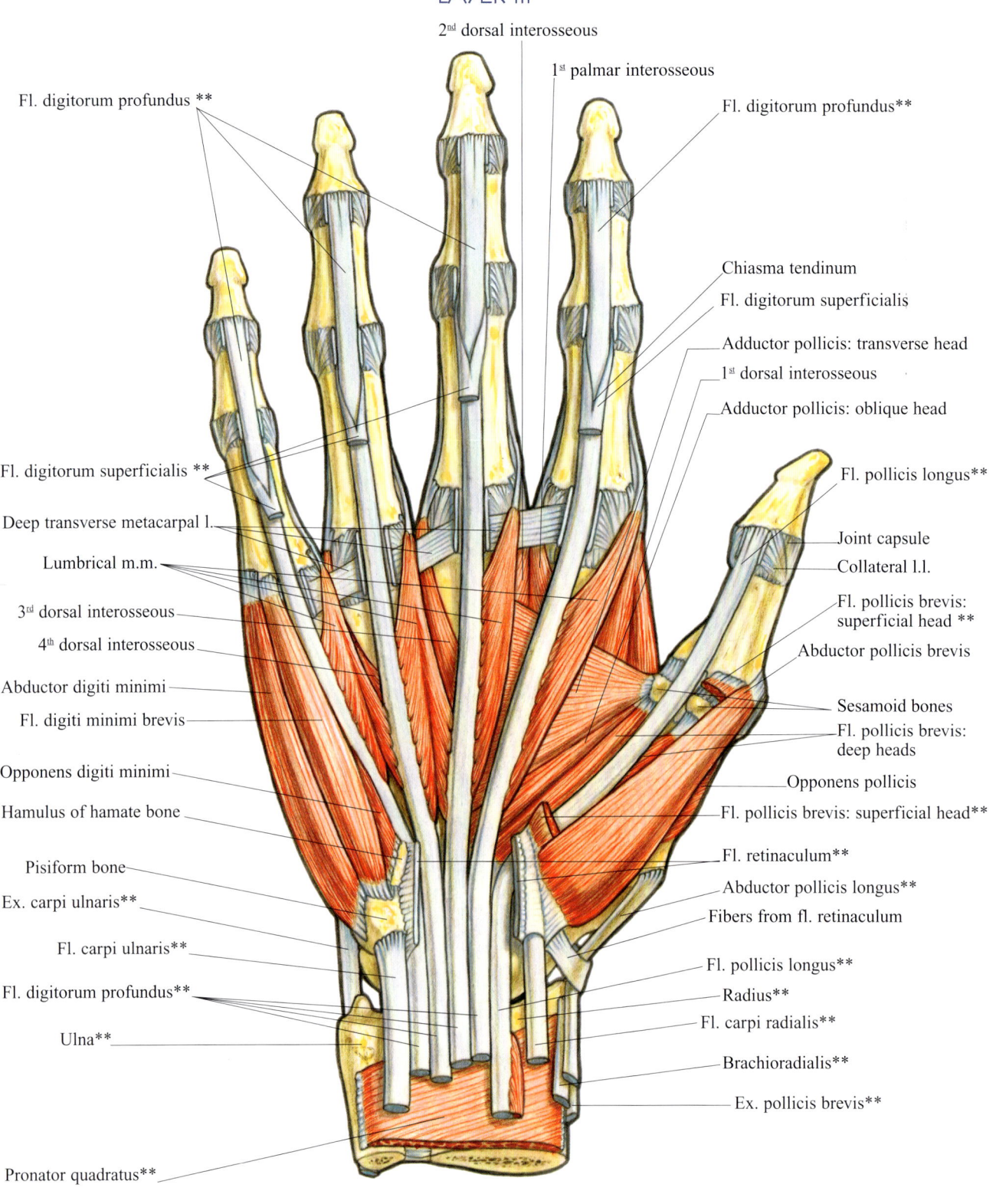

MUSCULAR SYSTEM

PALMAR HAND
LAYER IV

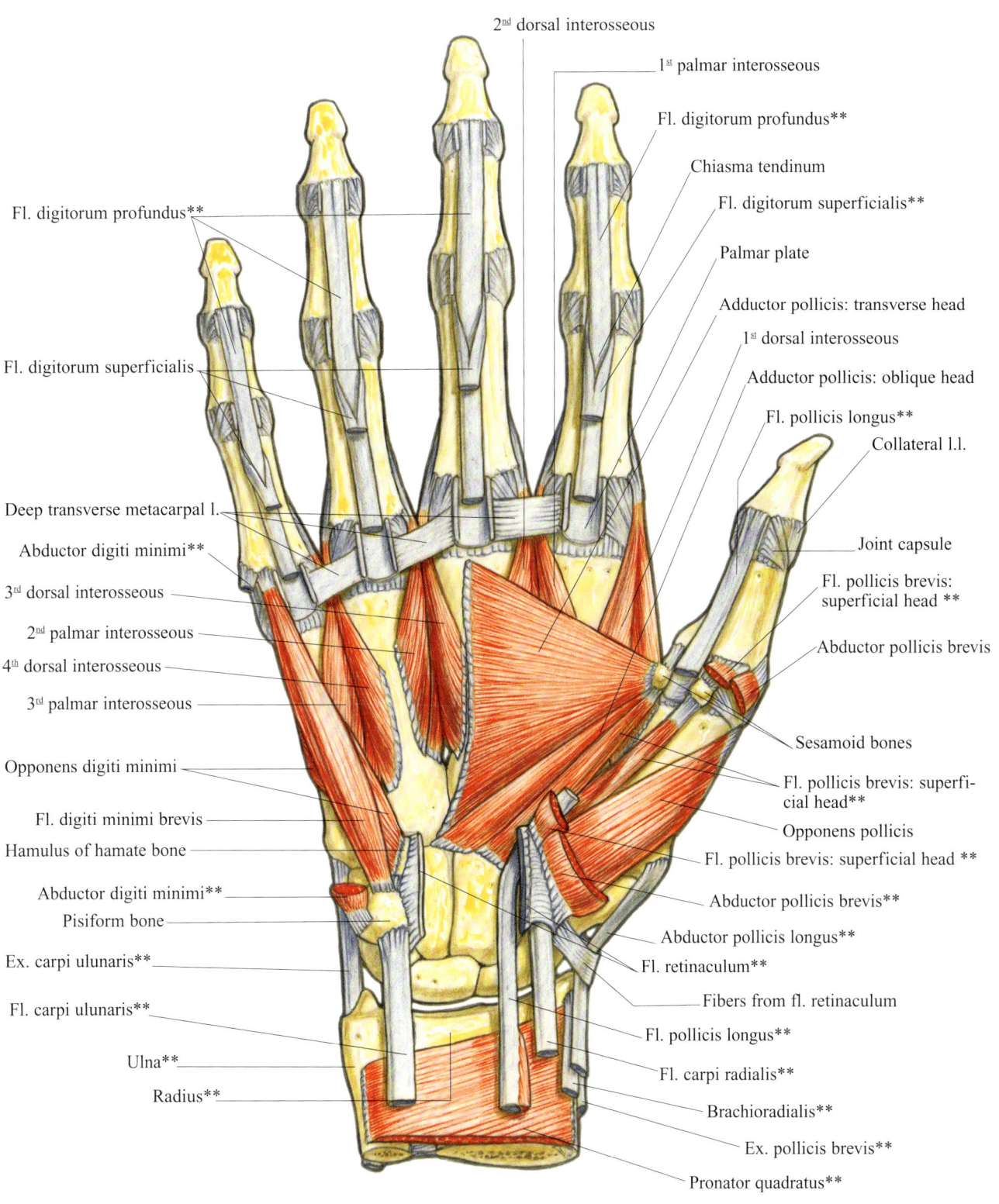

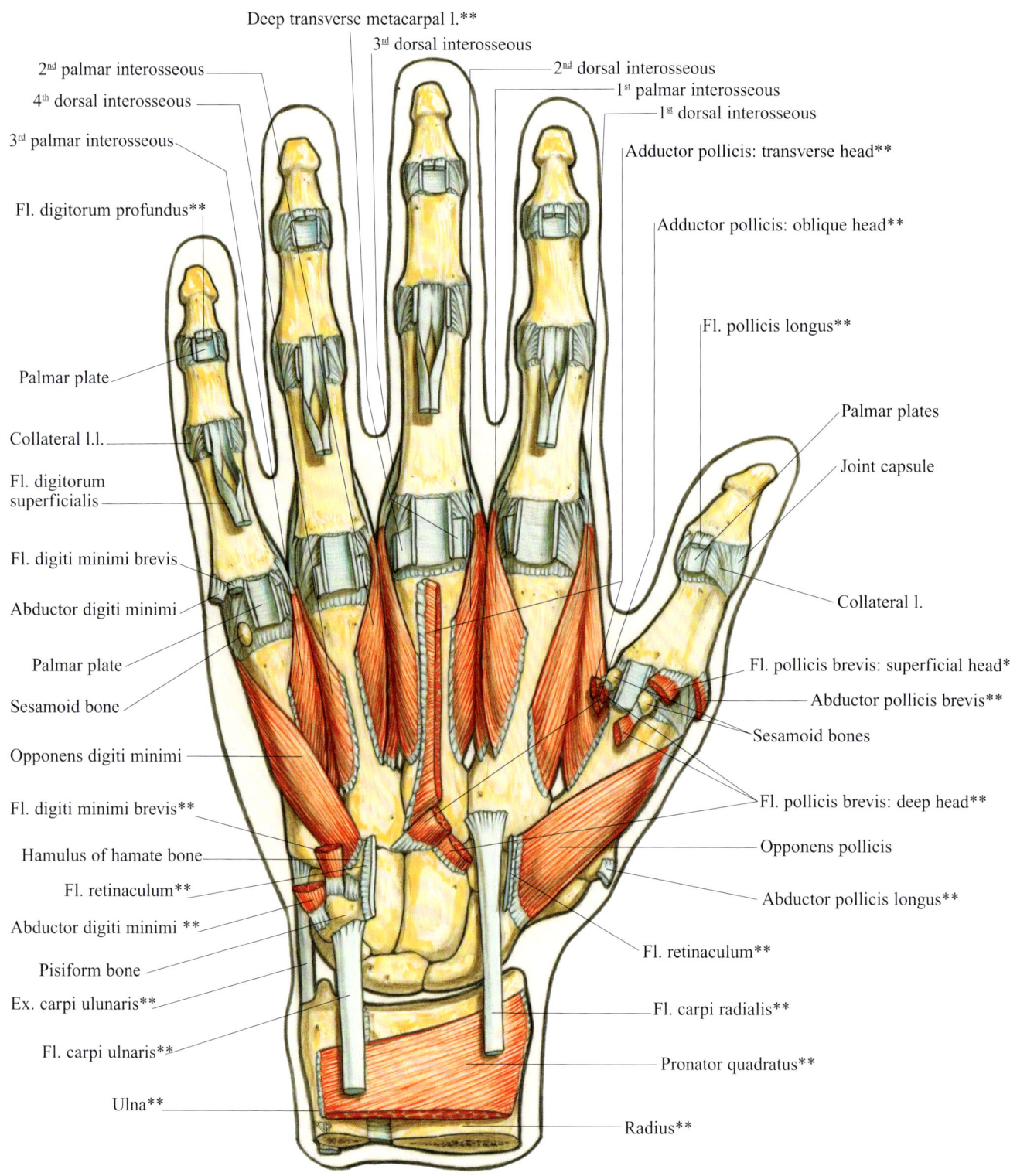

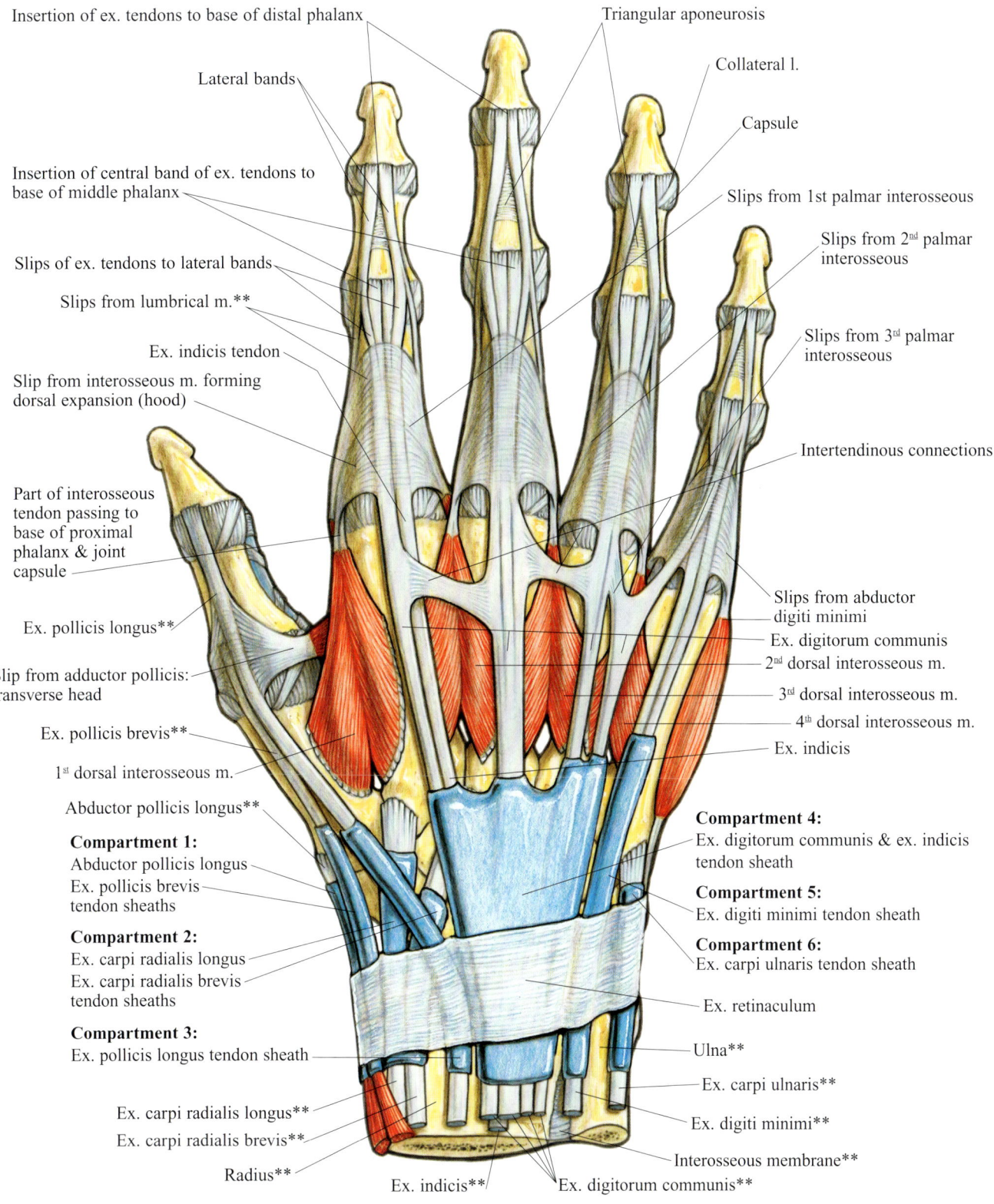

MUSCULAR SYSTEM

DORSAL HAND
LAYER II

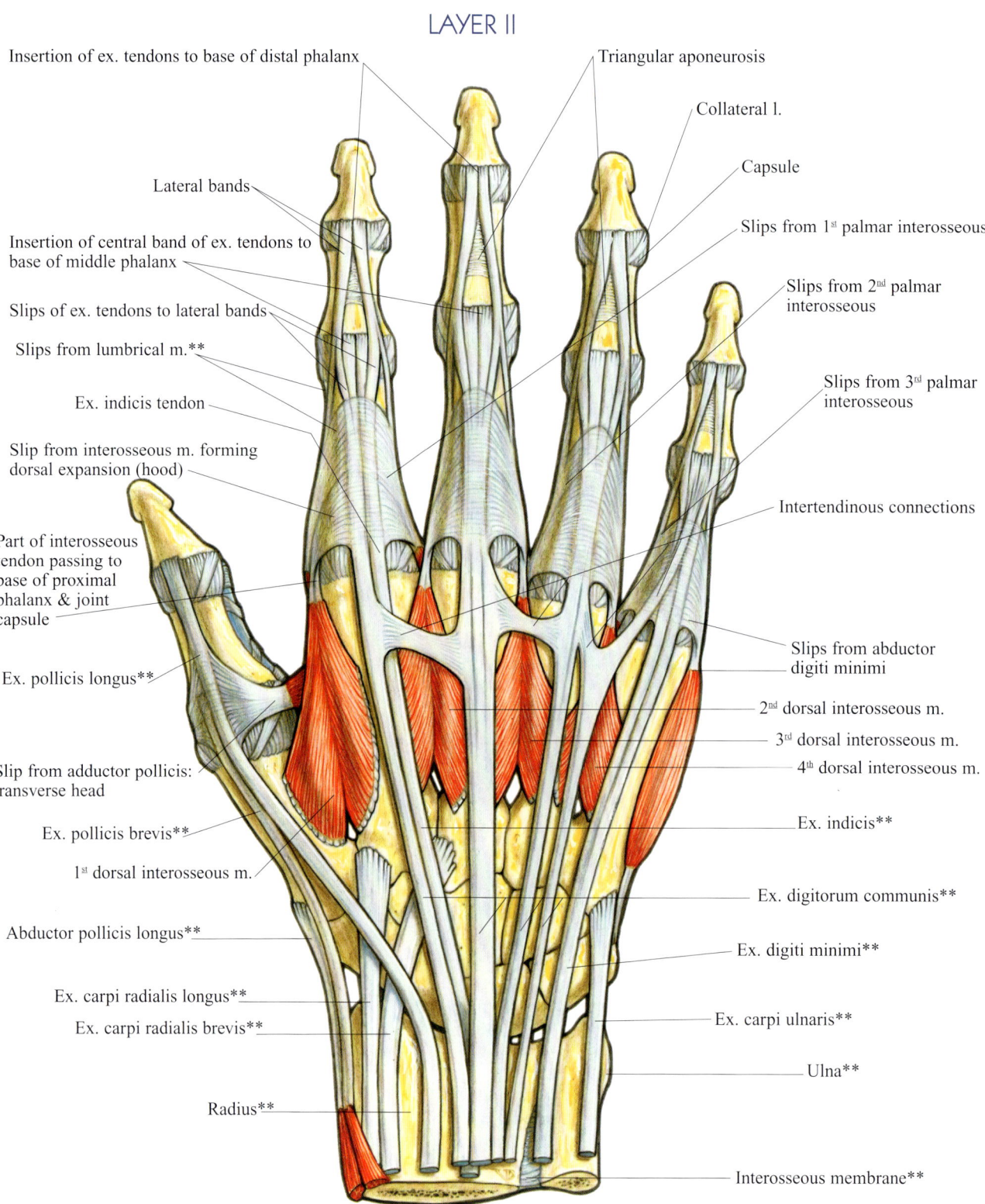

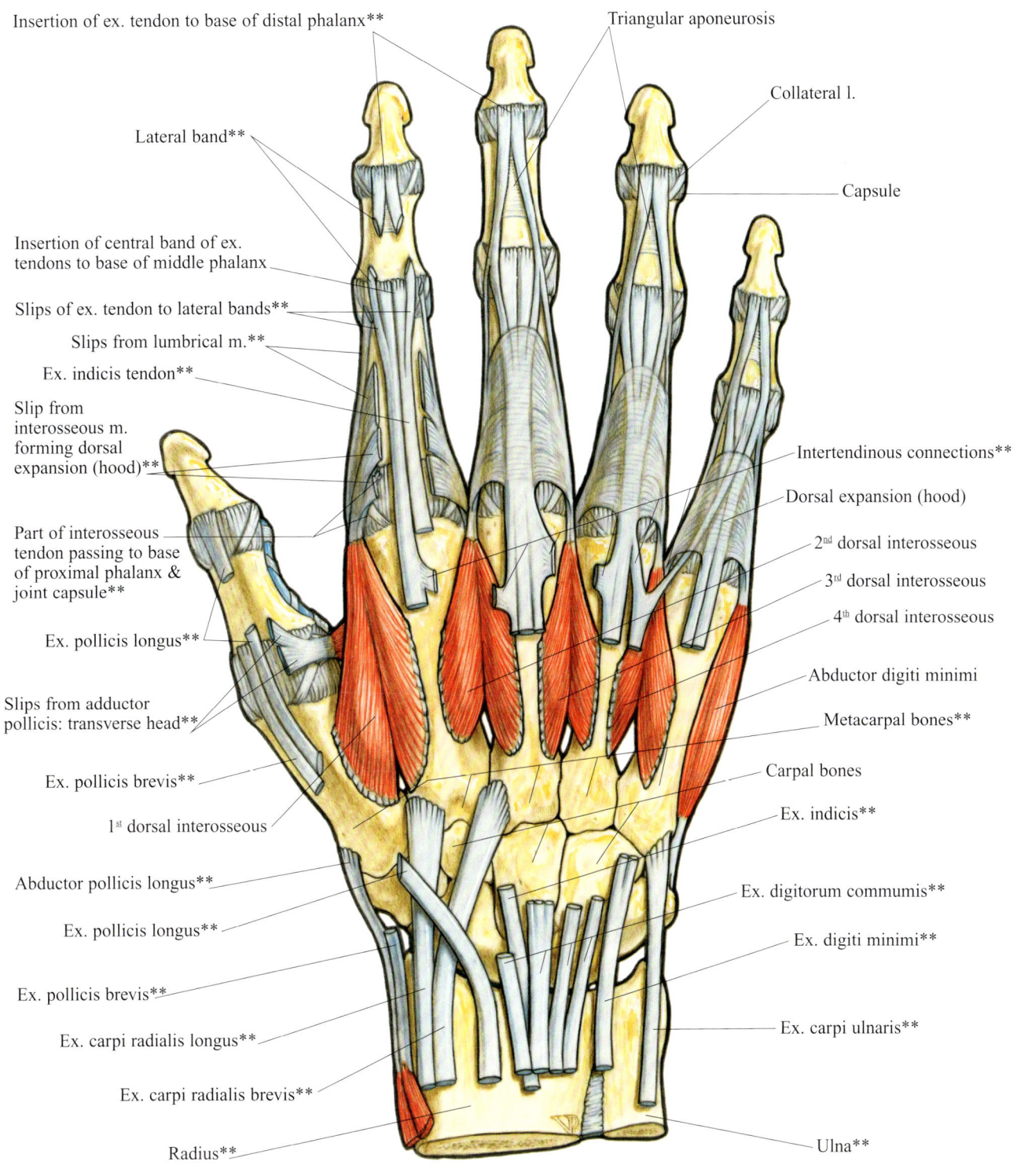

MUSCULAR SYSTEM

MEDIAL HAND

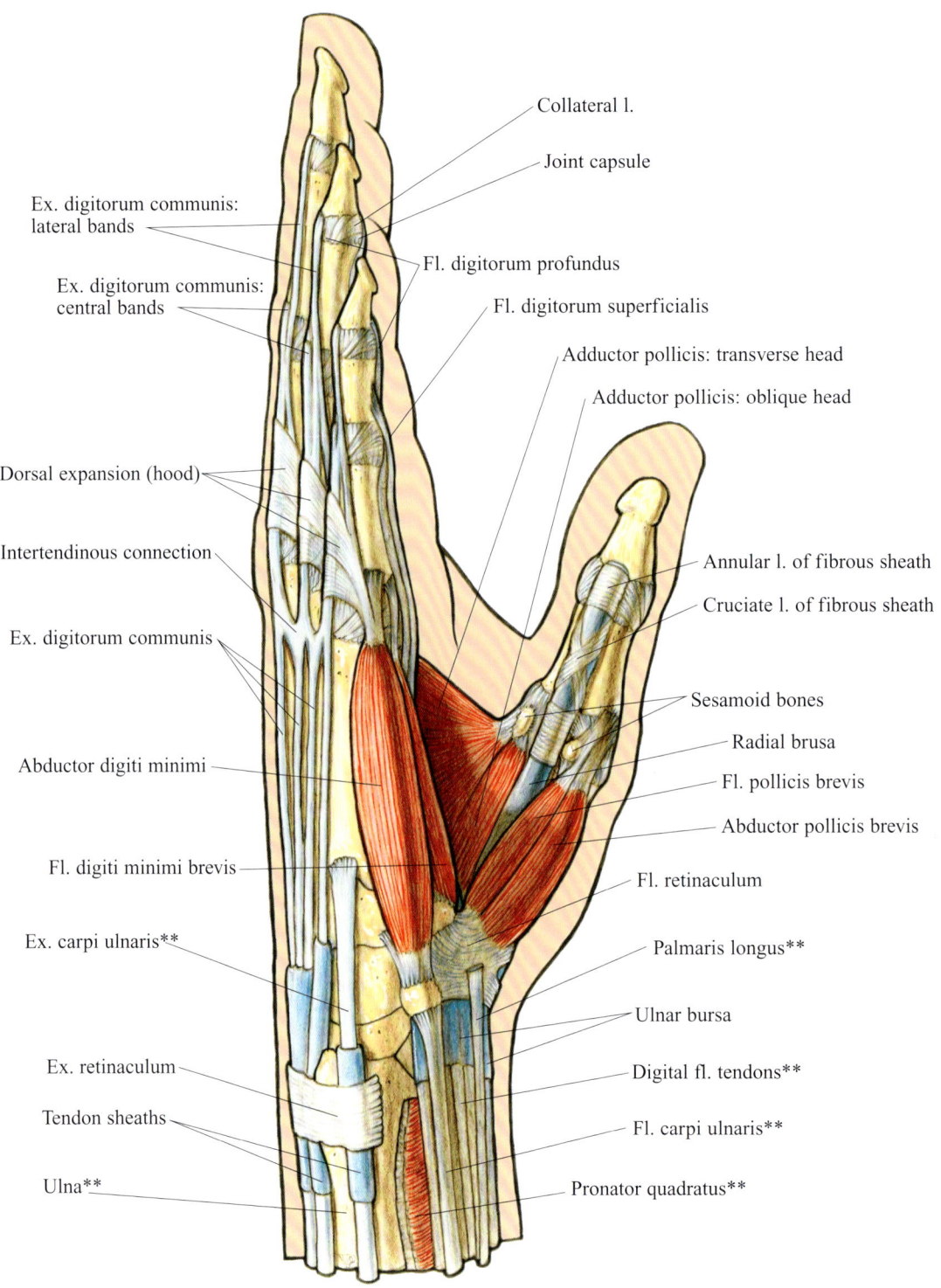

MUSCULAR SYSTEM

LATERAL HAND

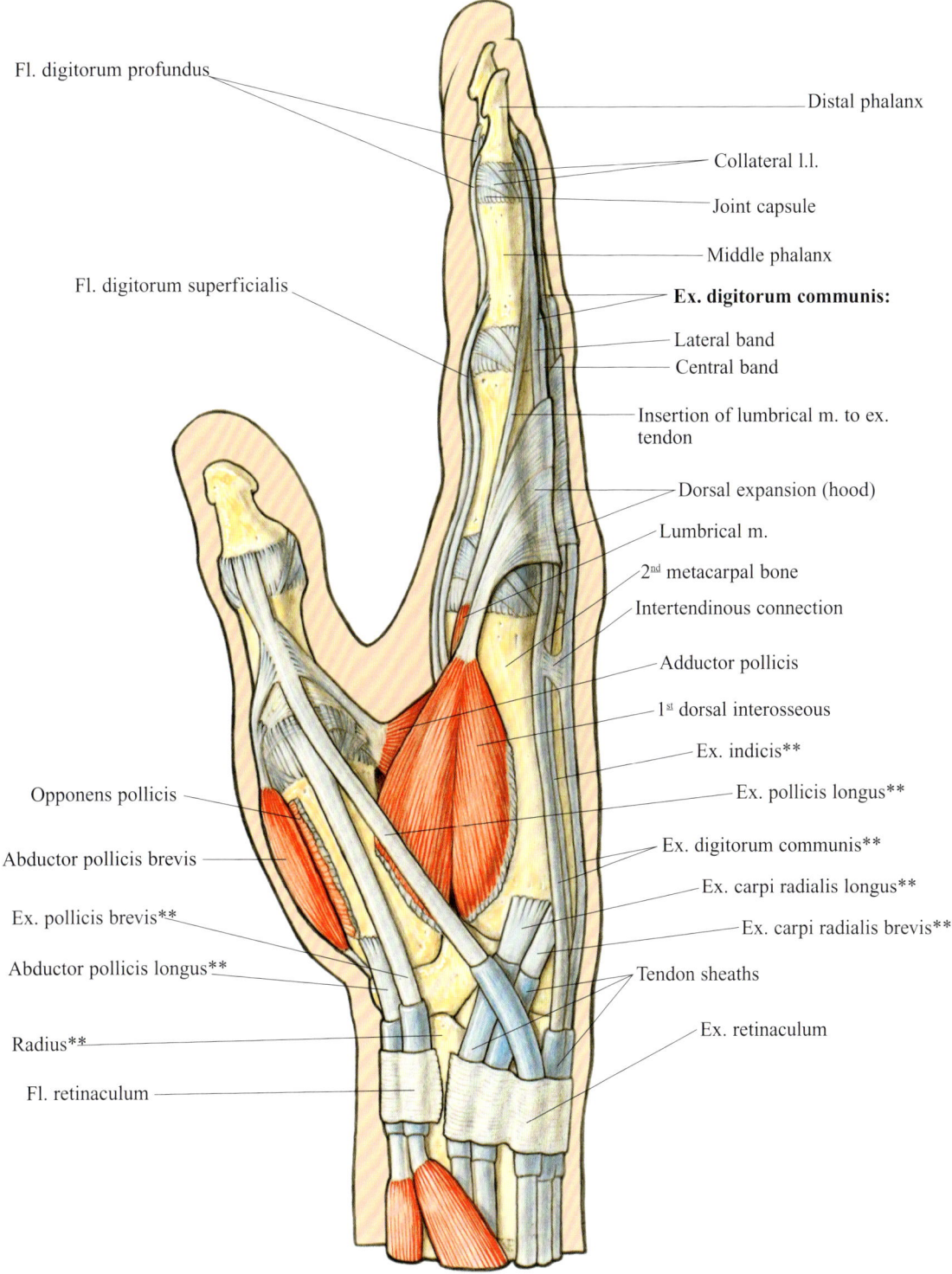

- Fl. digitorum profundus
- Fl. digitorum superficialis
- Opponens pollicis
- Abductor pollicis brevis
- Ex. pollicis brevis**
- Abductor pollicis longus**
- Radius**
- Fl. retinaculum
- Distal phalanx
- Collateral l.l.
- Joint capsule
- Middle phalanx
- **Ex. digitorum communis:**
  - Lateral band
  - Central band
- Insertion of lumbrical m. to ex. tendon
- Dorsal expansion (hood)
- Lumbrical m.
- 2nd metacarpal bone
- Intertendinous connection
- Adductor pollicis
- 1st dorsal interosseous
- Ex. indicis**
- Ex. pollicis longus**
- Ex. digitorum communis**
- Ex. carpi radialis longus**
- Ex. carpi radialis brevis**
- Tendon sheaths
- Ex. retinaculum

87

MUSCULAR SYSTEM

LEG & FOOT SURFACE MUSCLES

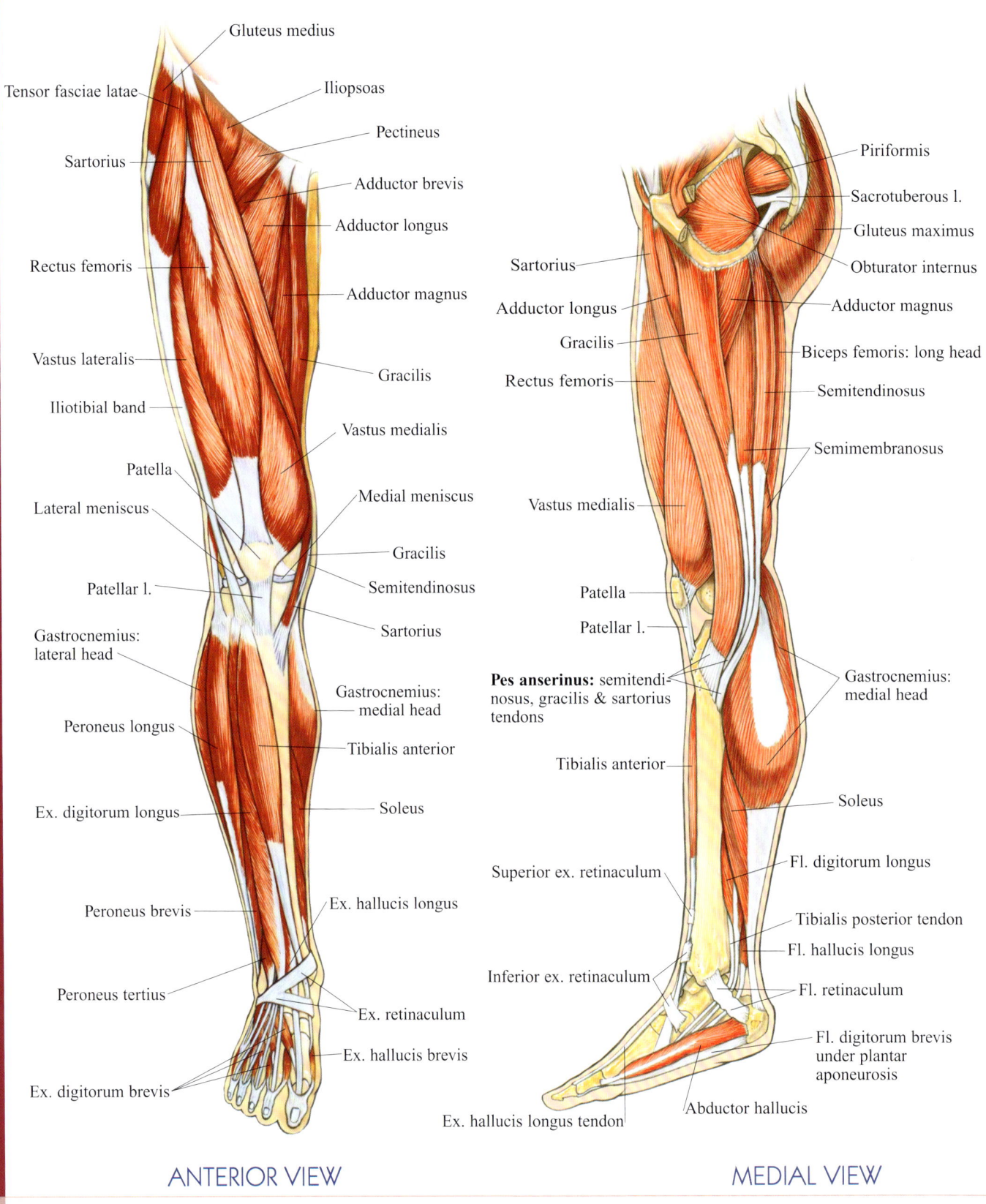

ANTERIOR VIEW

MEDIAL VIEW

88

MUSCULAR SYSTEM

LEG & FOOT SURFACE MUSCLES

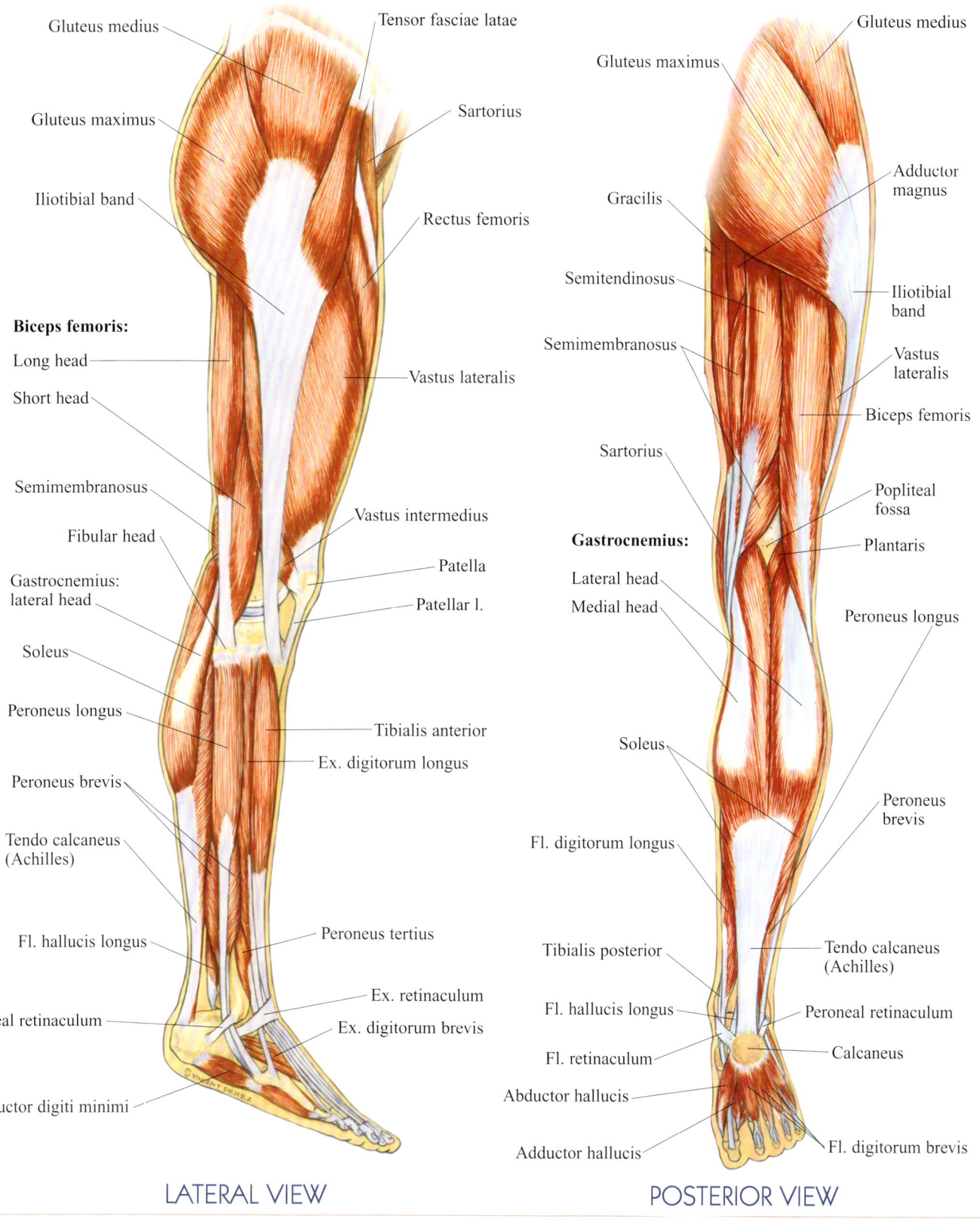

LATERAL VIEW

POSTERIOR VIEW

89

MUSCULAR SYSTEM

DORSAL FOOT
LAYER I

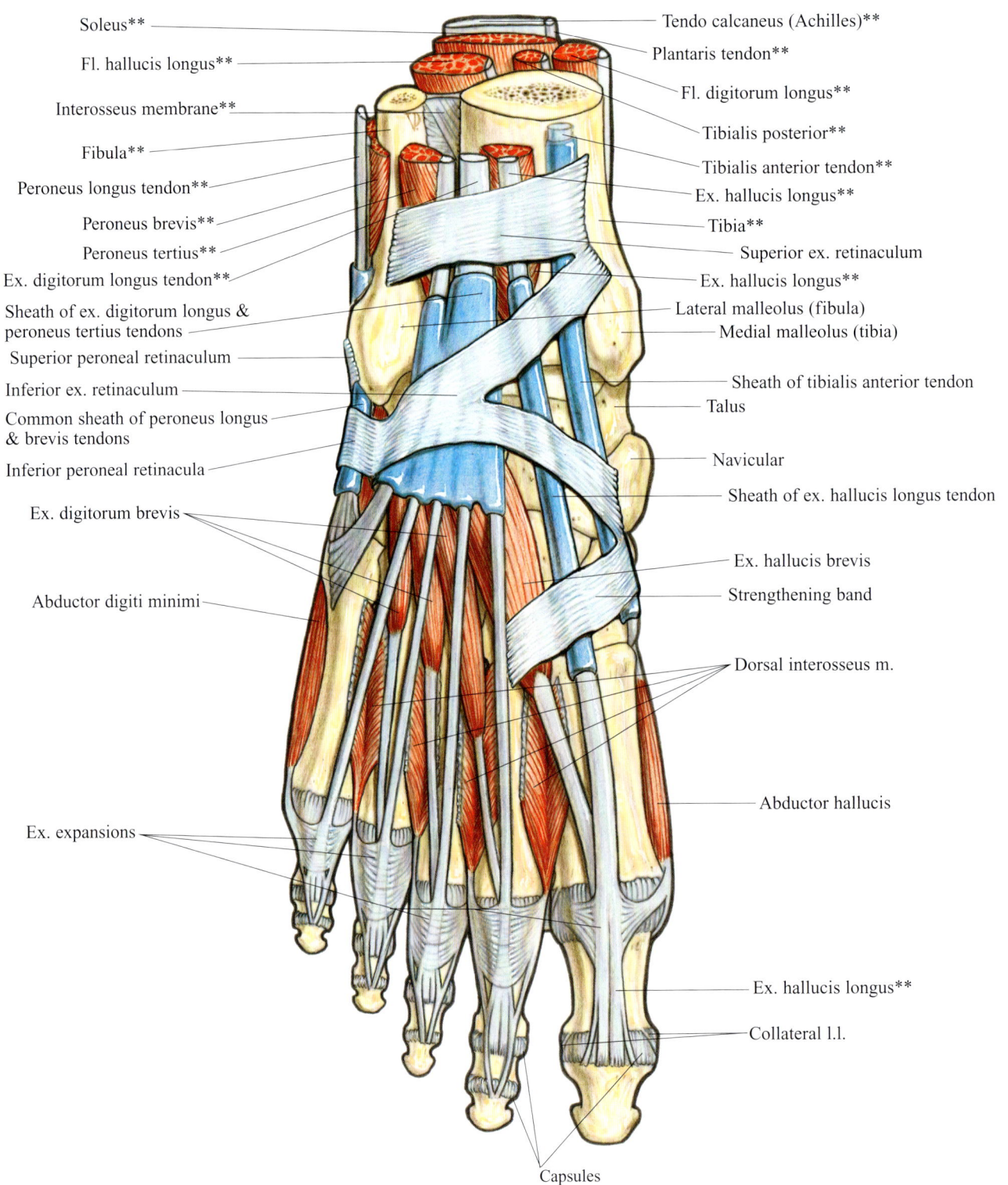

MUSCULAR SYSTEM

DORSAL FOOT
LAYER II

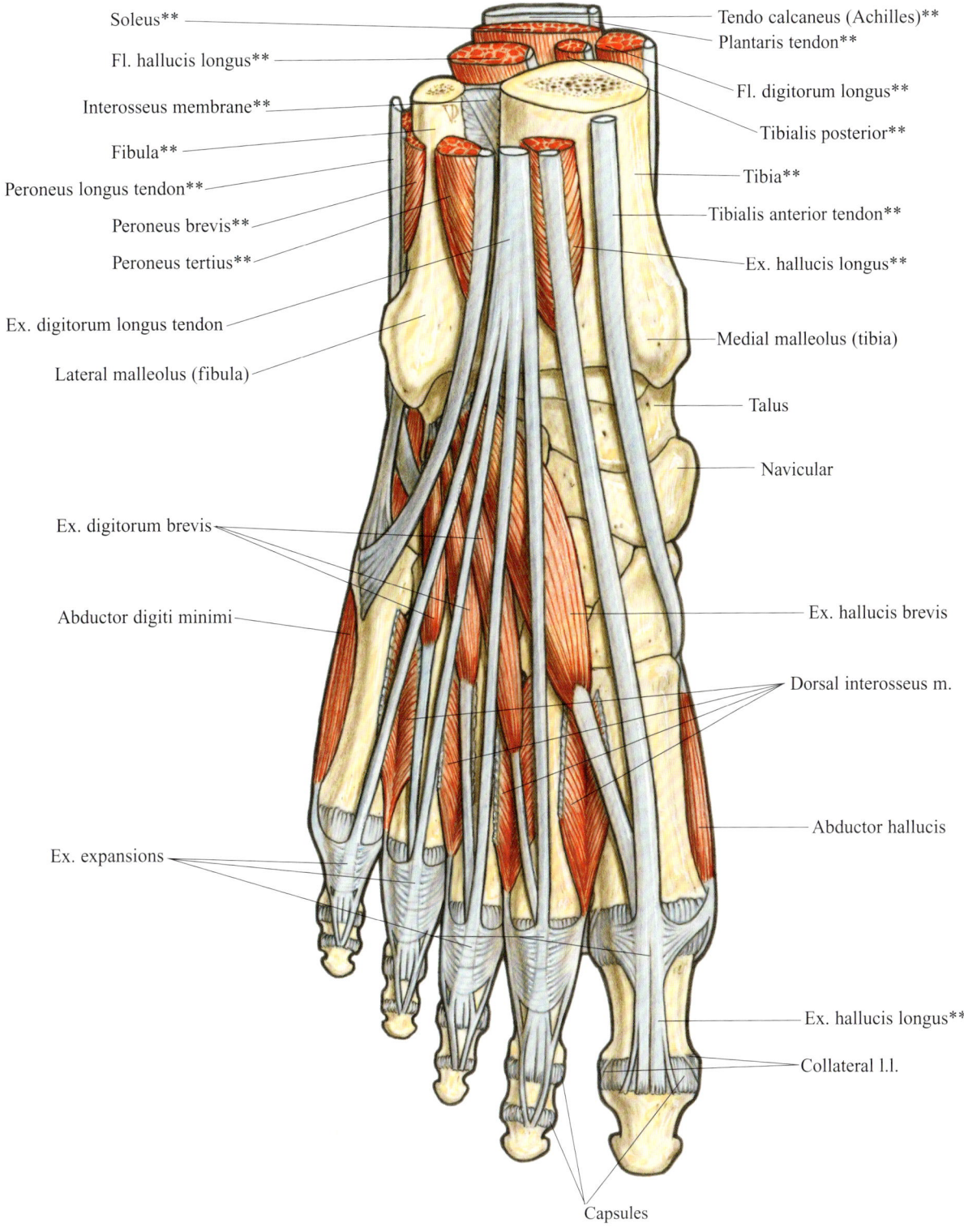

MUSCULAR SYSTEM

DORSAL FOOT
LAYER III

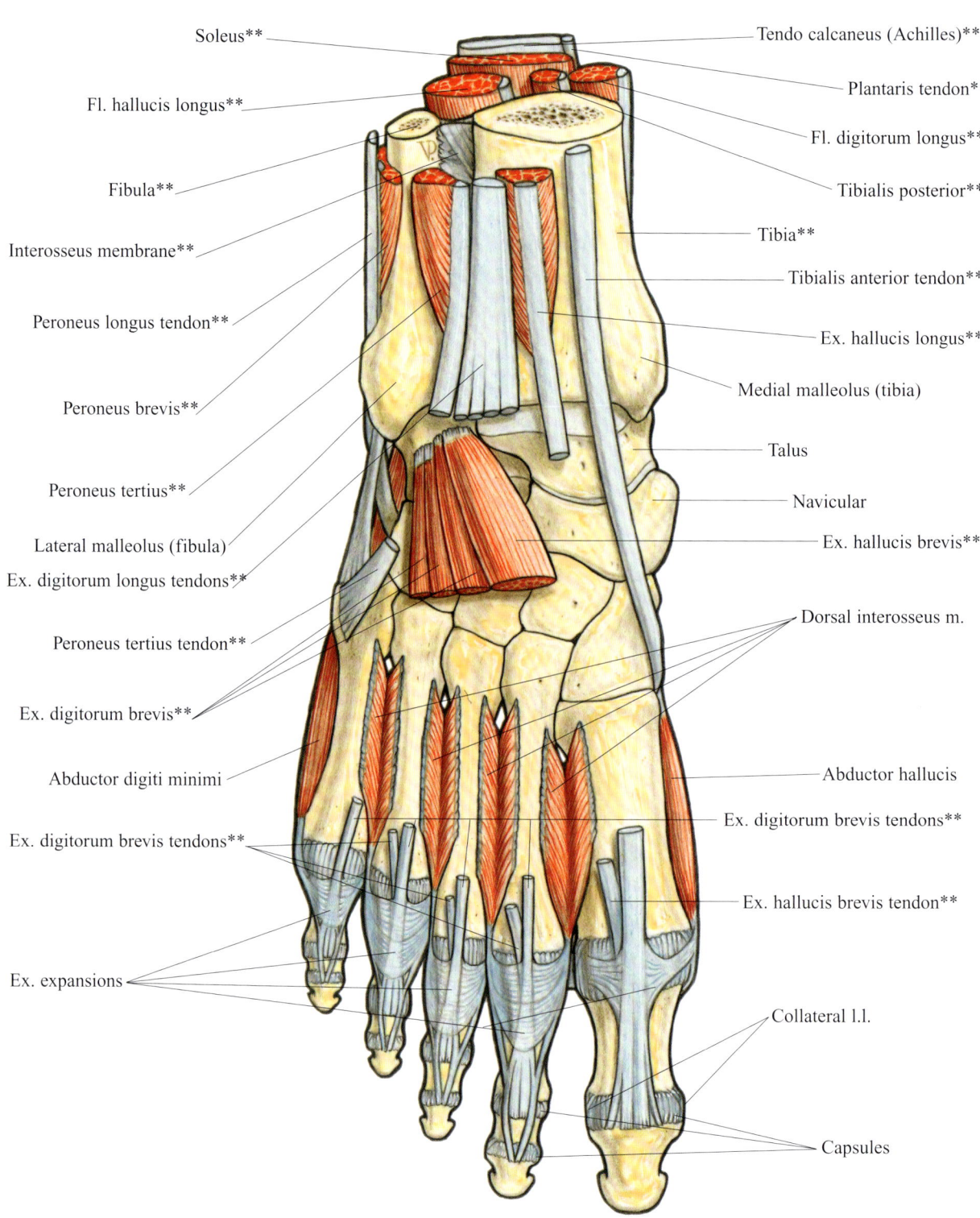

MUSCULAR SYSTEM

PLANTAR FOOT
LAYER I

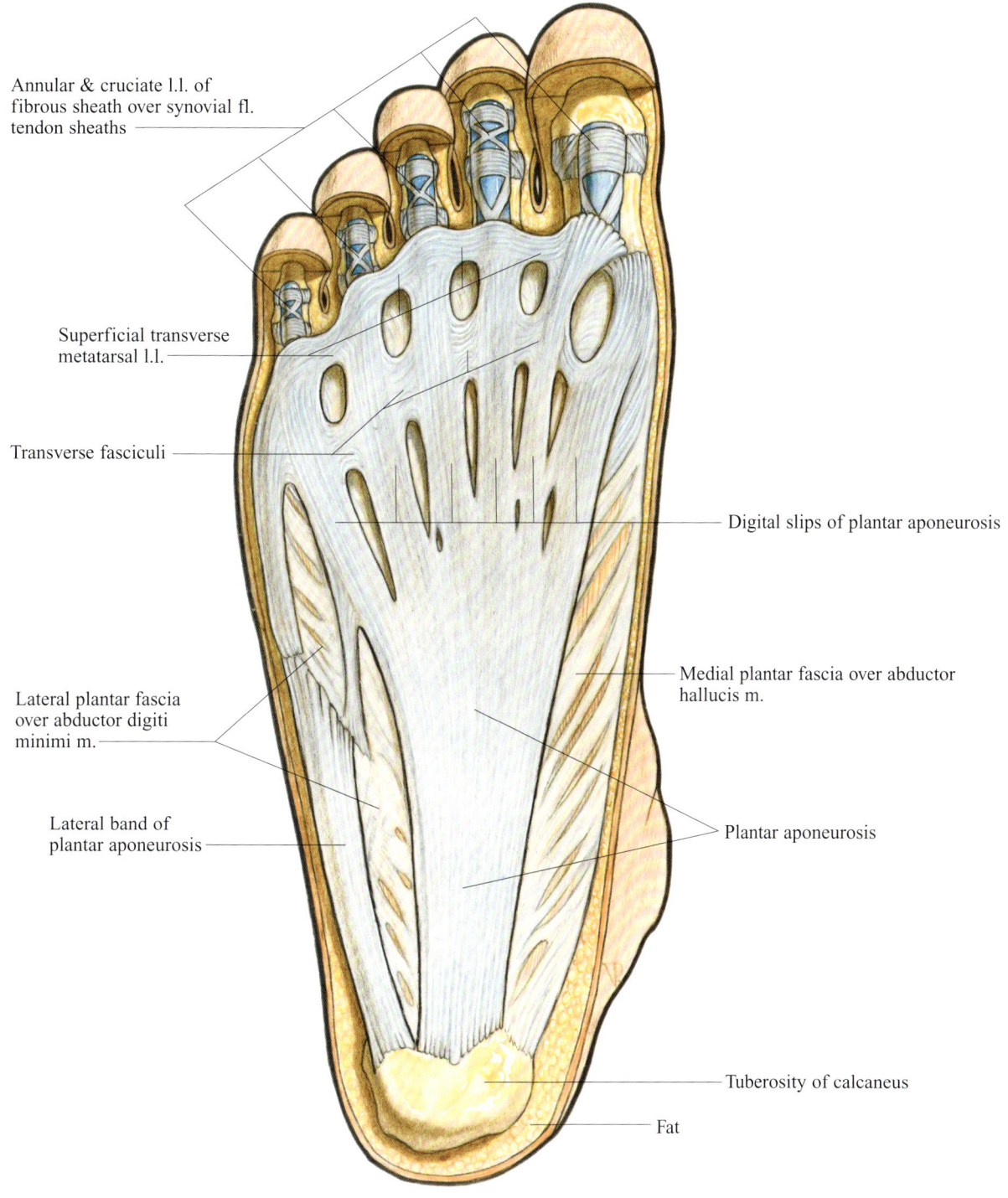

MUSCULAR SYSTEM

PLANTAR FOOT
LAYER II

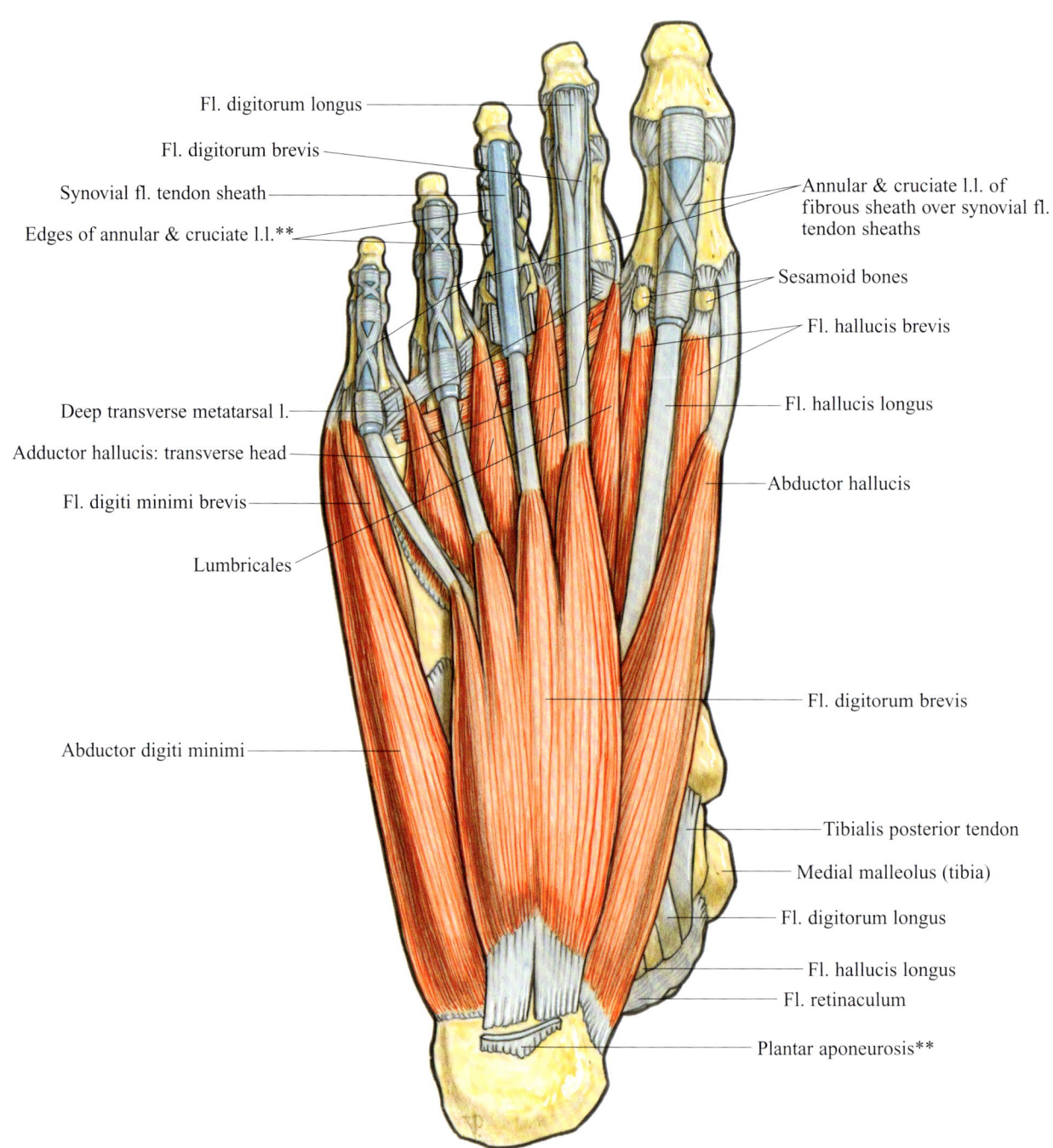

MUSCULAR SYSTEM

PLANTAR FOOT
LAYER III

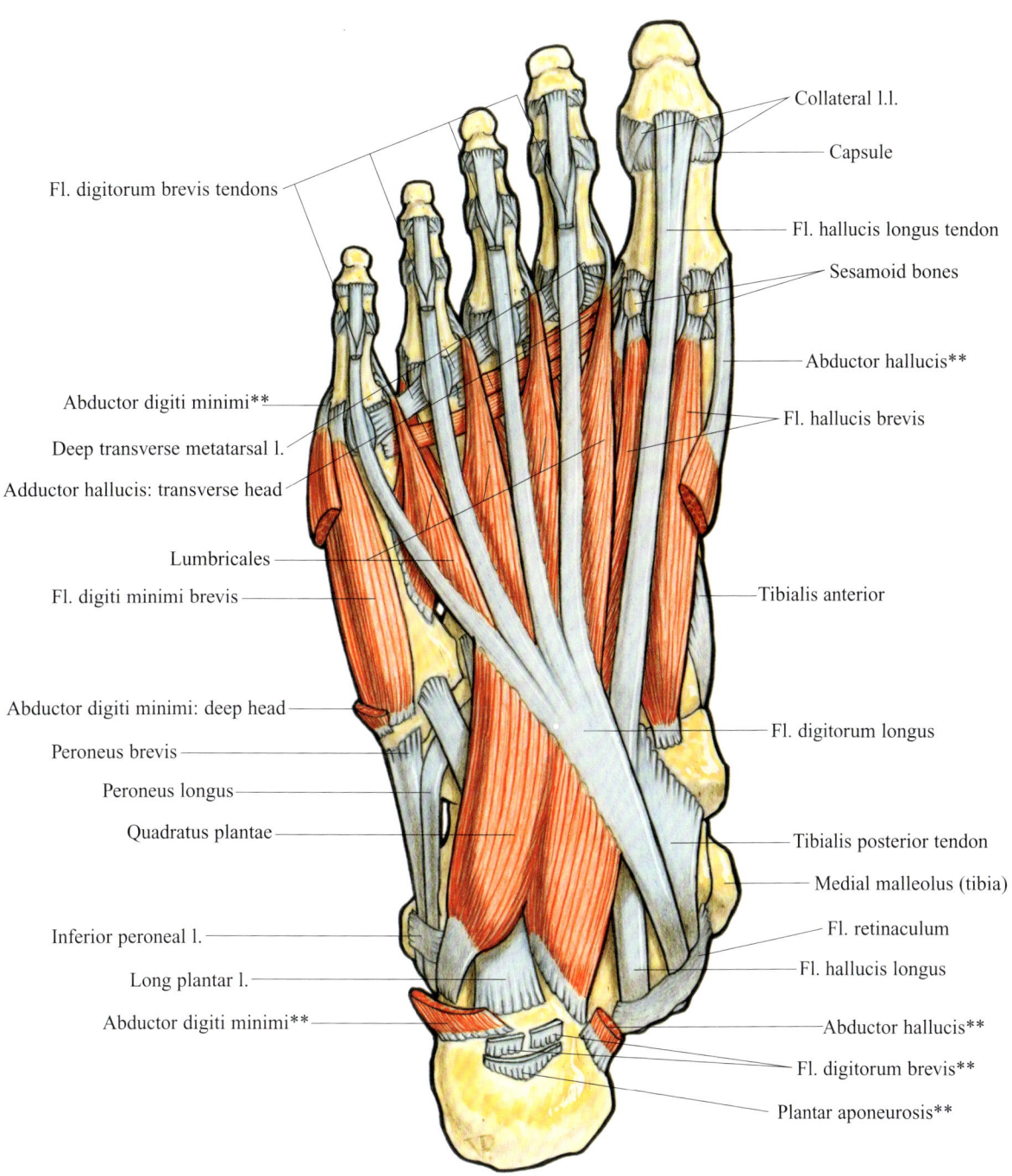

MUSCULAR SYSTEM

PLANTAR FOOT
LAYER IV

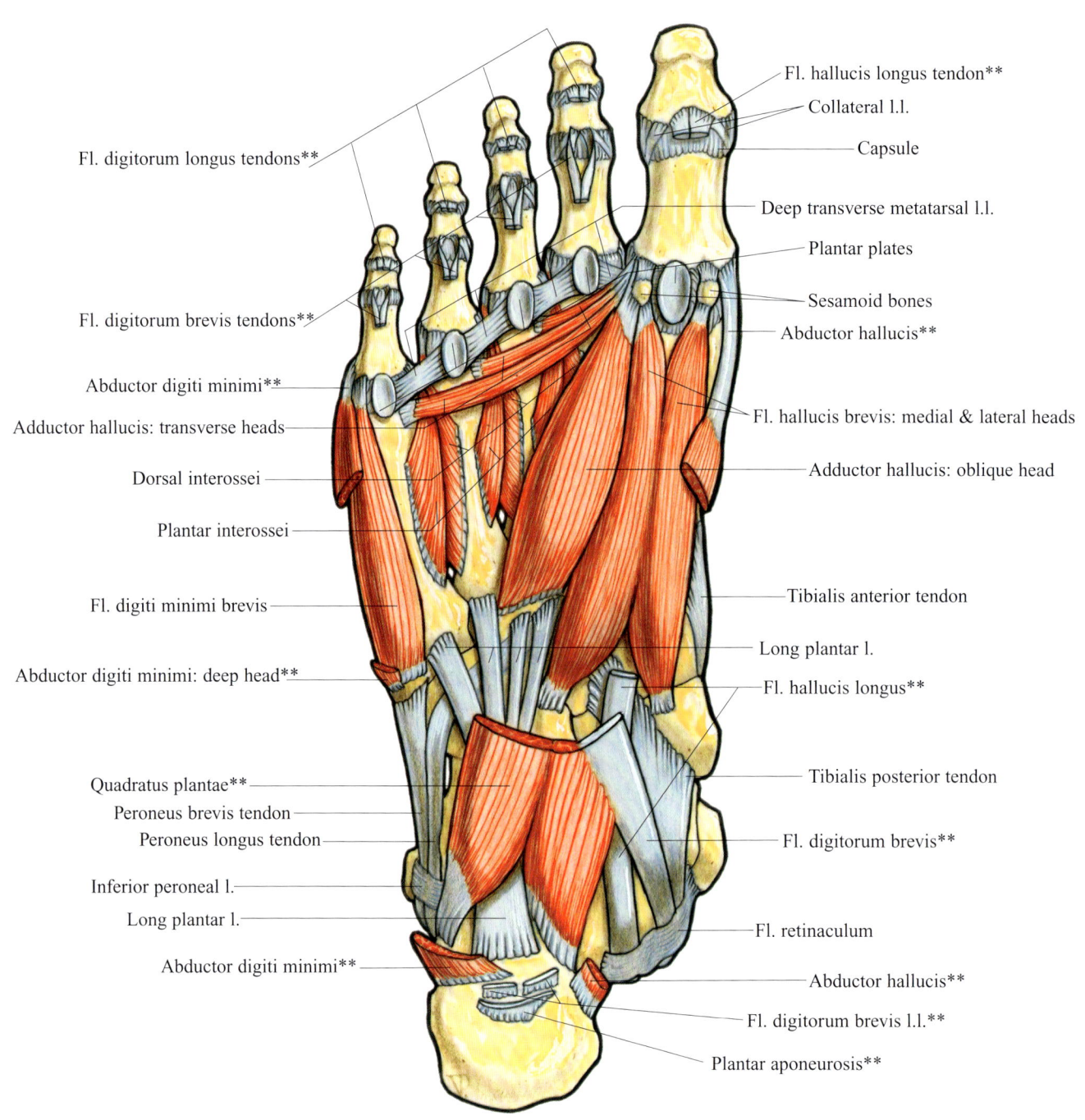

MUSCULAR SYSTEM

PLANTAR FOOT
LAYER V

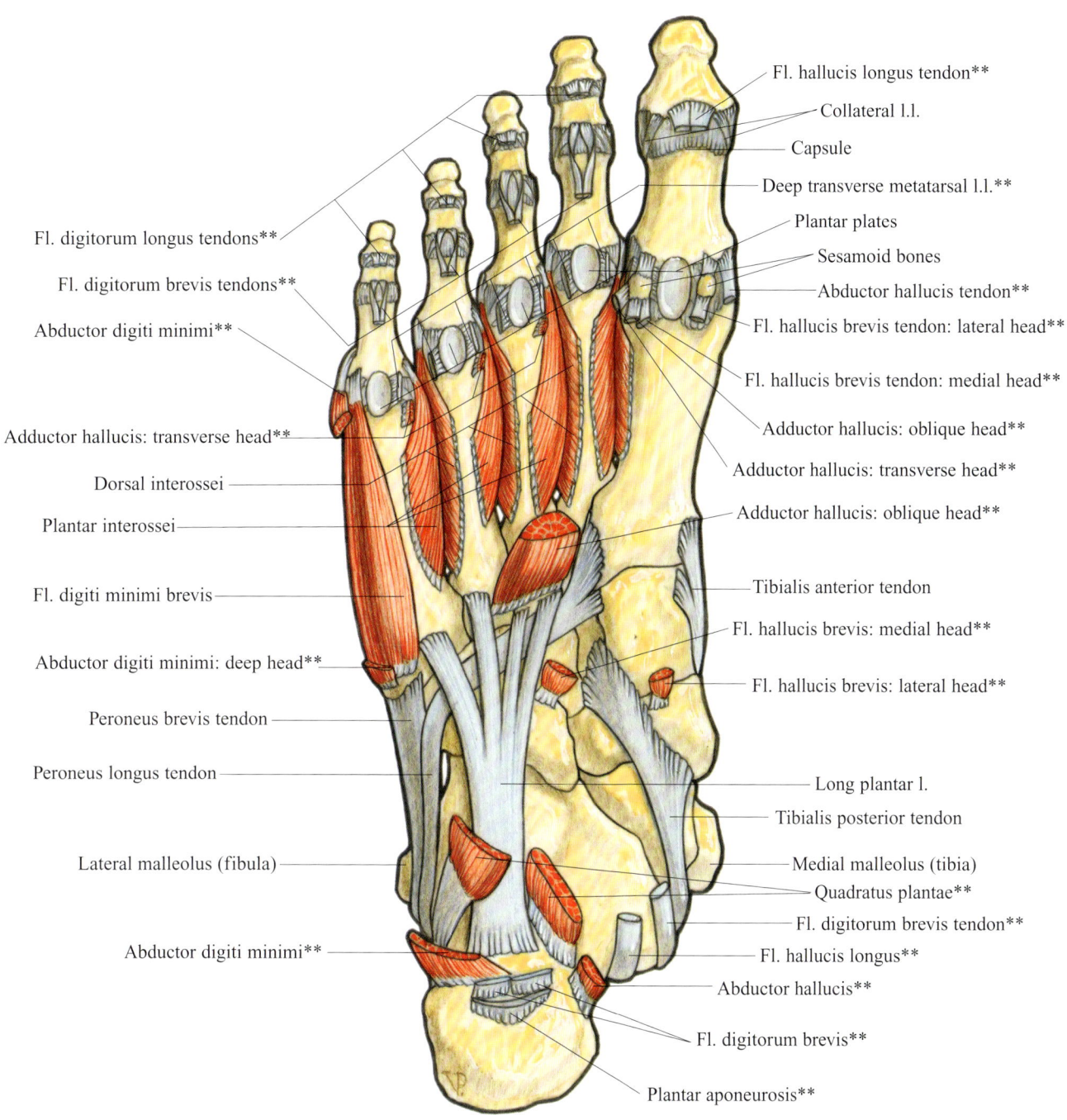

97

MUSCULAR SYSTEM

LATERAL FOOT

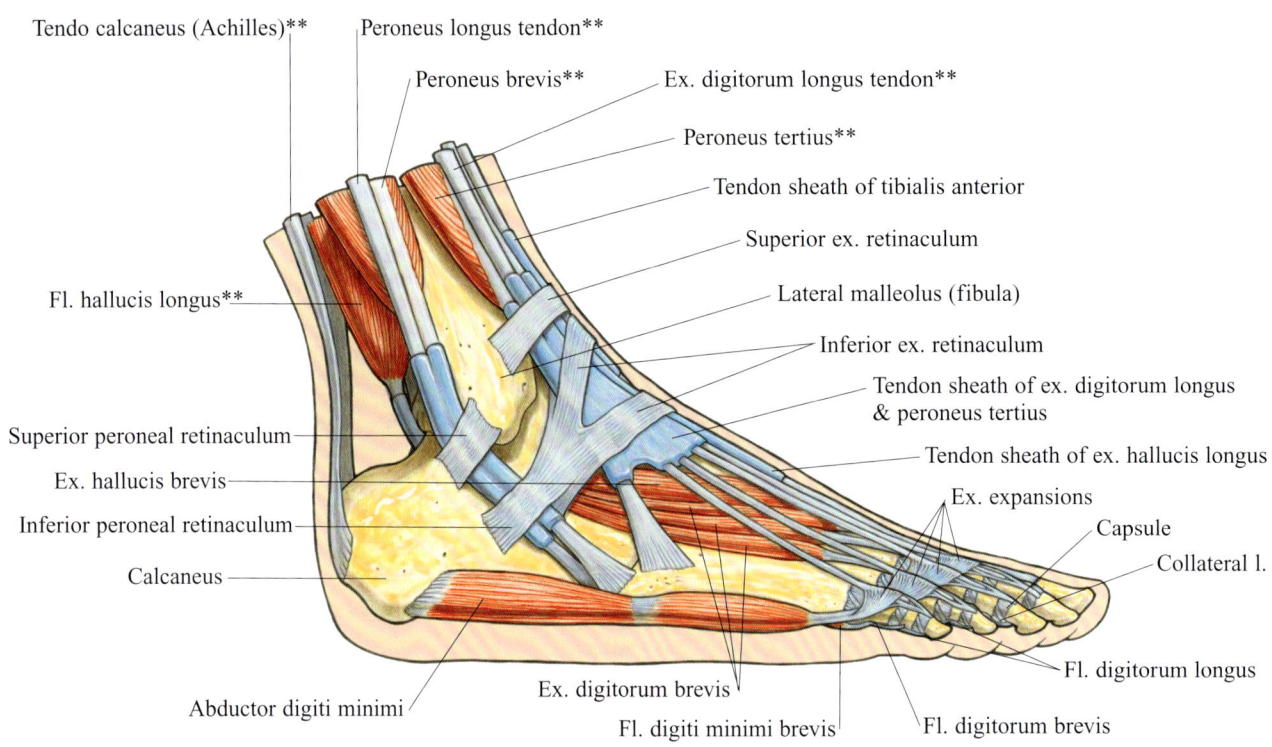

MEDIAL FOOT

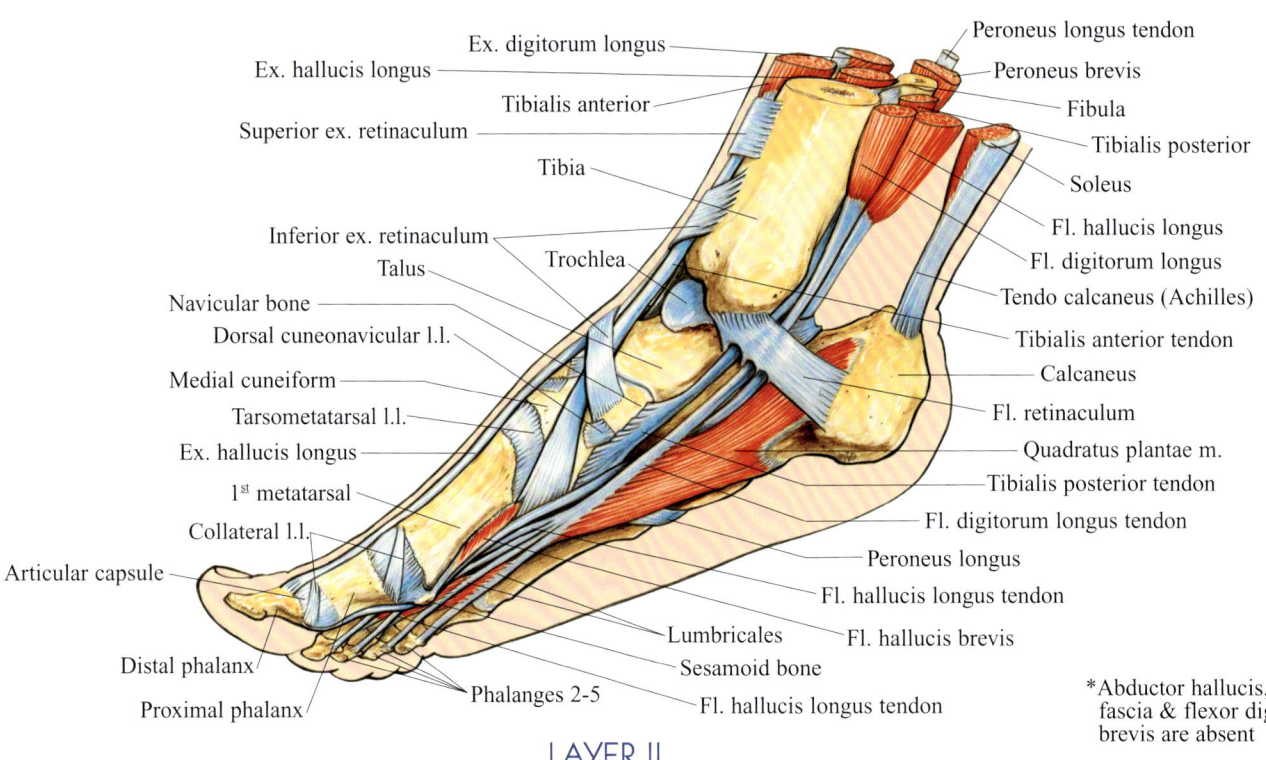

LAYER II

MUSCULAR SYSTEM

MUSCLE MICROSTRUCTURE
EXTENSOR INDICIS

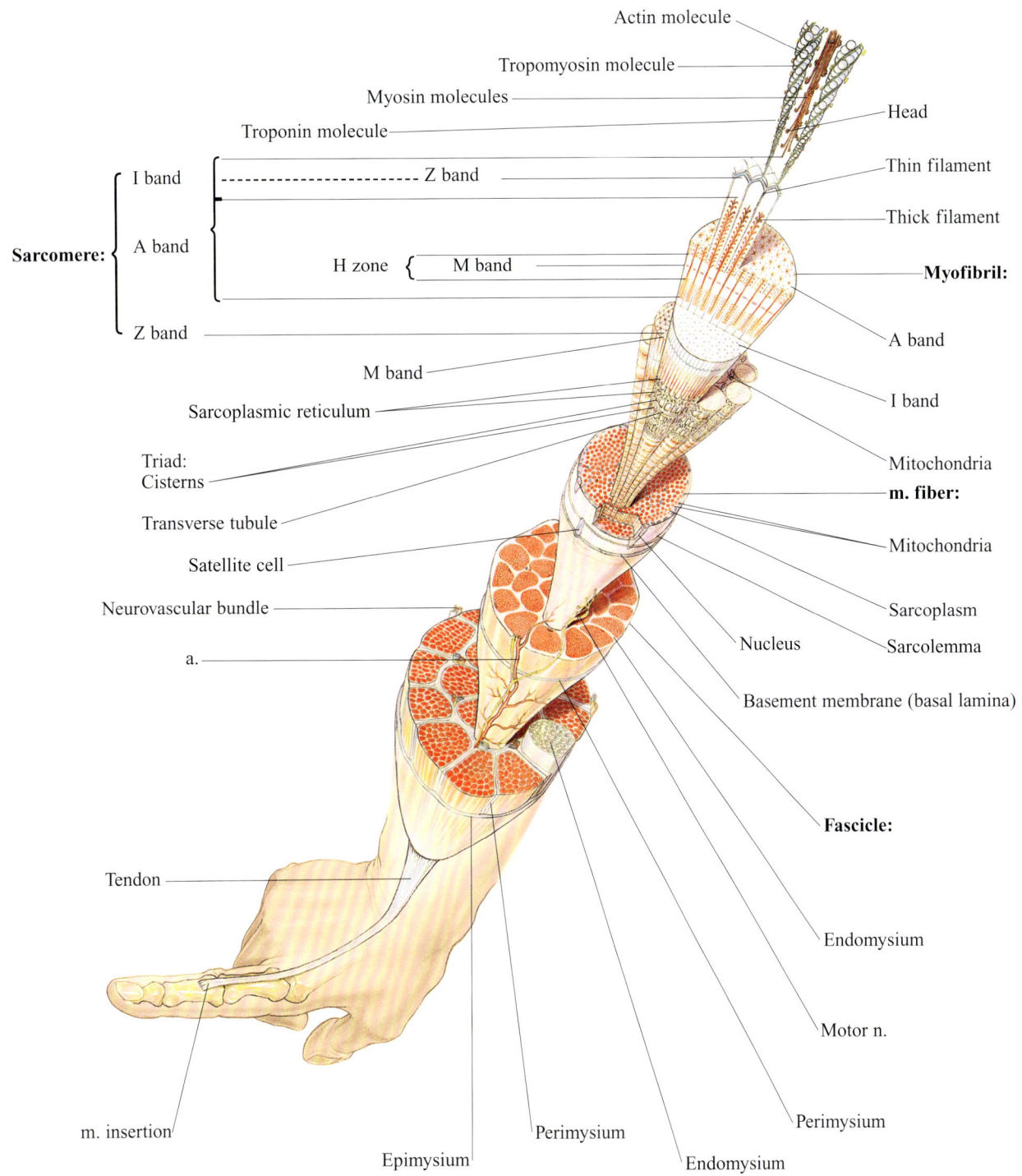

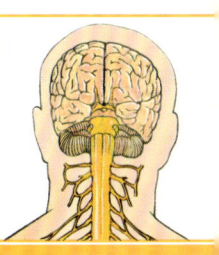

6

NERVOUS SYSTEM

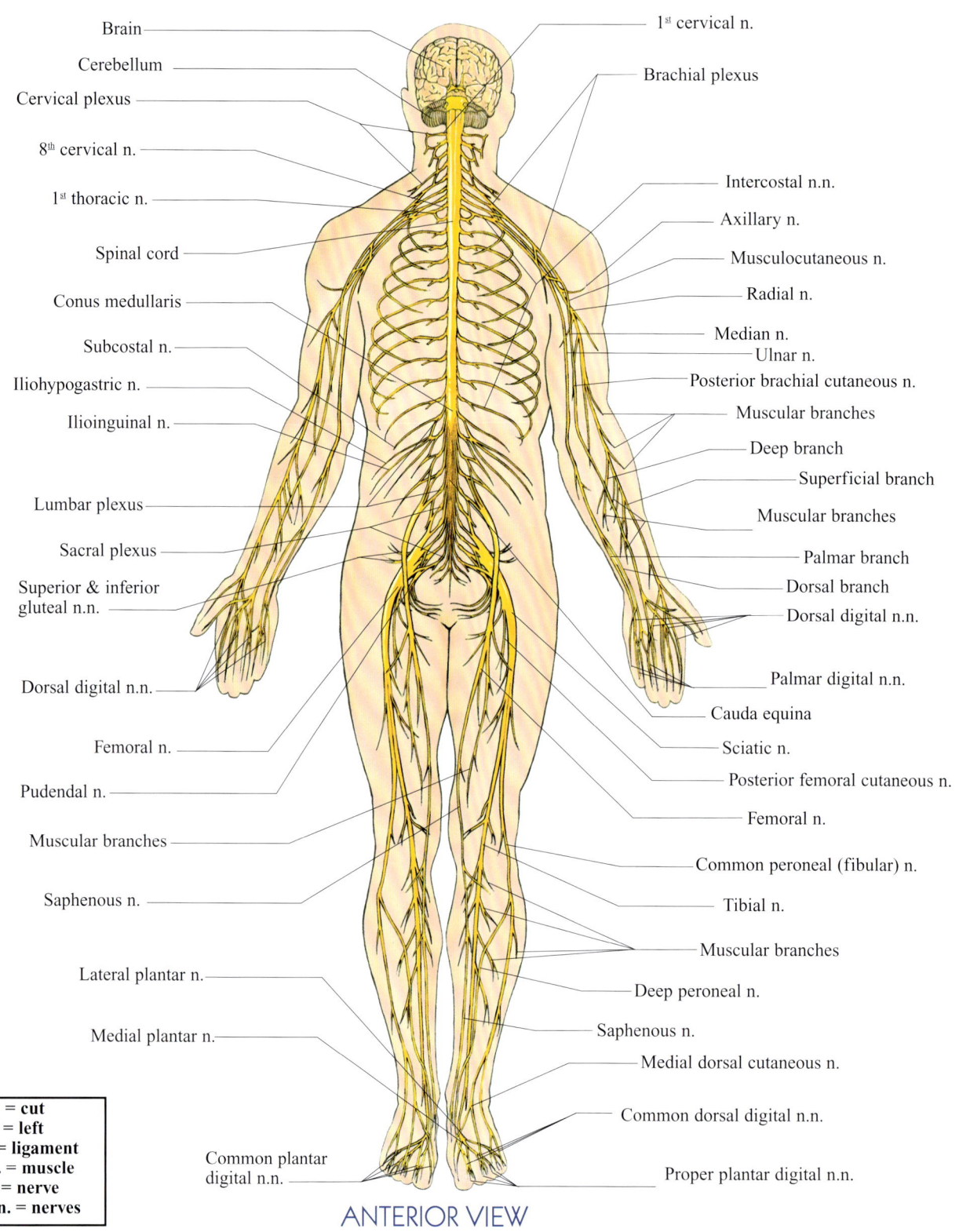

NERVOUS SYSTEM

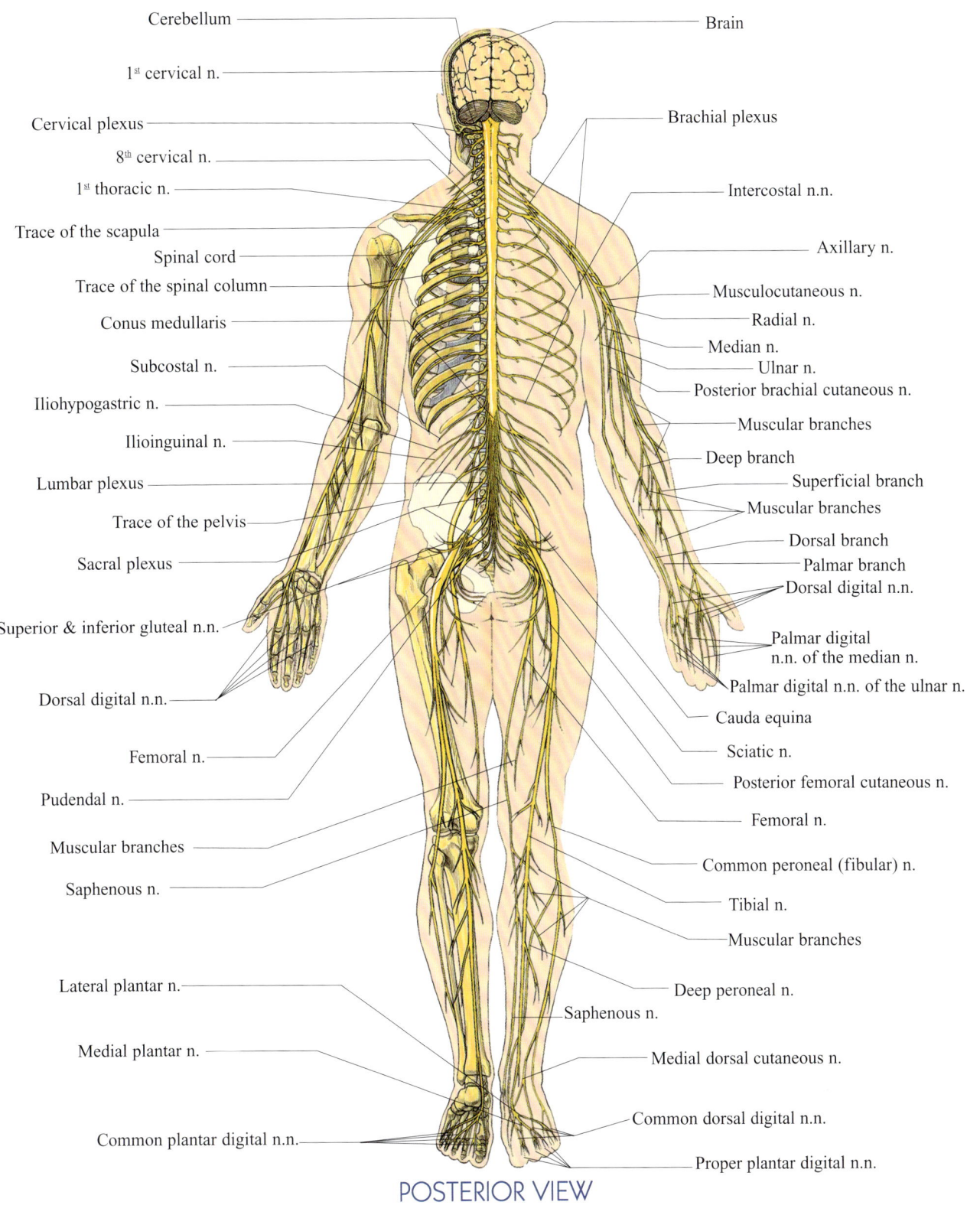

POSTERIOR VIEW

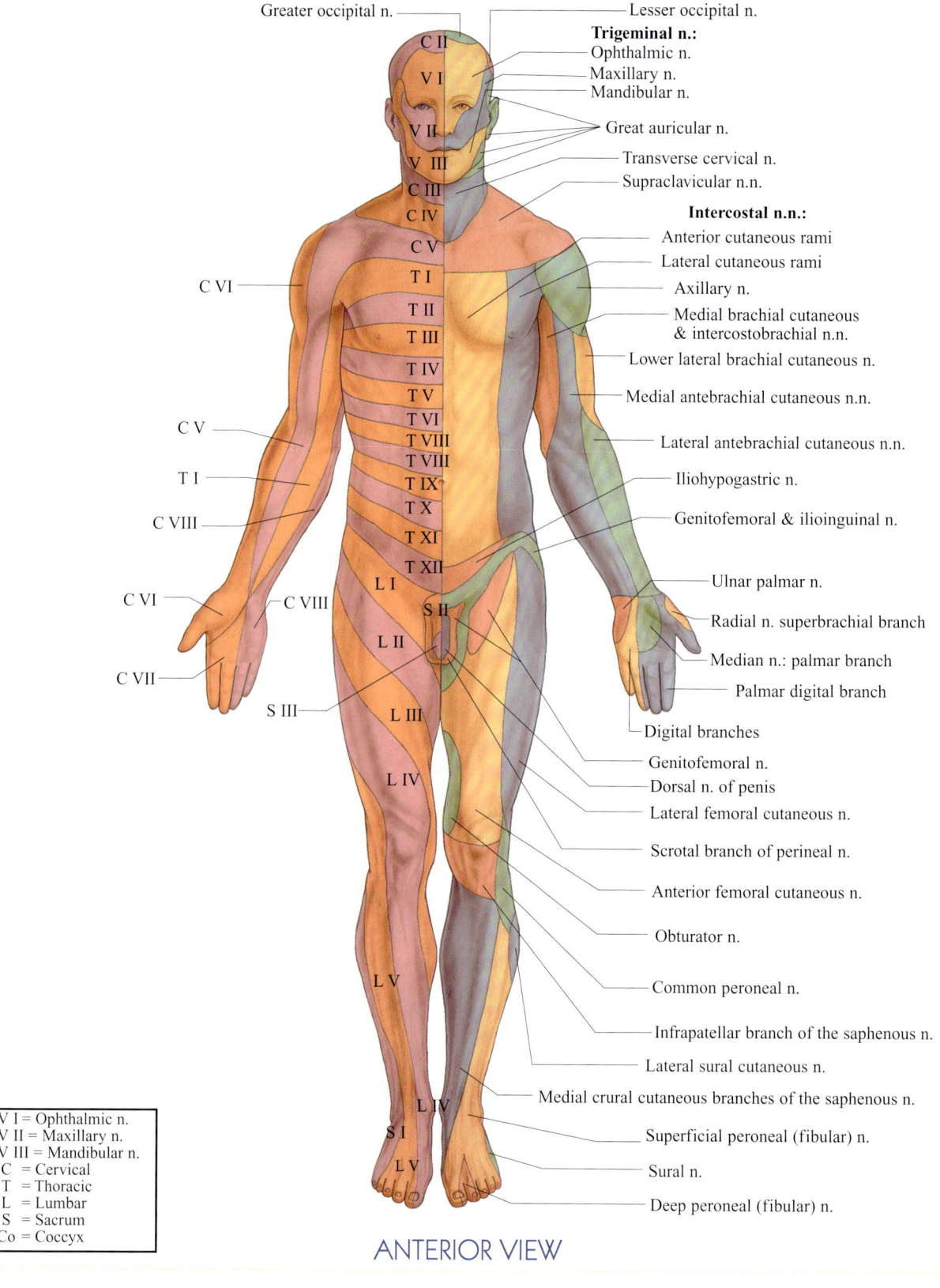

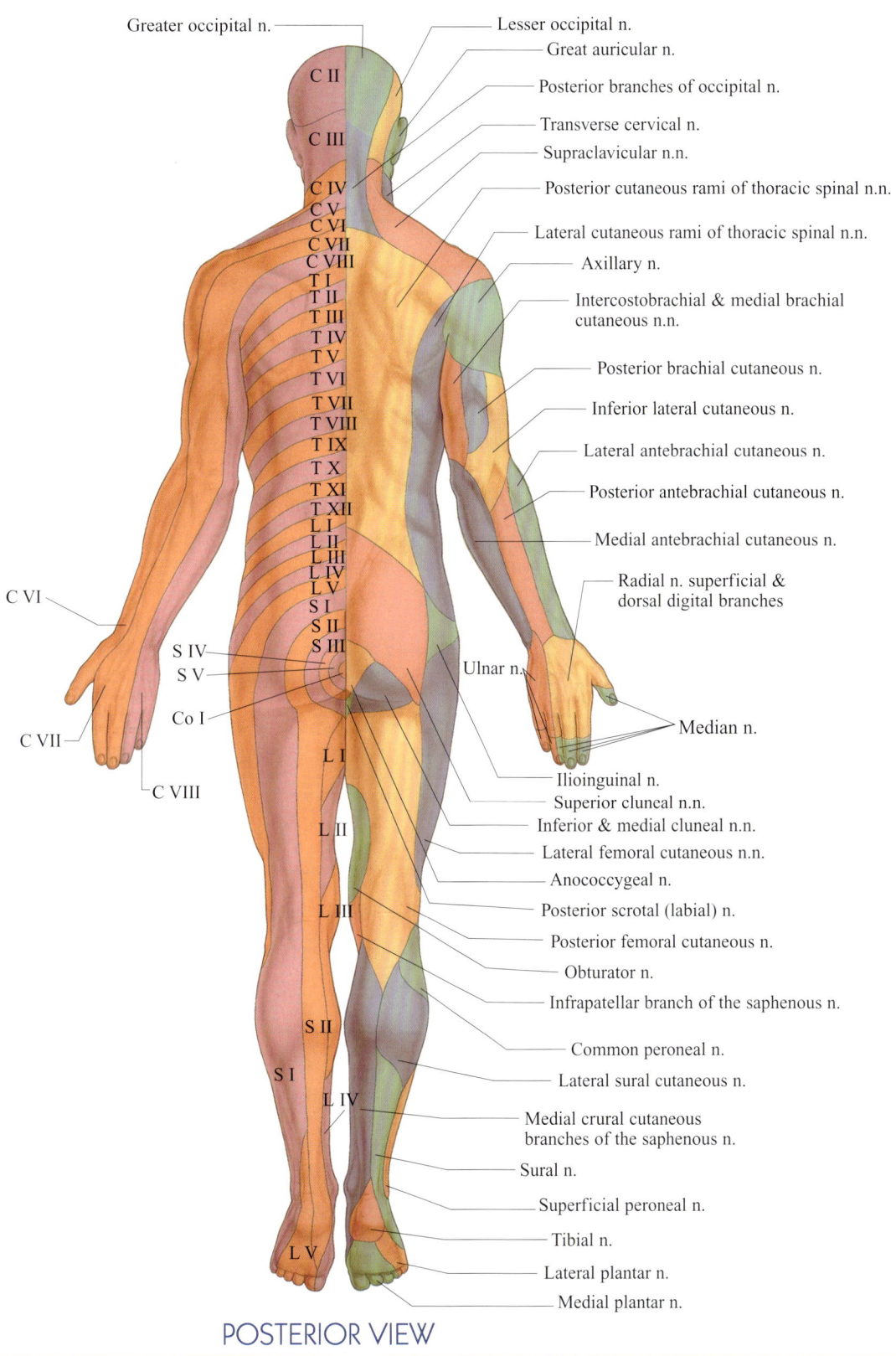

CUTANEOUS INNERVATION
DERMATOMES & PERIPHERAL NERVE DISTRIBUTIONS

POSTERIOR VIEW

CERVICOBRACHIAL PLEXUS

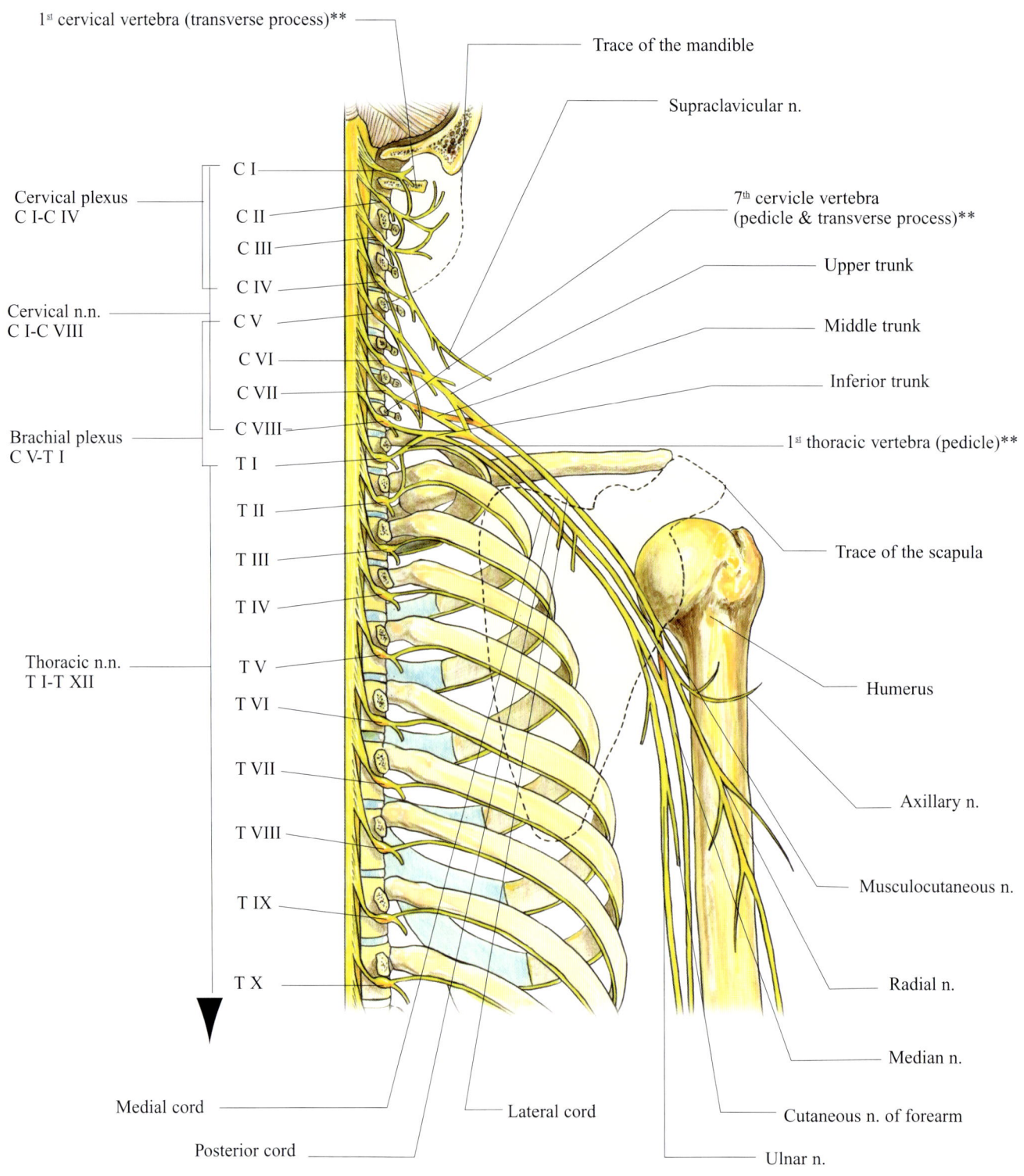

POSTERIOR VIEW

LUMBOSACRAL PLEXUS

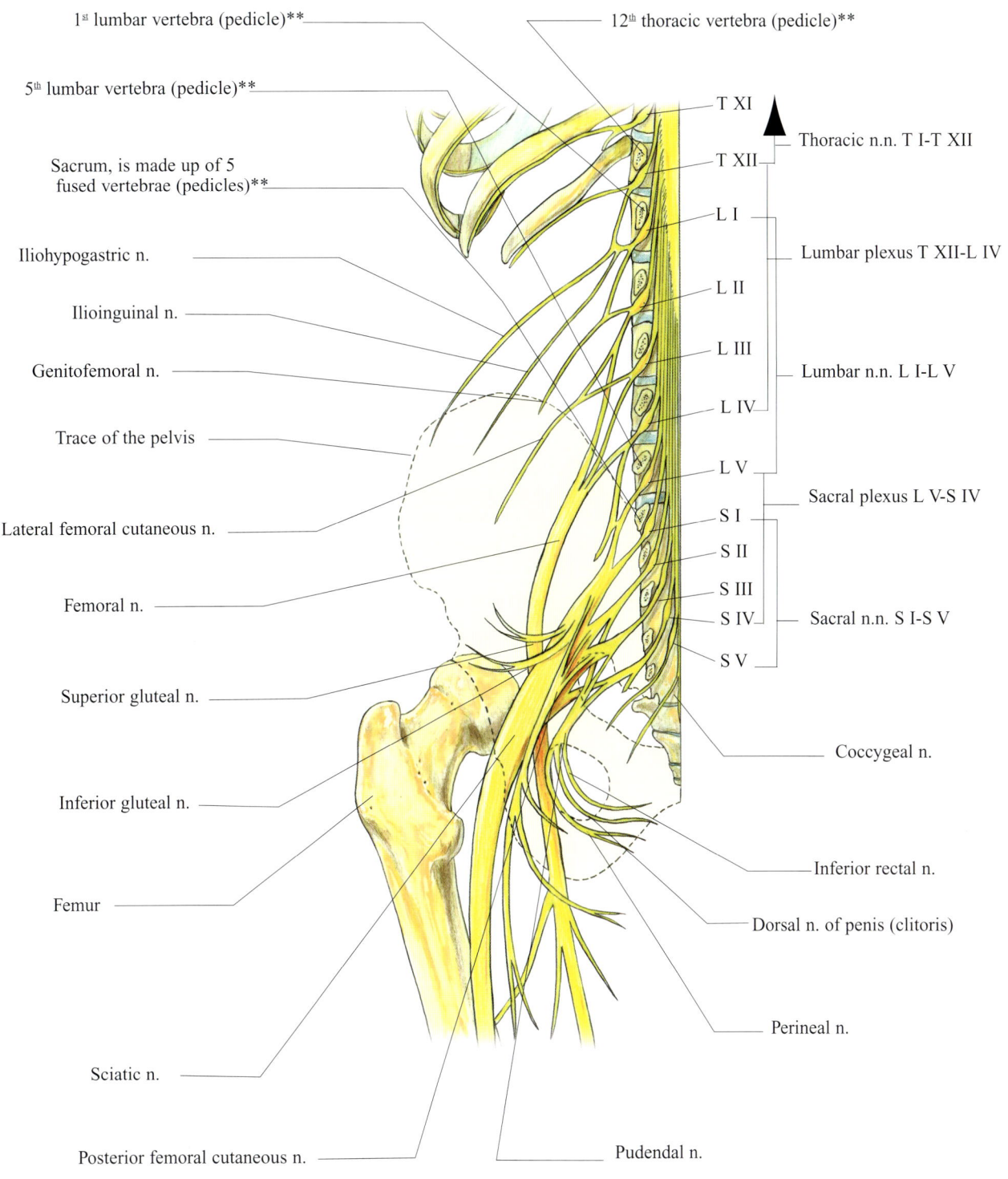

POSTERIOR VIEW

NERVOUS SYSTEM

SPINAL CORD

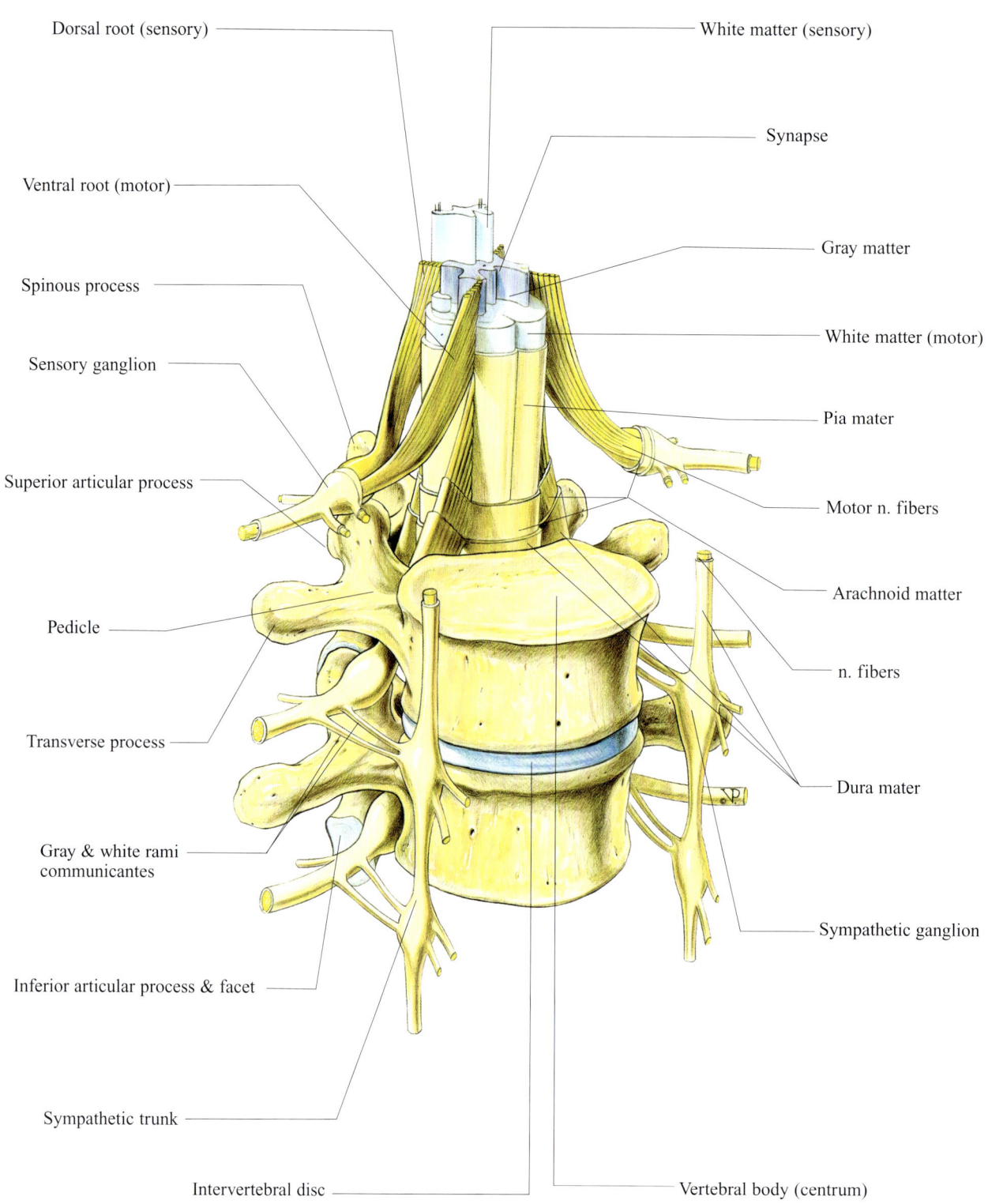

108

SCIATIC NERVE

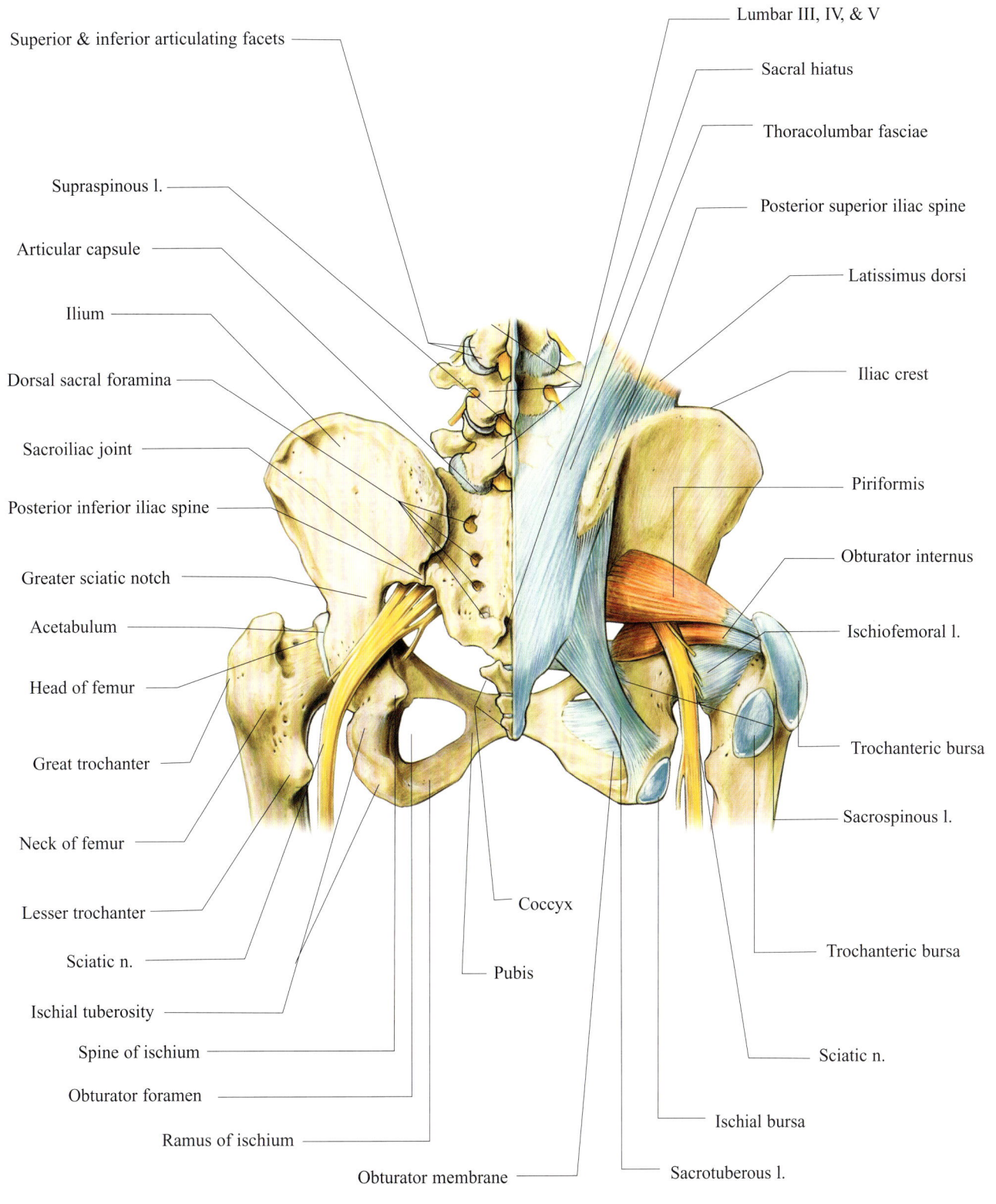

TRIGEMINAL NERVE

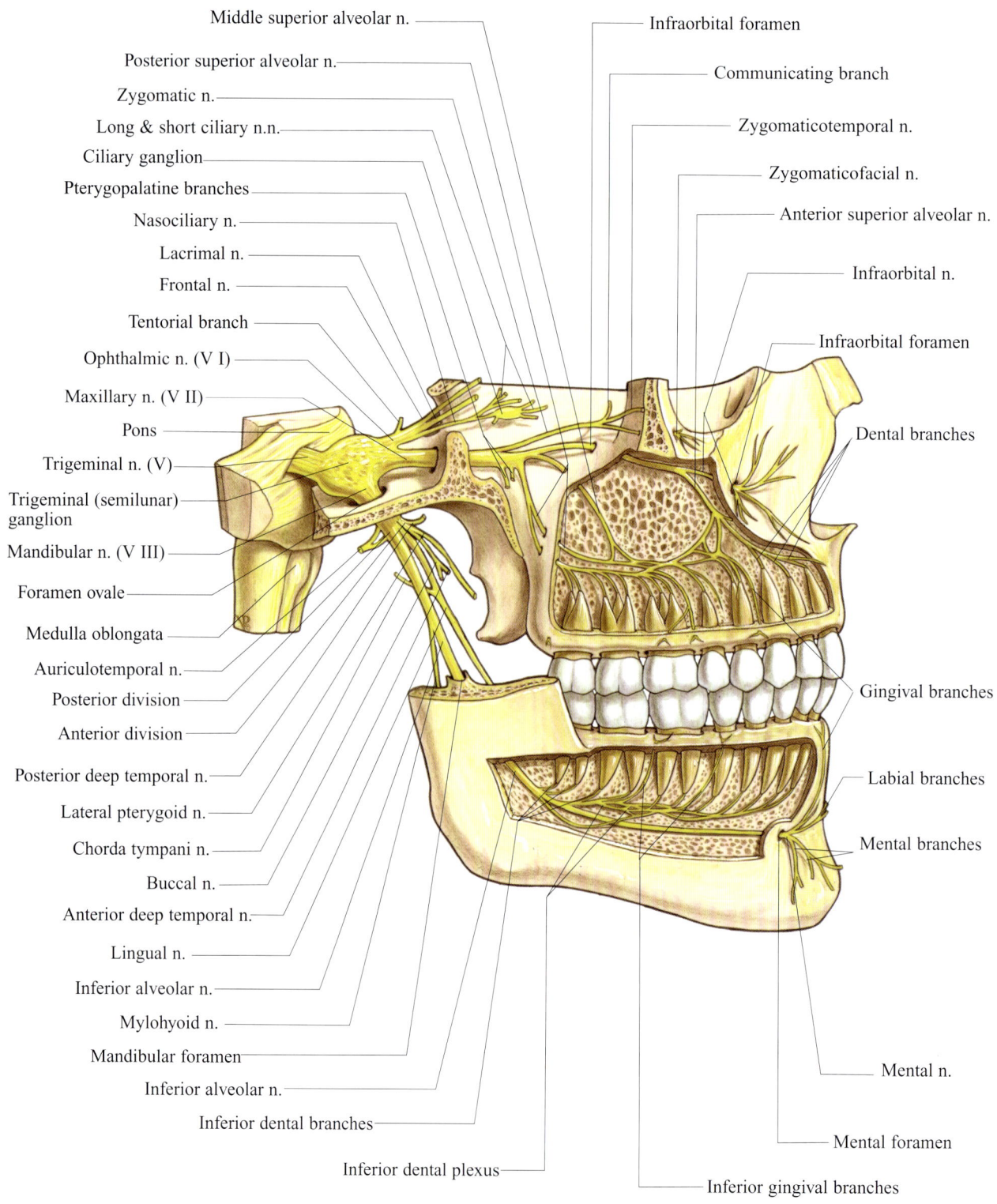

NERVOUS SYSTEM

NERVES OF THE FACE & HEAD

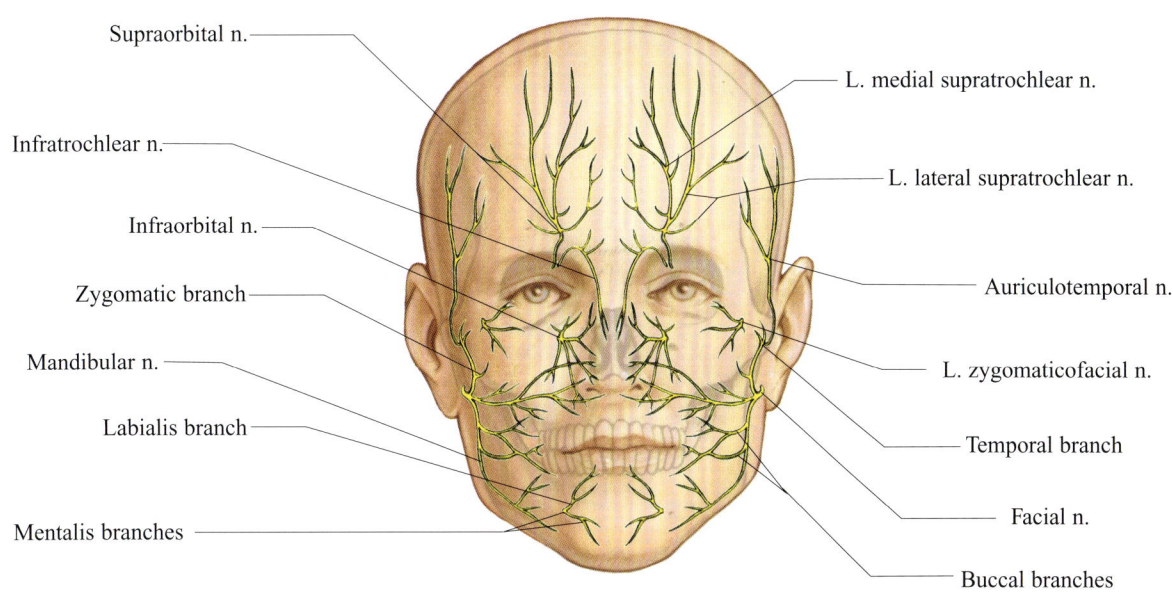

NERVE STRUCTURE

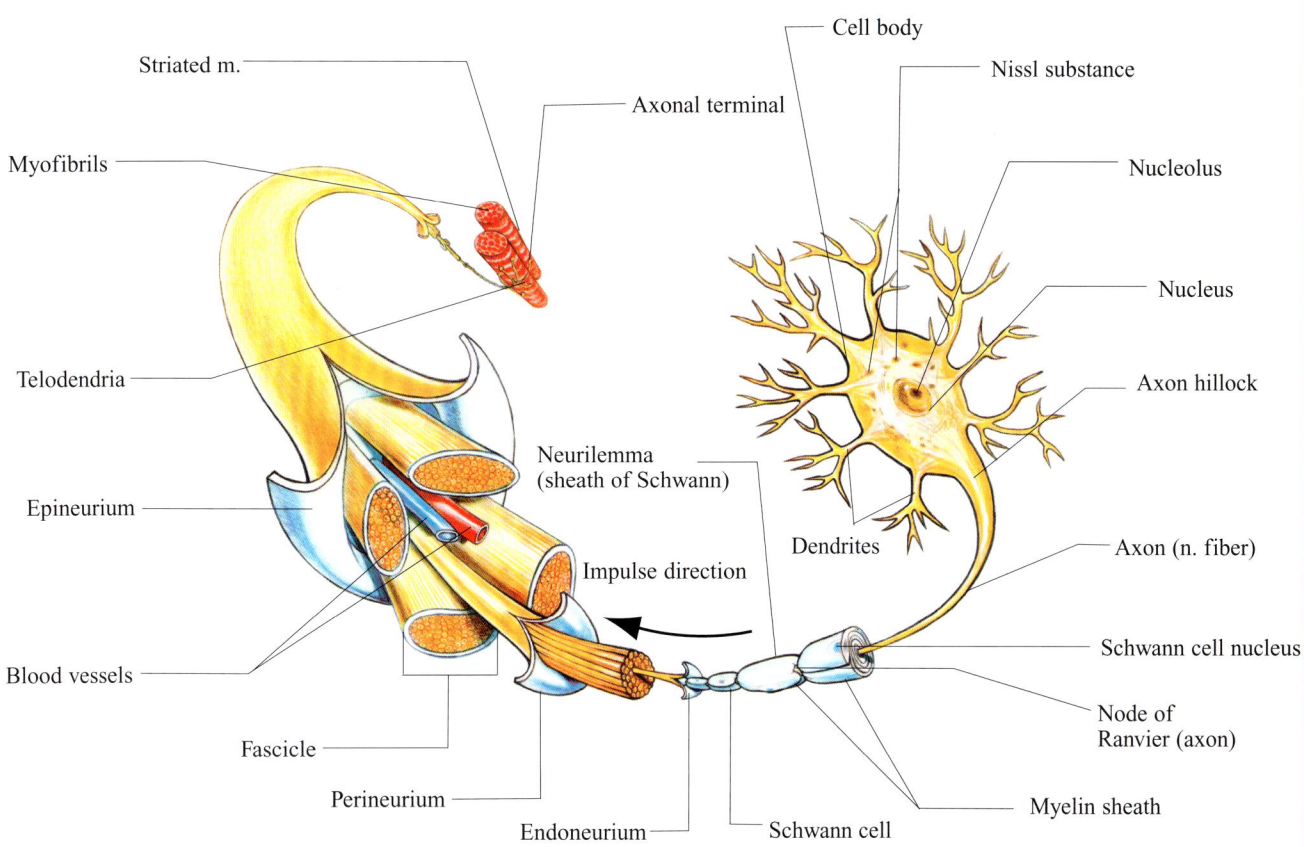

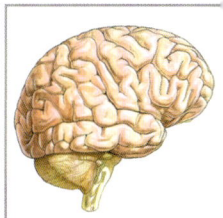

7
THE BRAIN

BRAIN IN PLACE

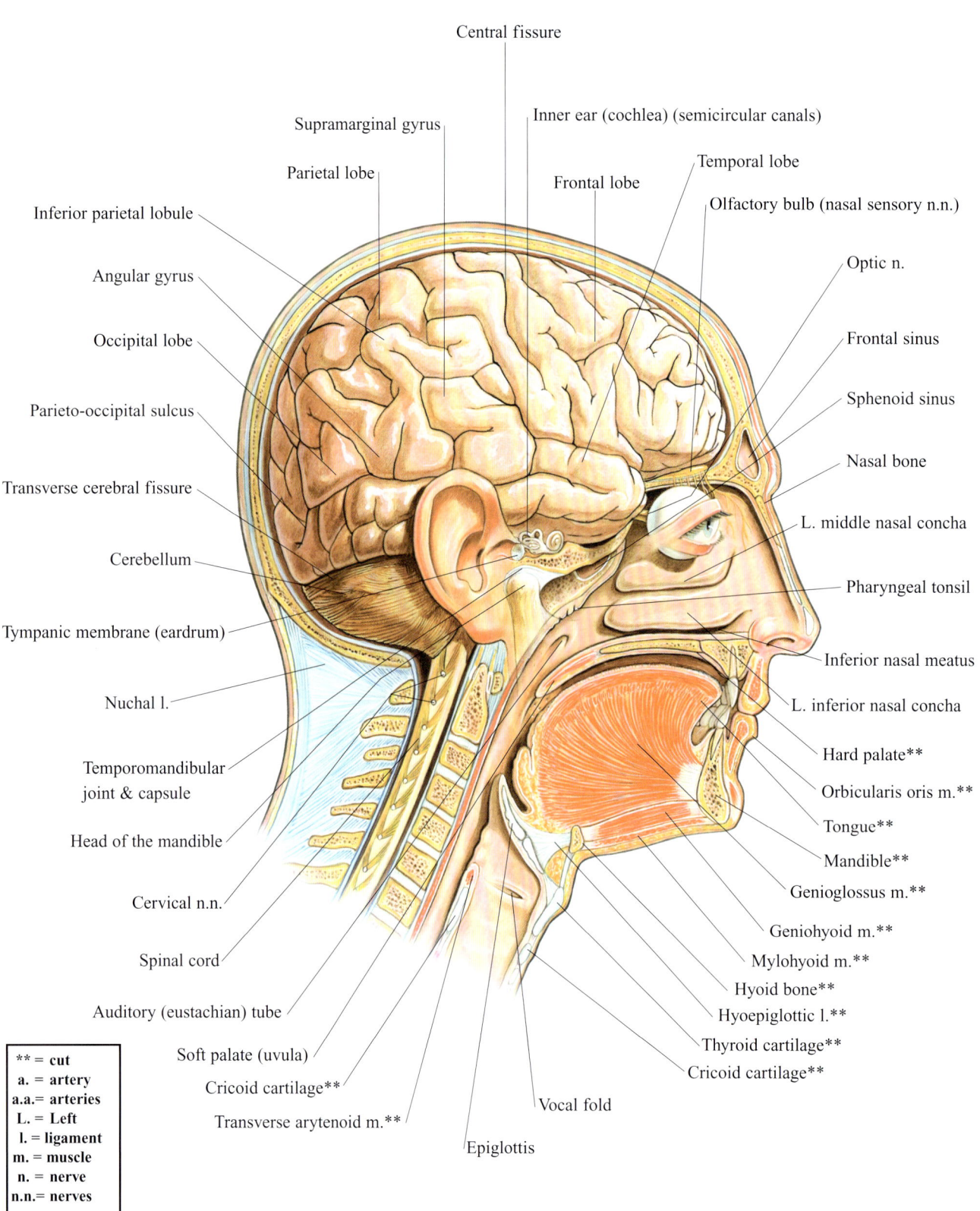

114

BRAIN

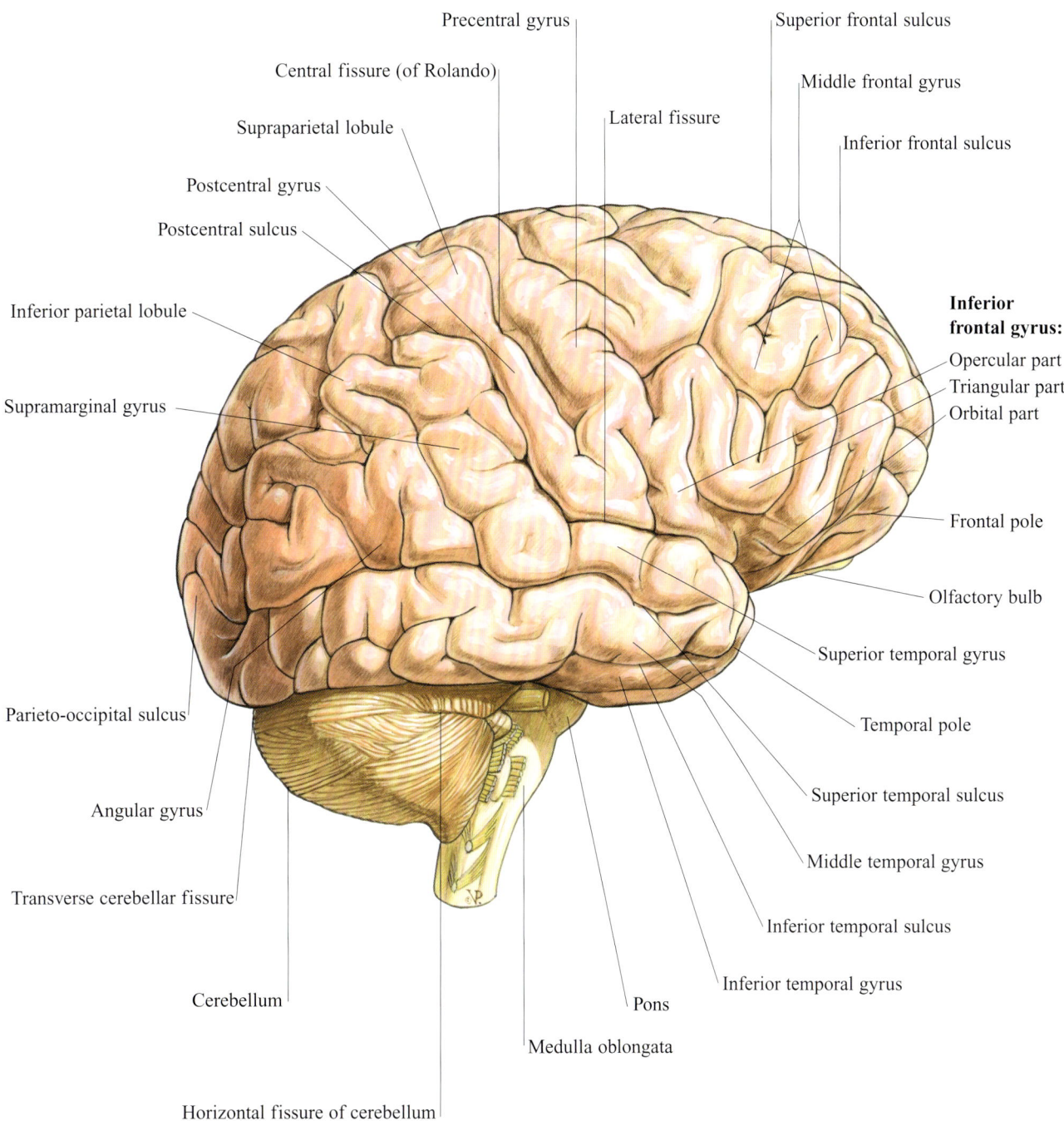

LATERAL VIEW

BRAIN

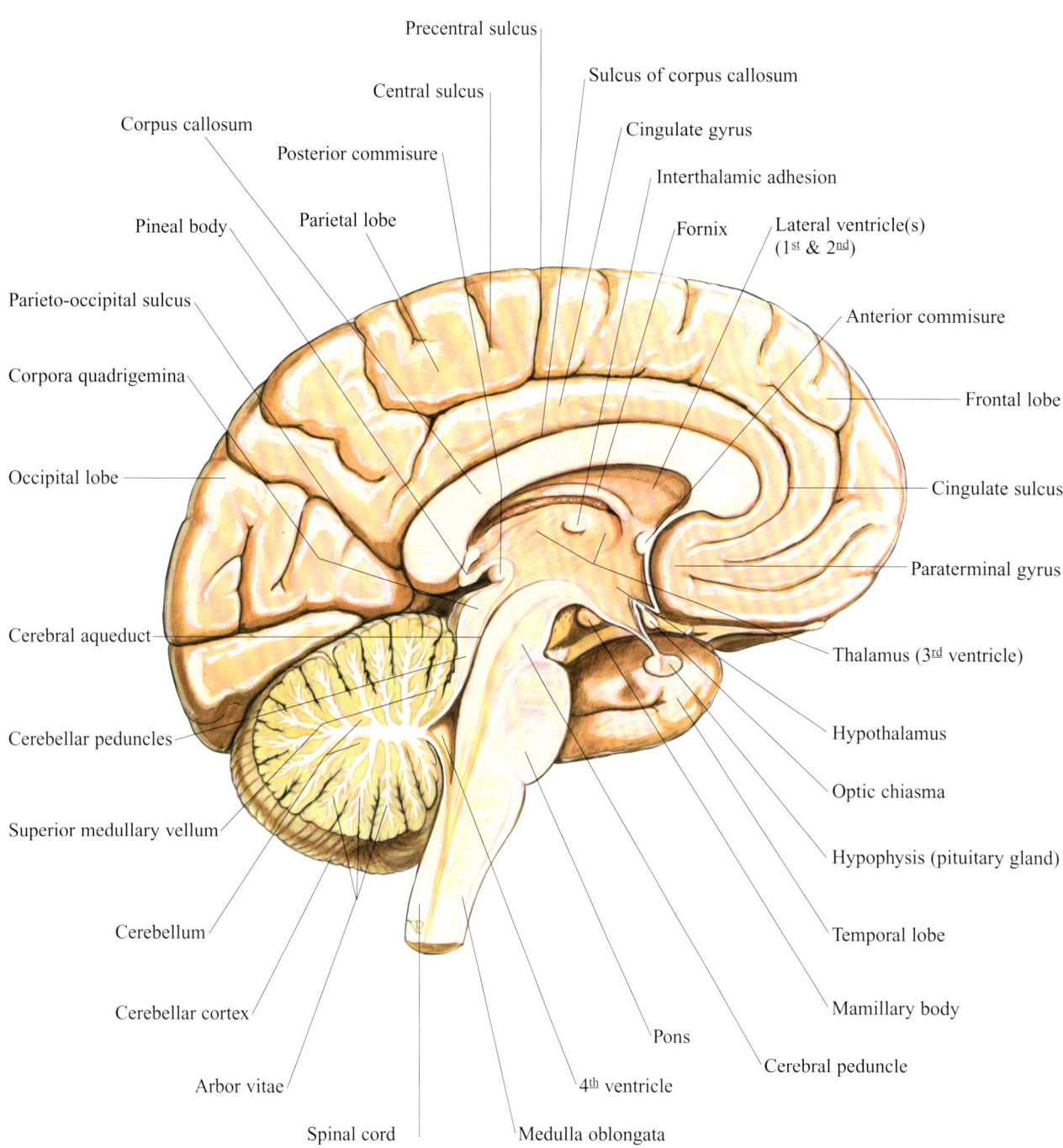

MEDIAL (SAGITTAL) VIEW

BRAIN

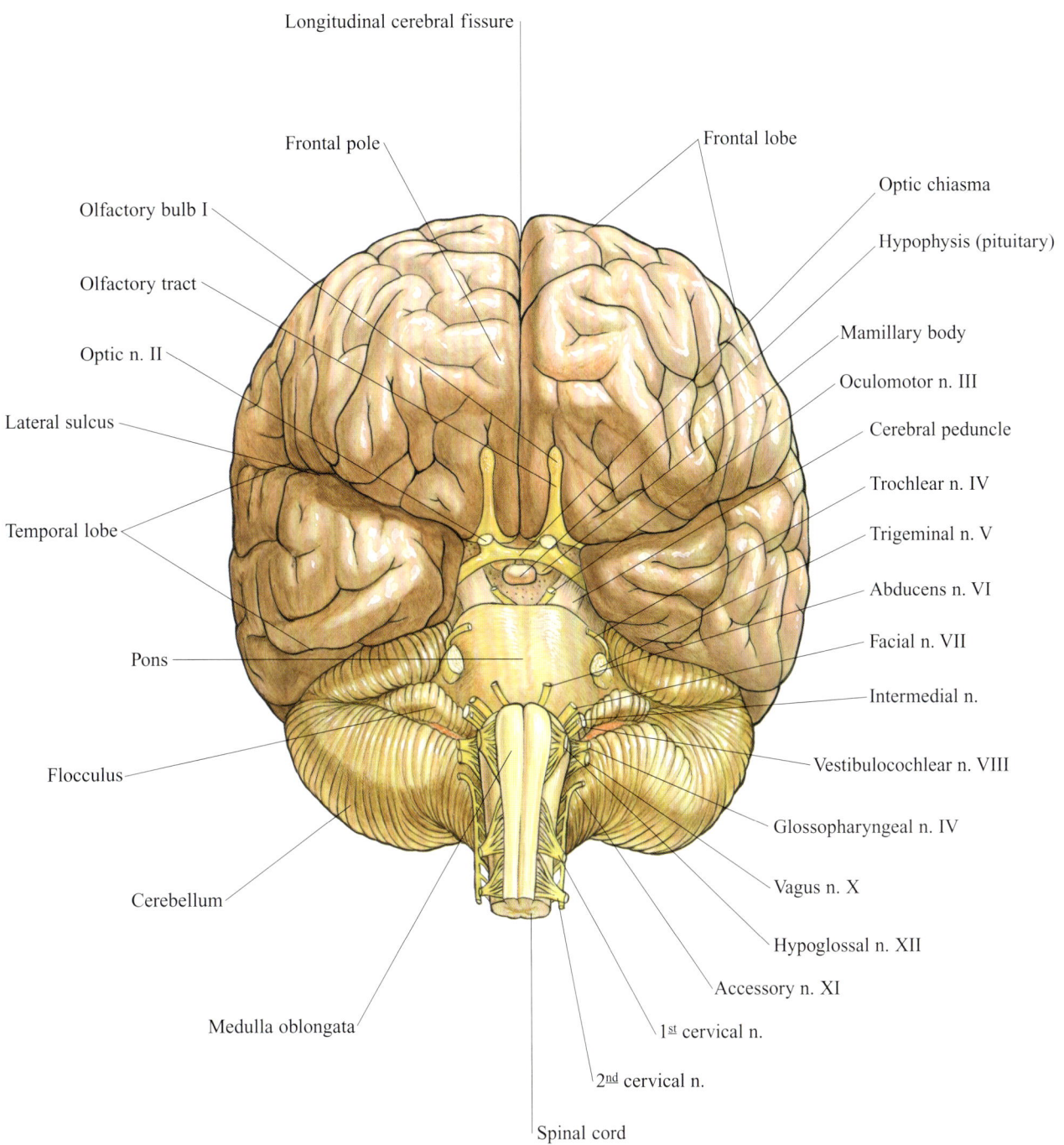

ANTERIOR VIEW

BRAIN

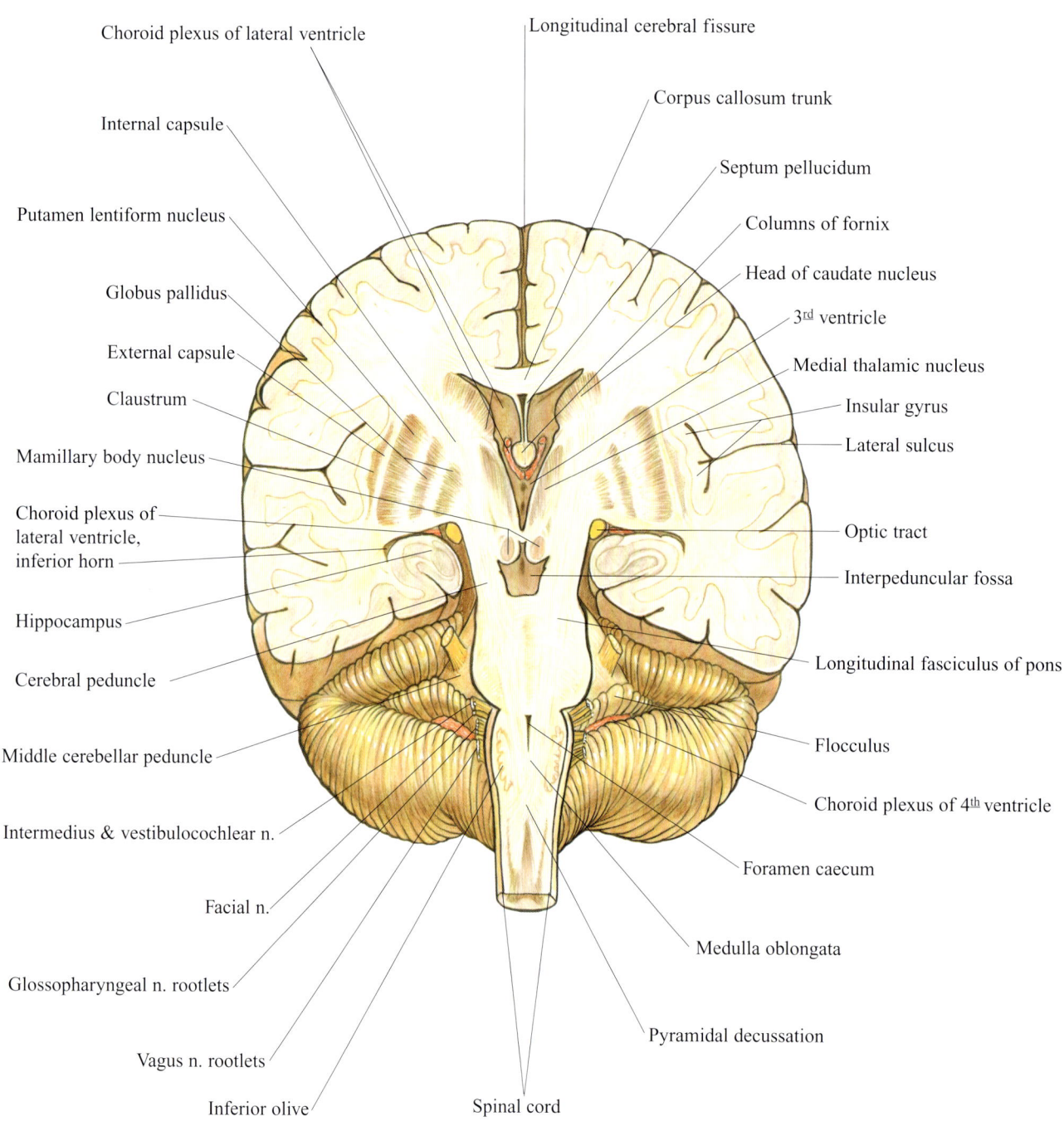

FRONTAL SECTION

BRAIN

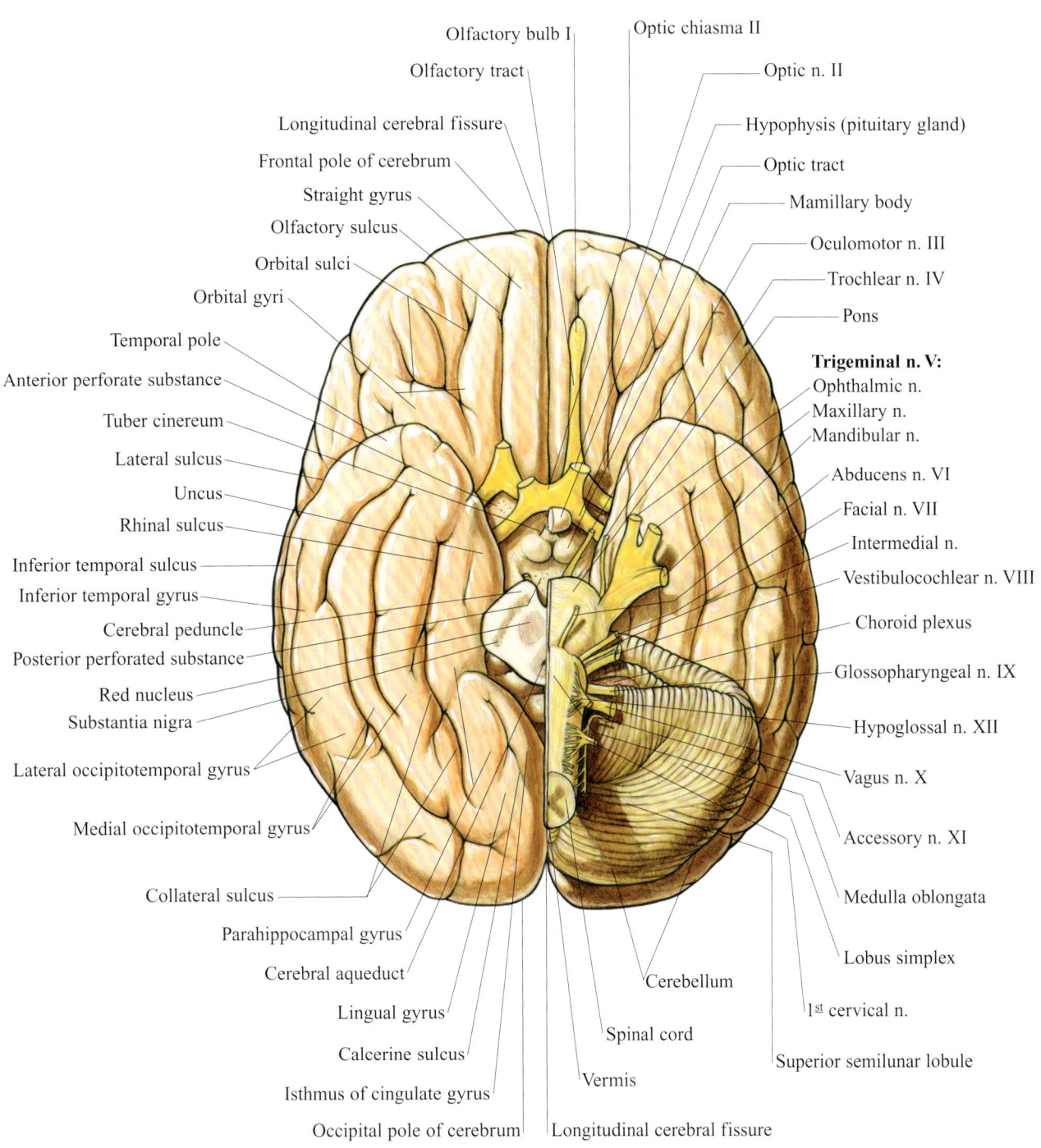

INFERIOR VIEW

BRAIN

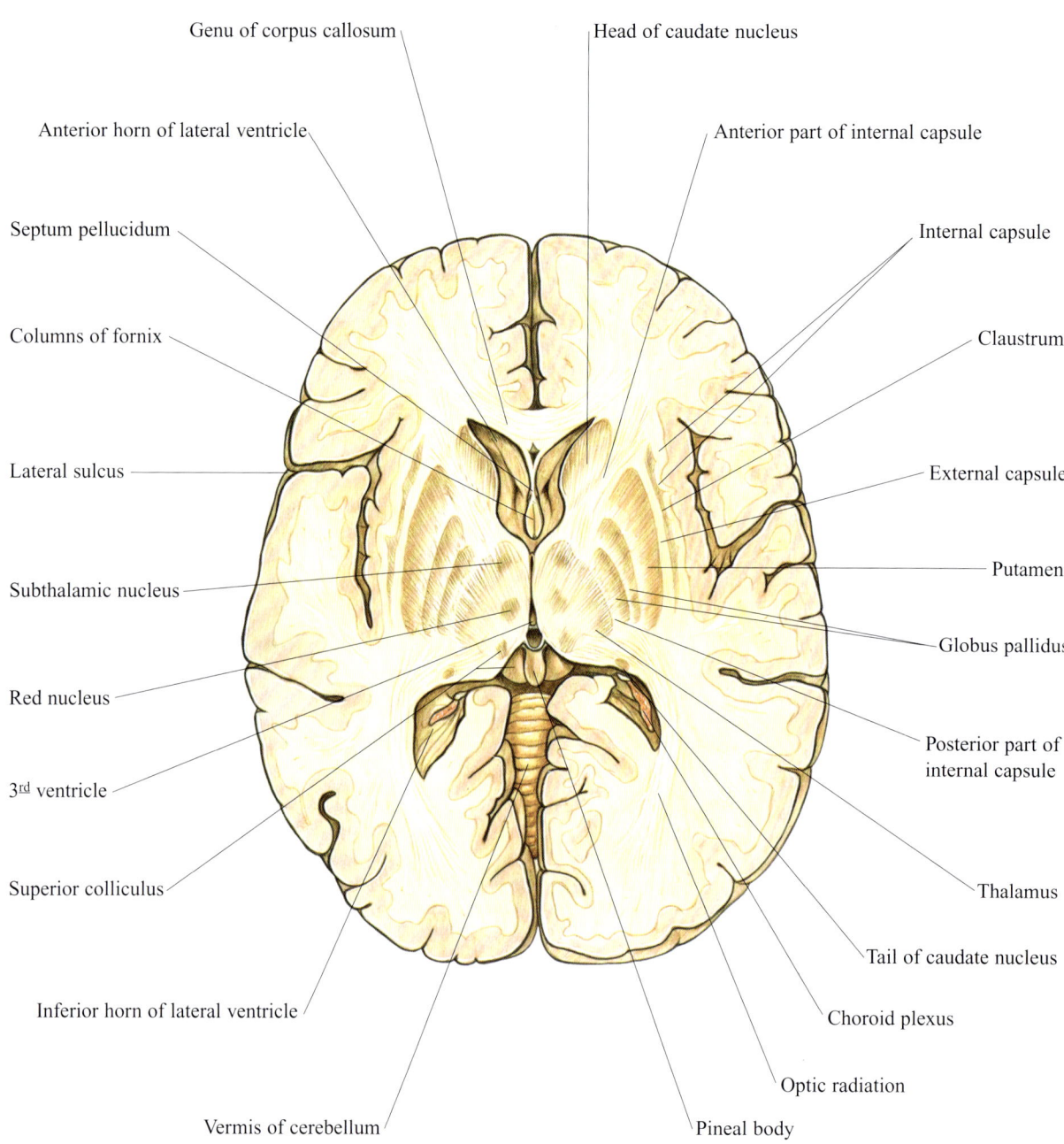

HORIZONTAL SECTION

BRAIN

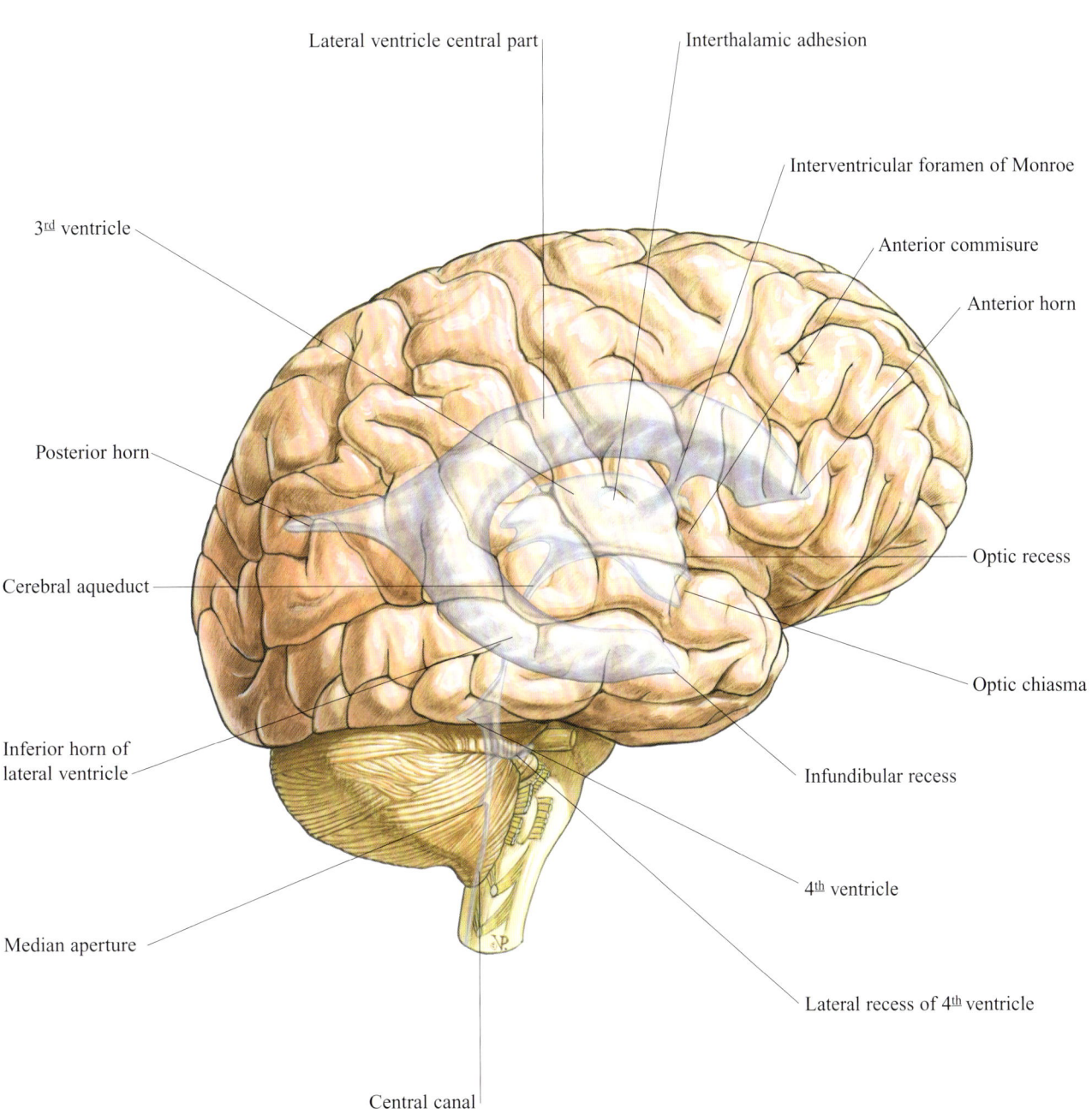

VENTRICLES

BRAIN

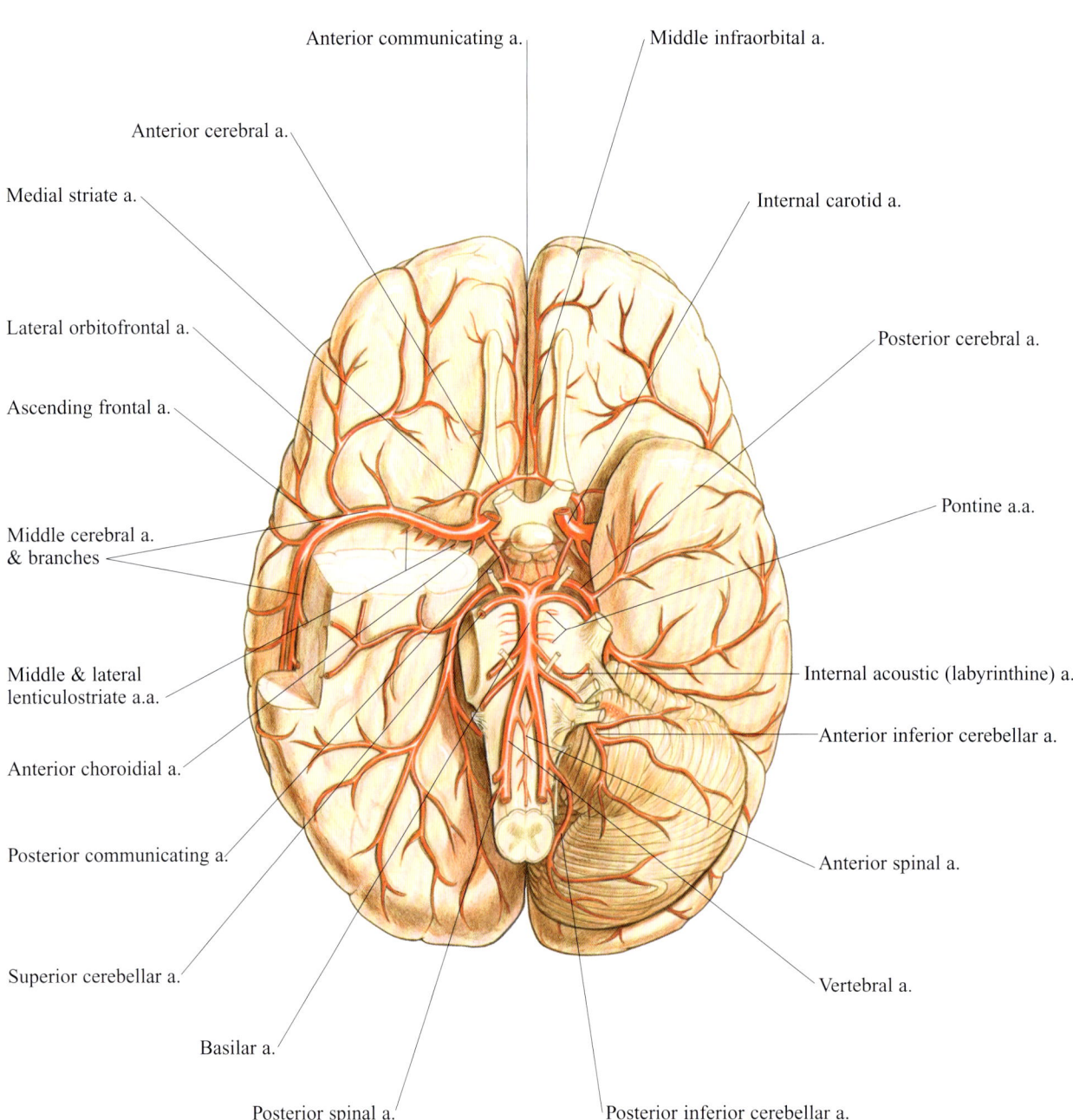

ARTERIES

NOTES

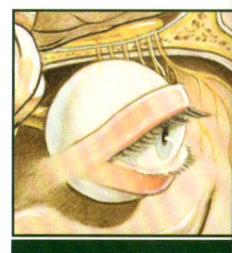

8
THE SENSES

THE SENSES

HEAD: EYE, EAR, NOSE & MOUTH

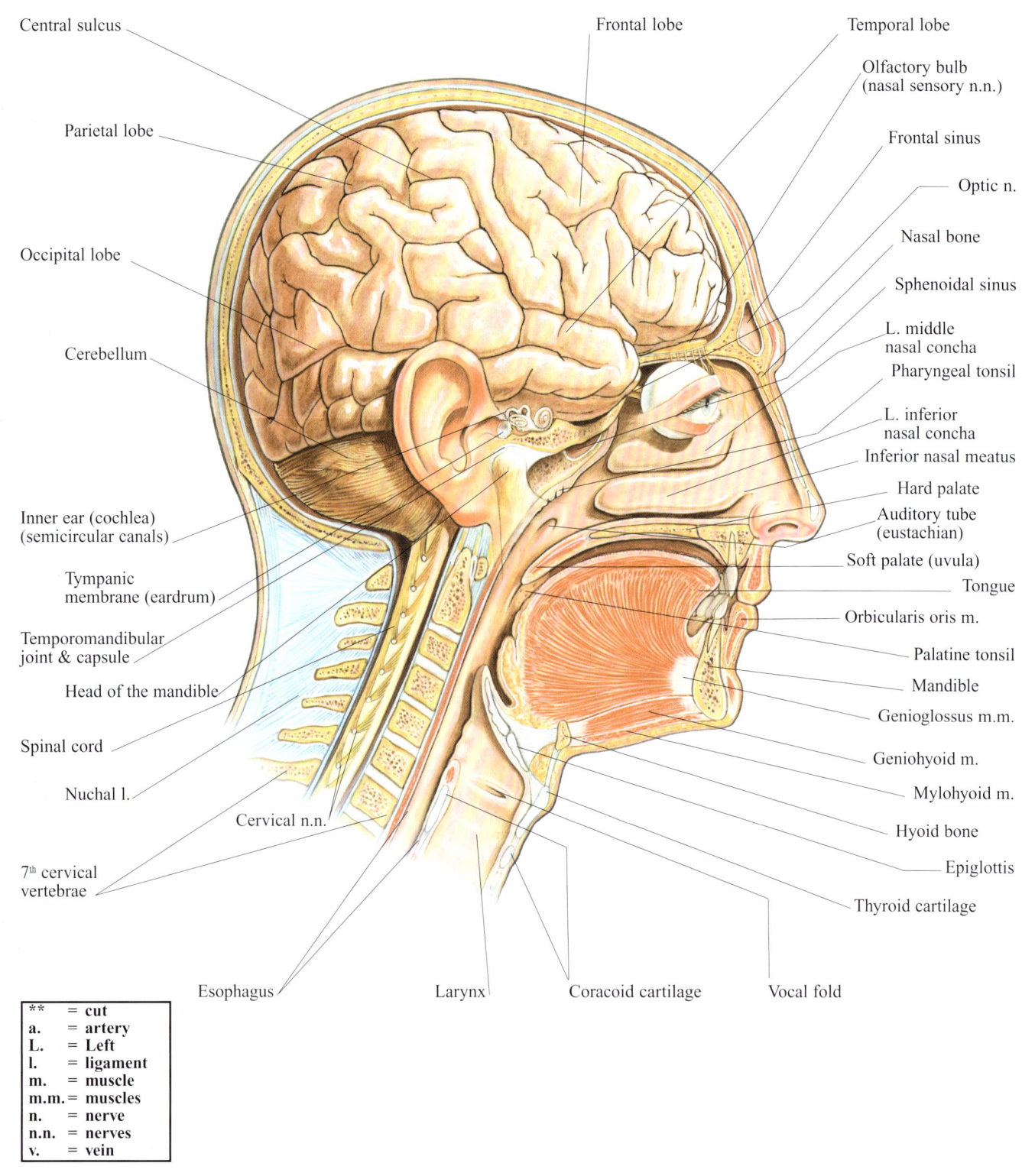

THE SENSES

SEEING

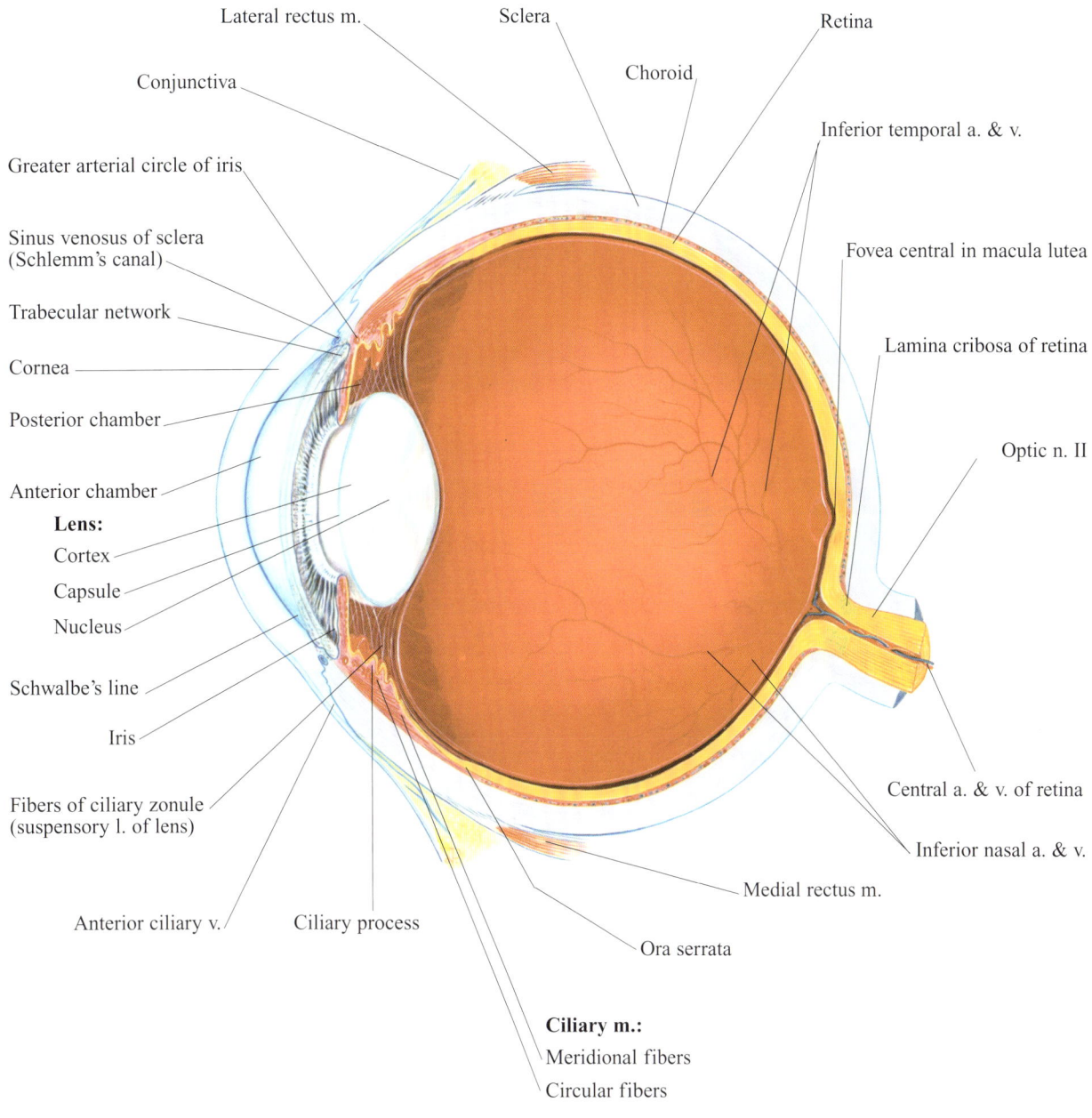

THE SENSES

HEARING

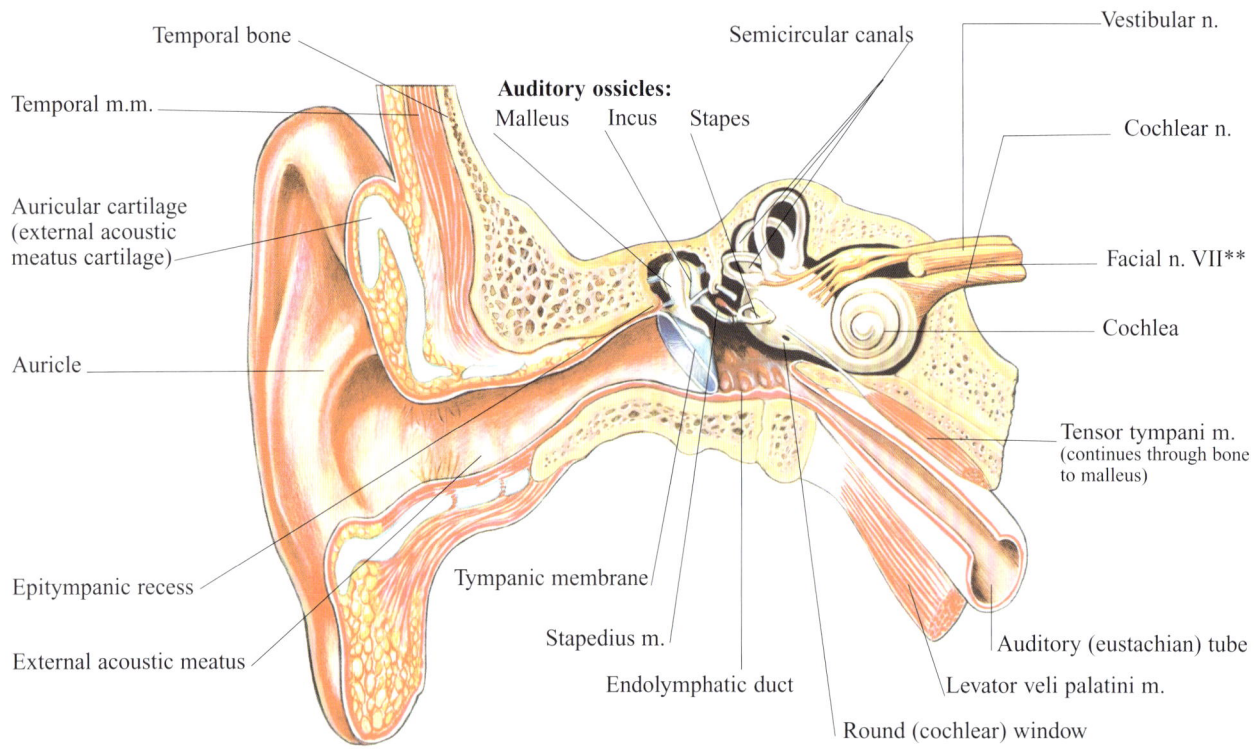

SEMICIRCULAR CANALS & DUCTS

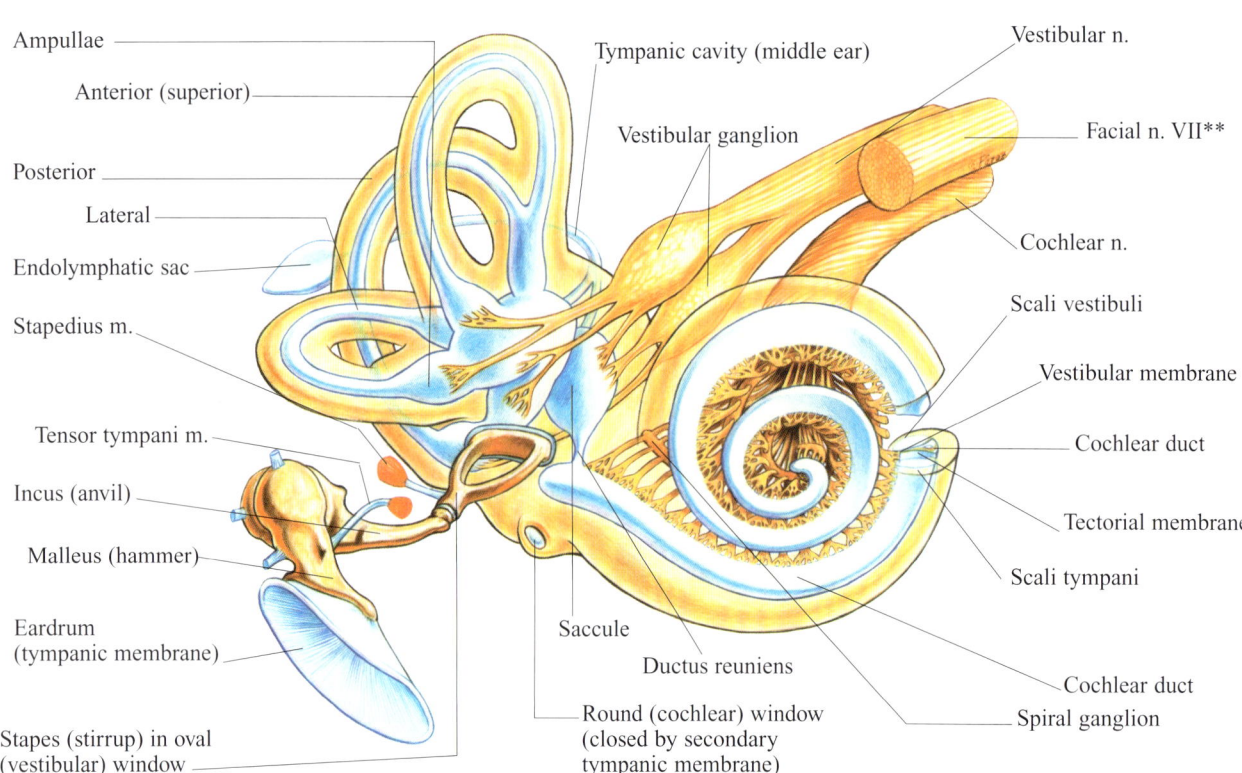

THE SENSES

SMELL

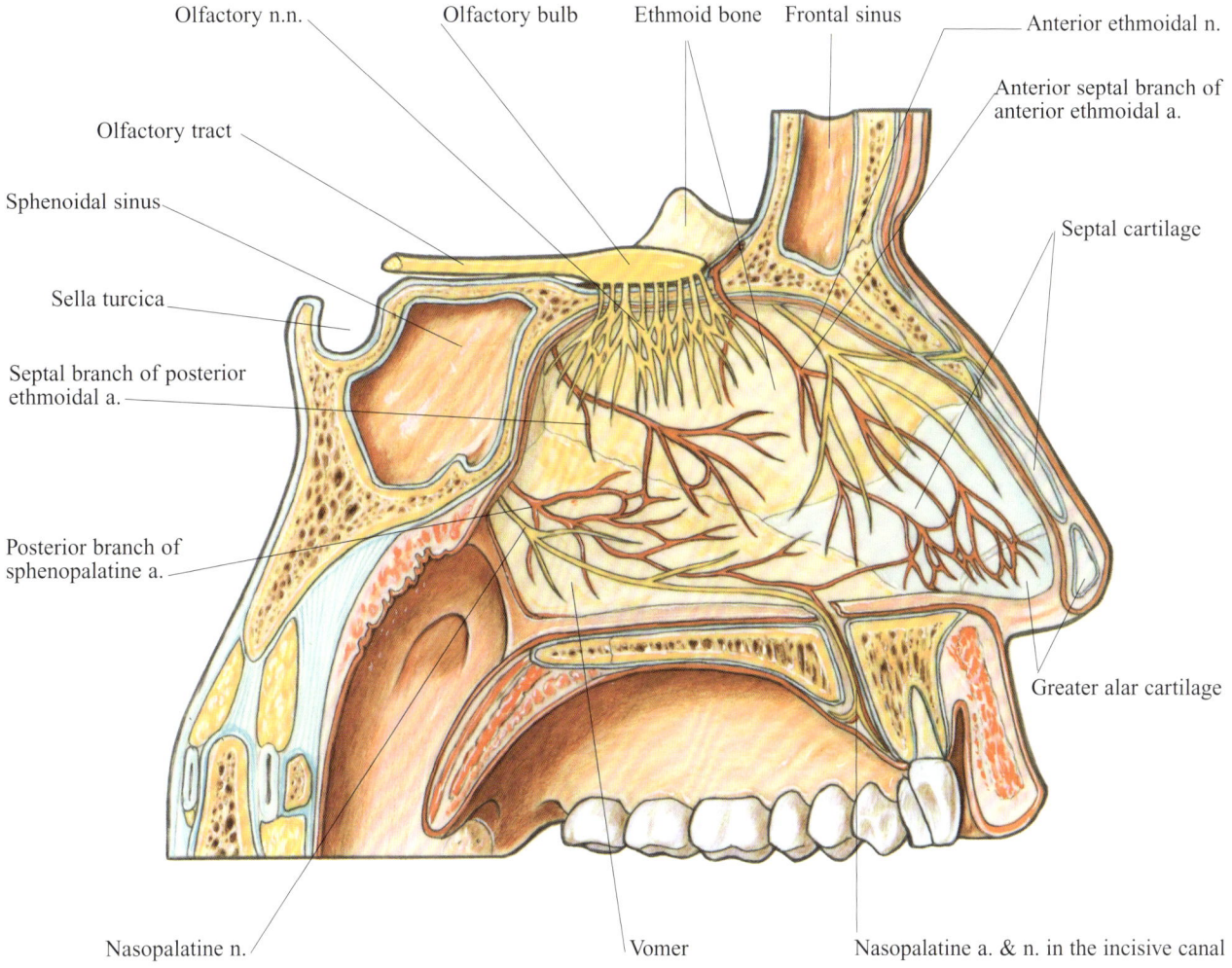

THE SENSES

TASTE

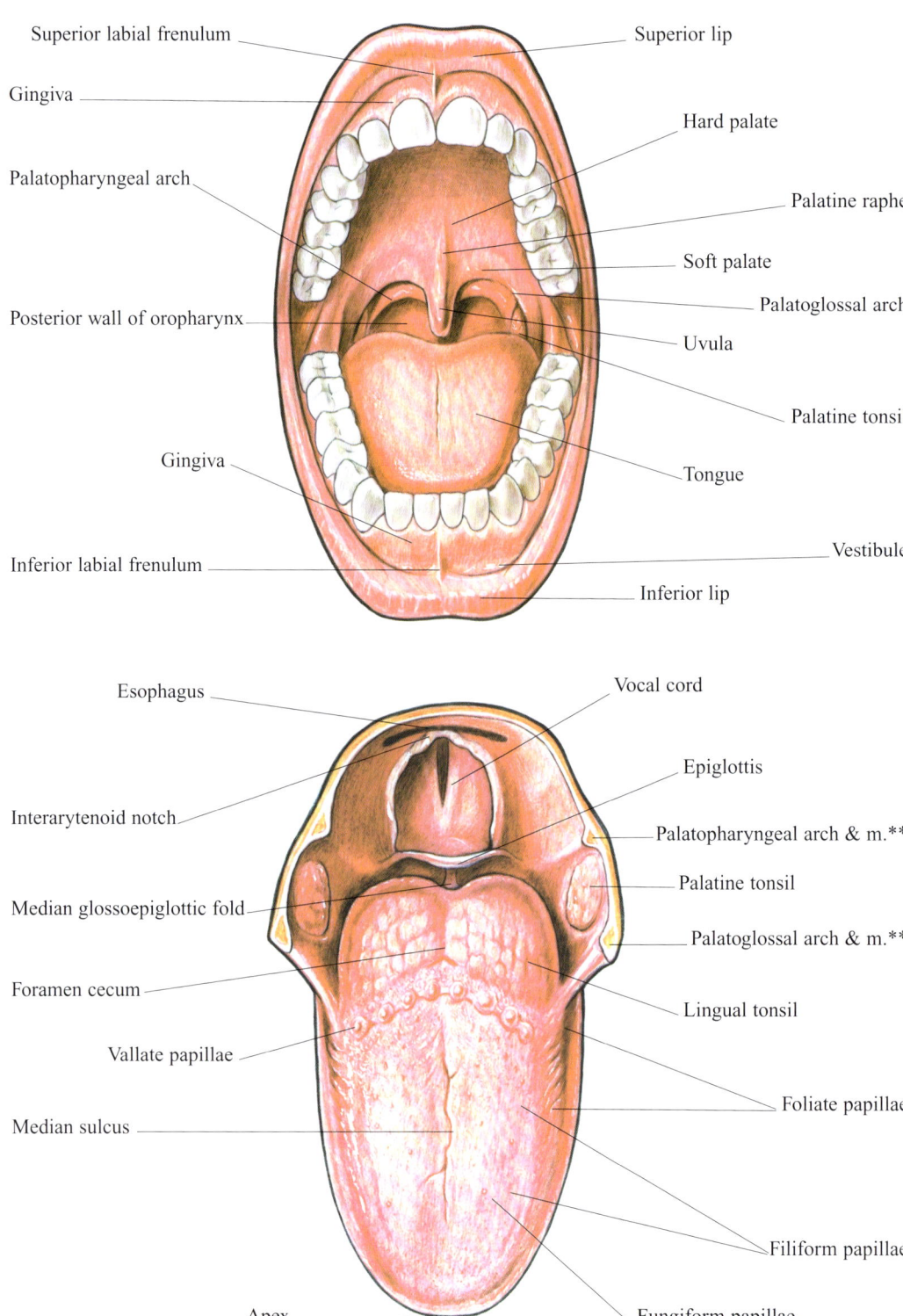

THE SENSES

TOUCH

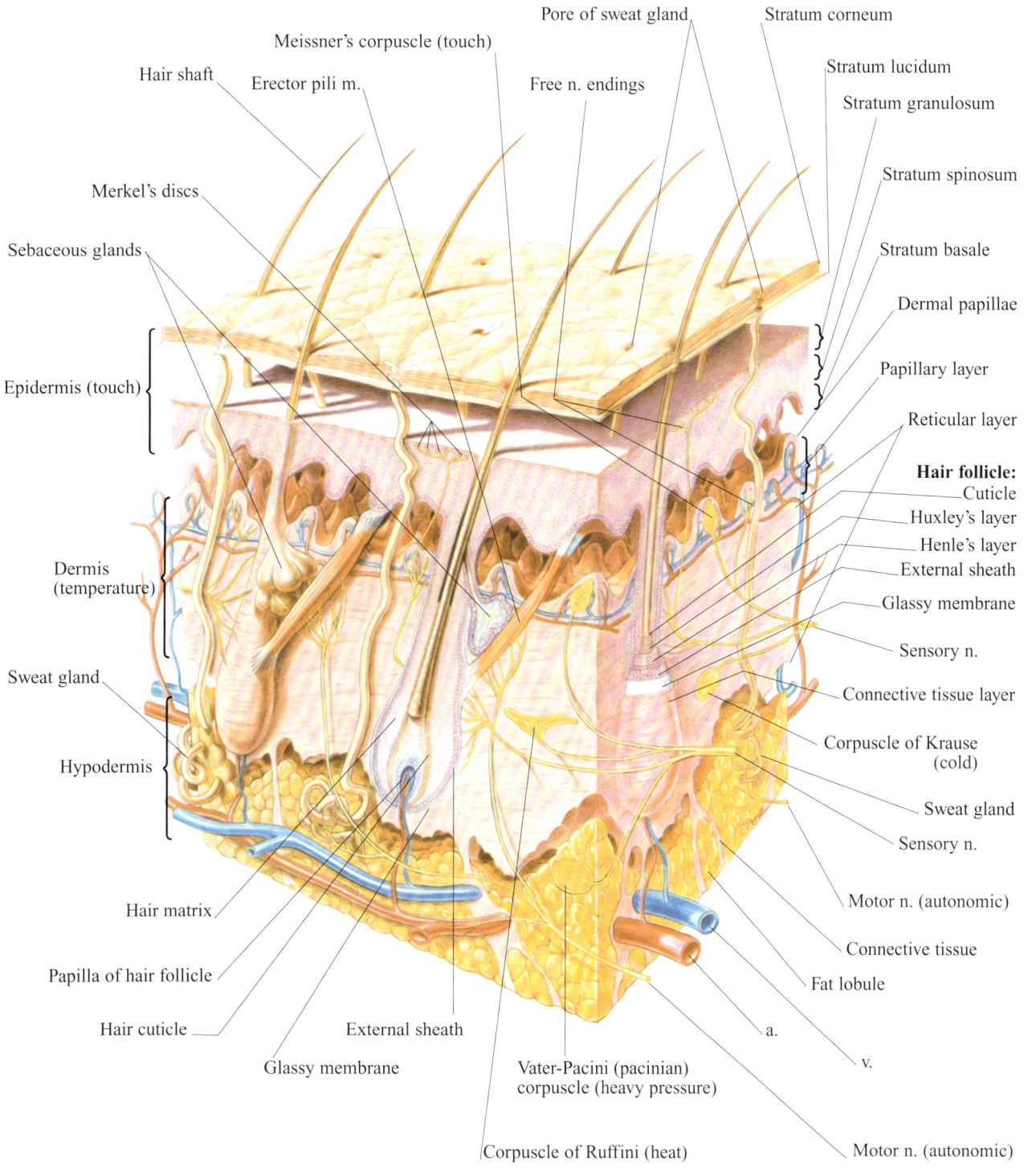

9
DIGESTIVE SYSTEM

DIGESTIVE SYSTEM

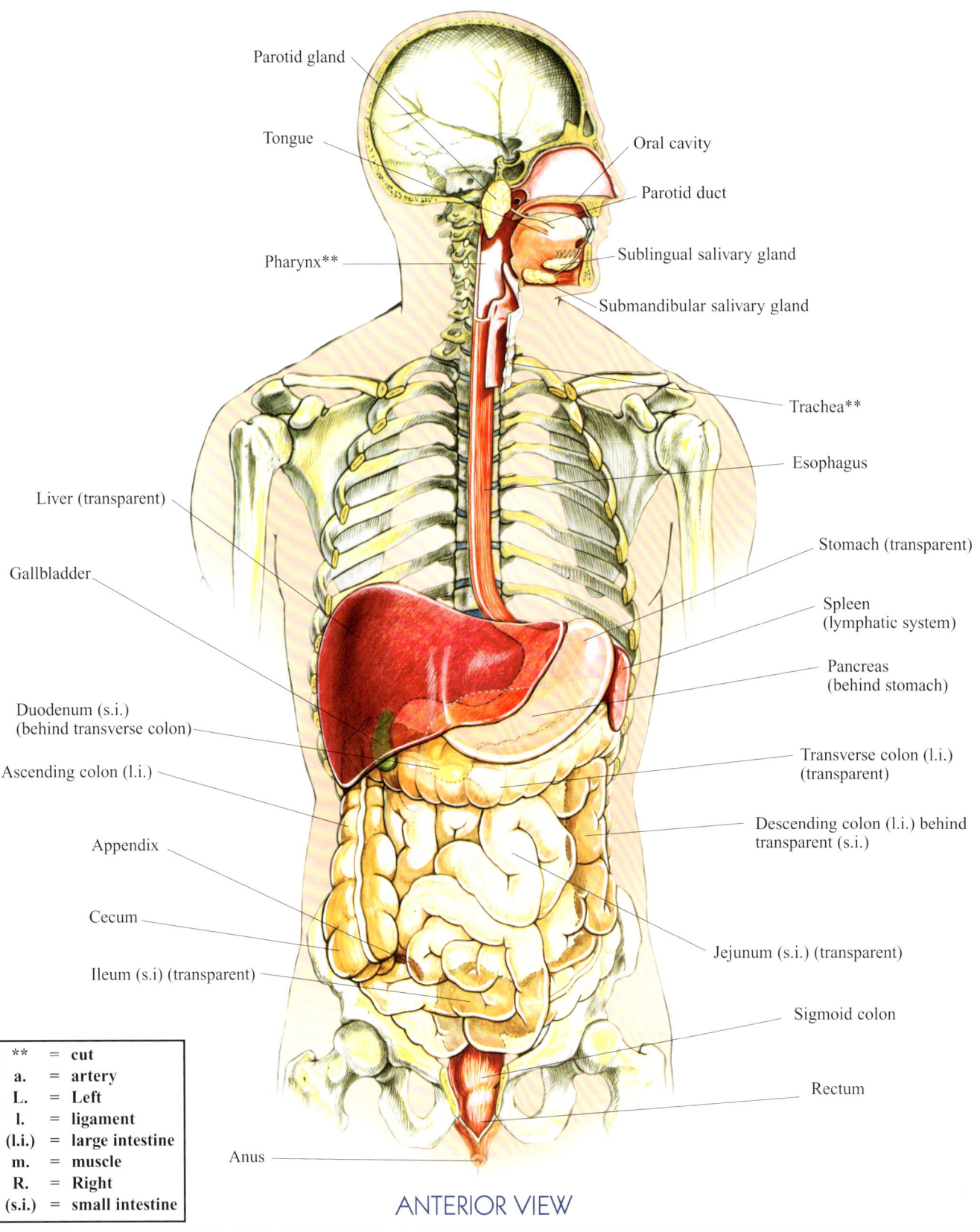

DIGESTIVE SYSTEM

DIGESTIVE SYSTEM

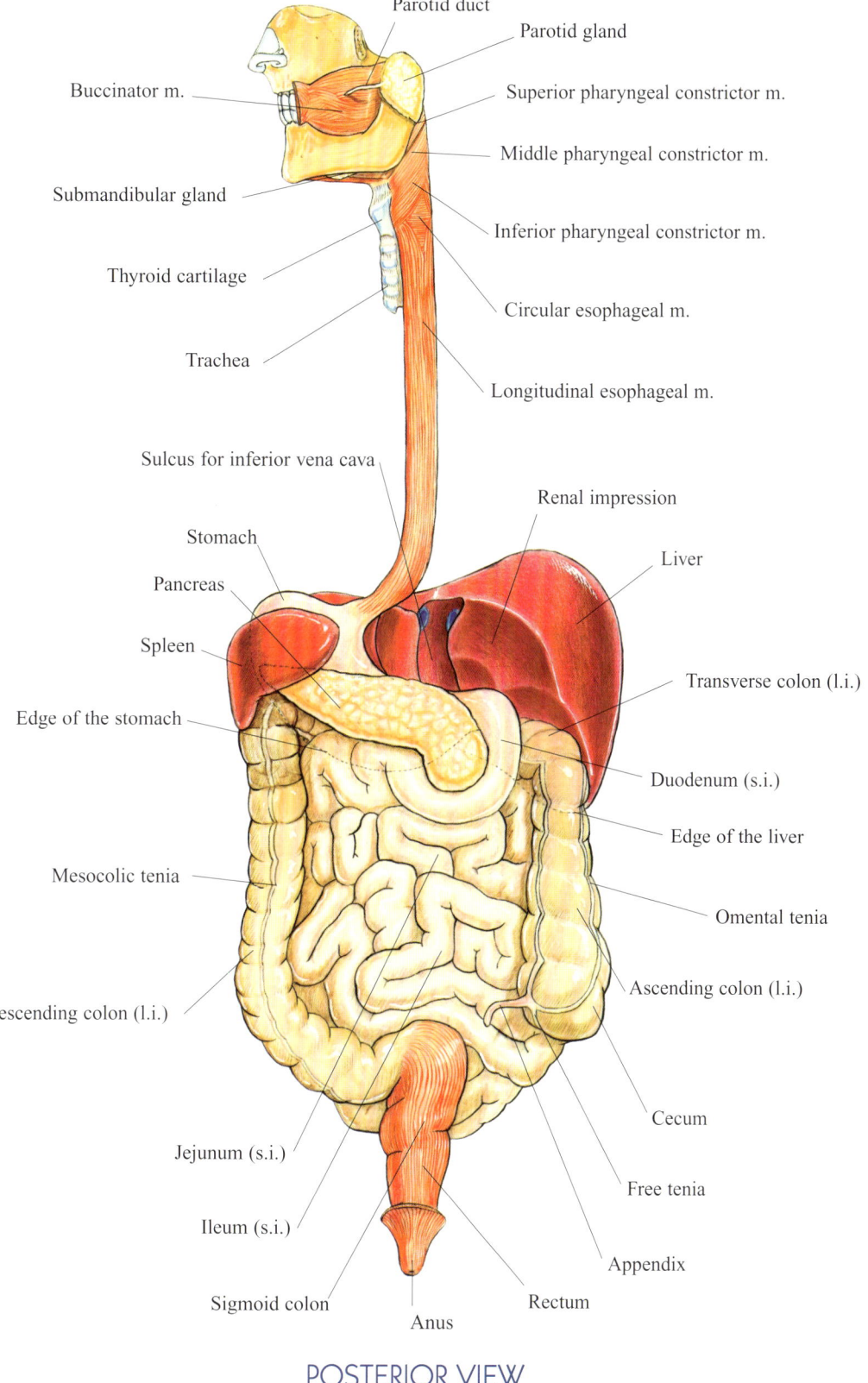

POSTERIOR VIEW

DIGESTIVE SYSTEM

MOUTH & SALIVARY GLANDS

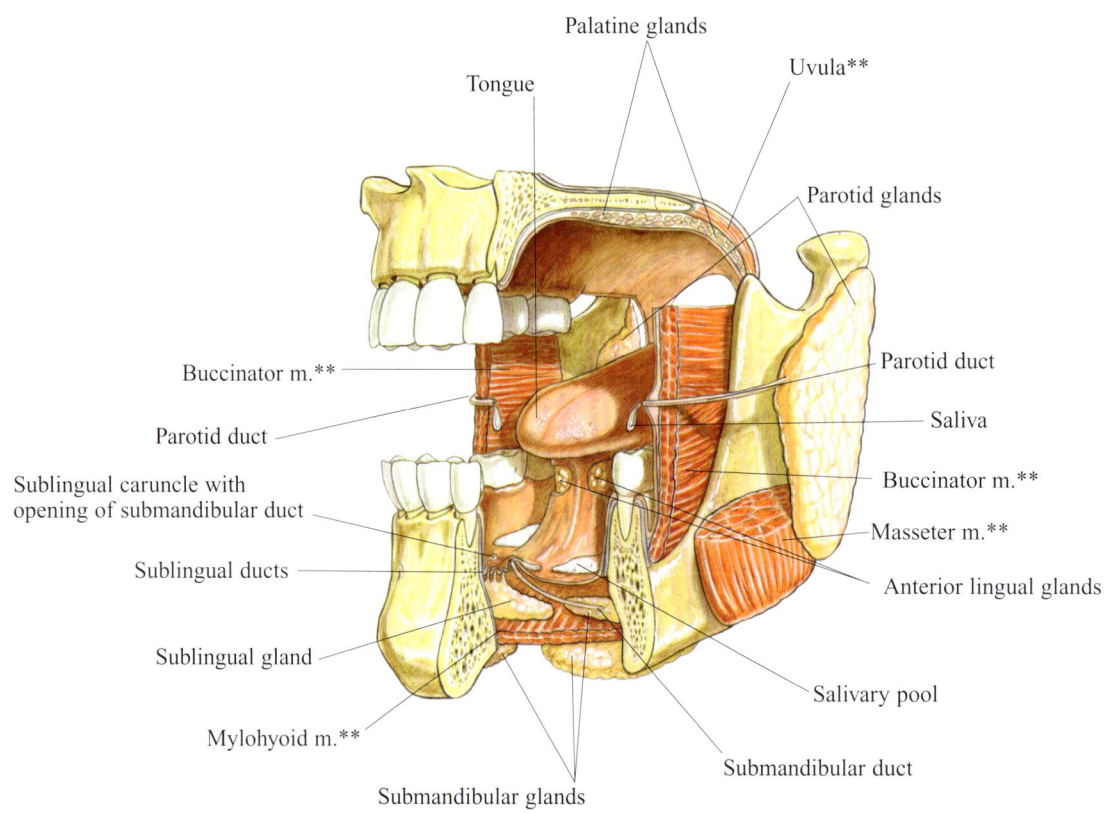

TONGUE

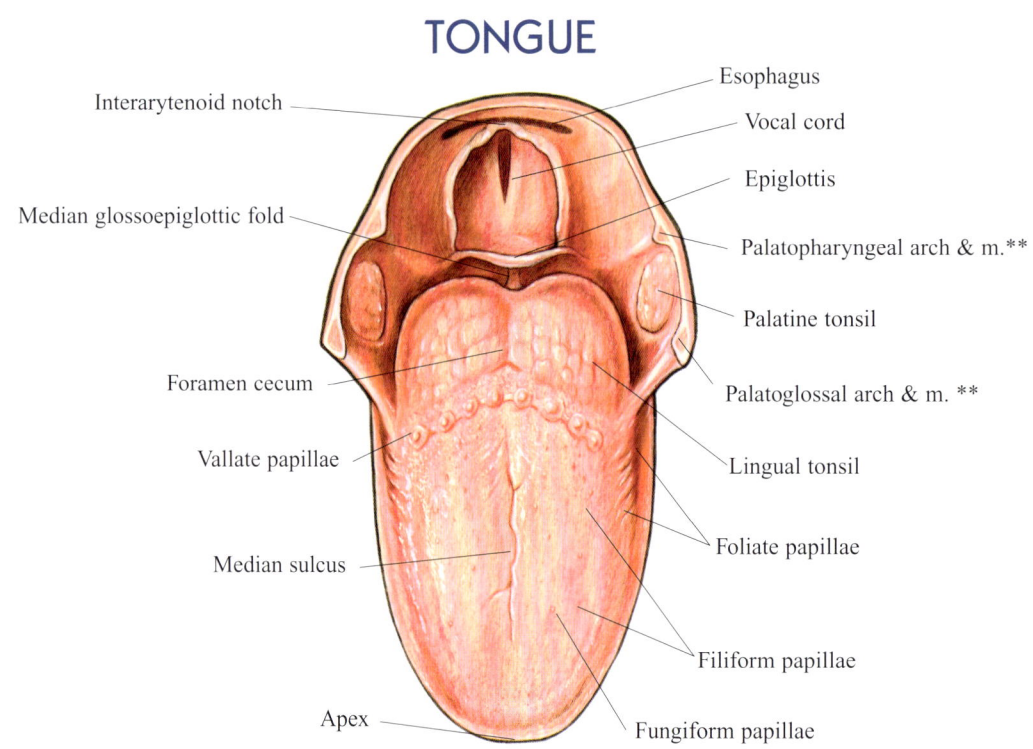

DIGESTIVE SYSTEM

STOMACH

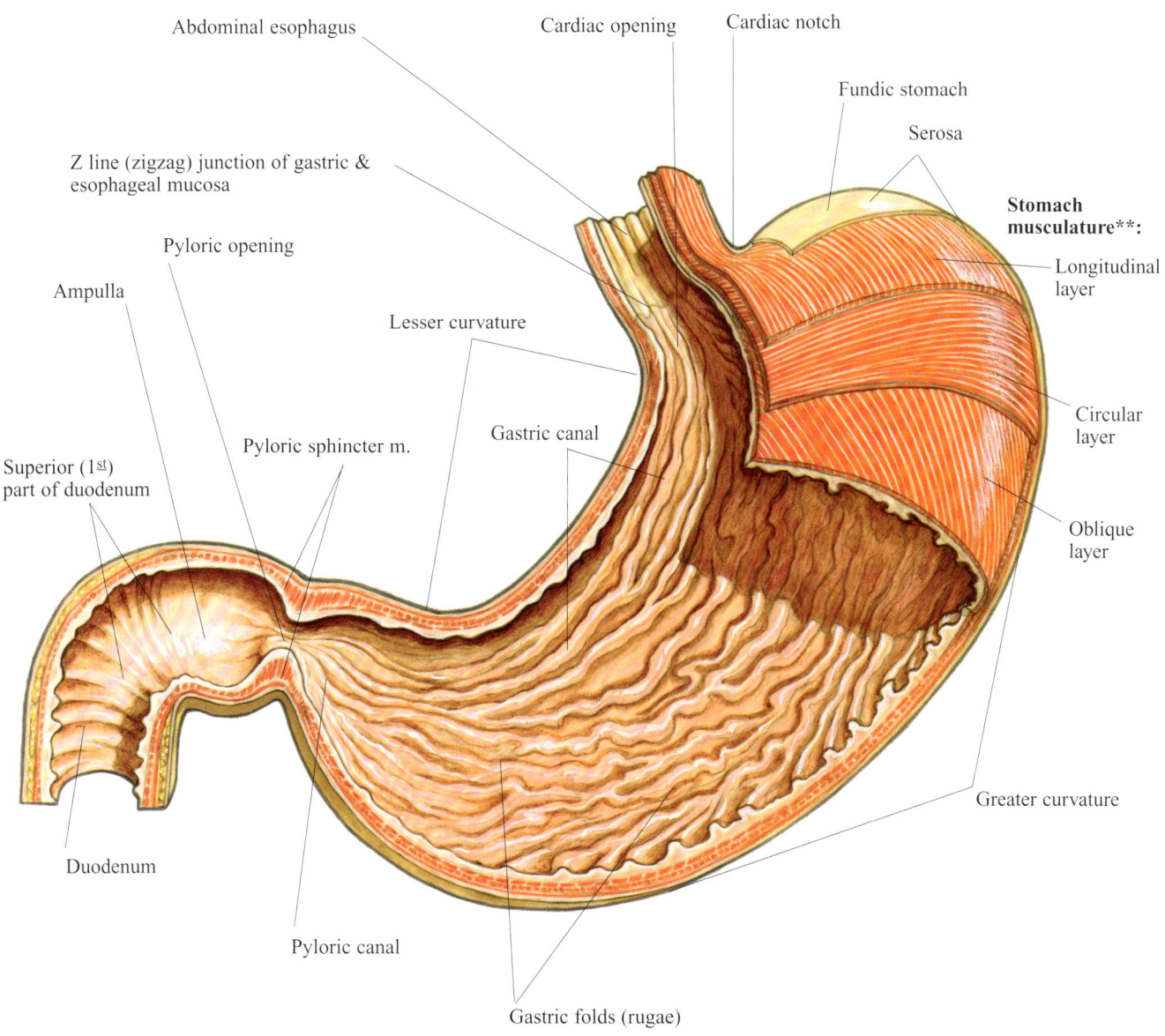

DIGESTIVE SYSTEM

BILE & PANCREATIC DUCT

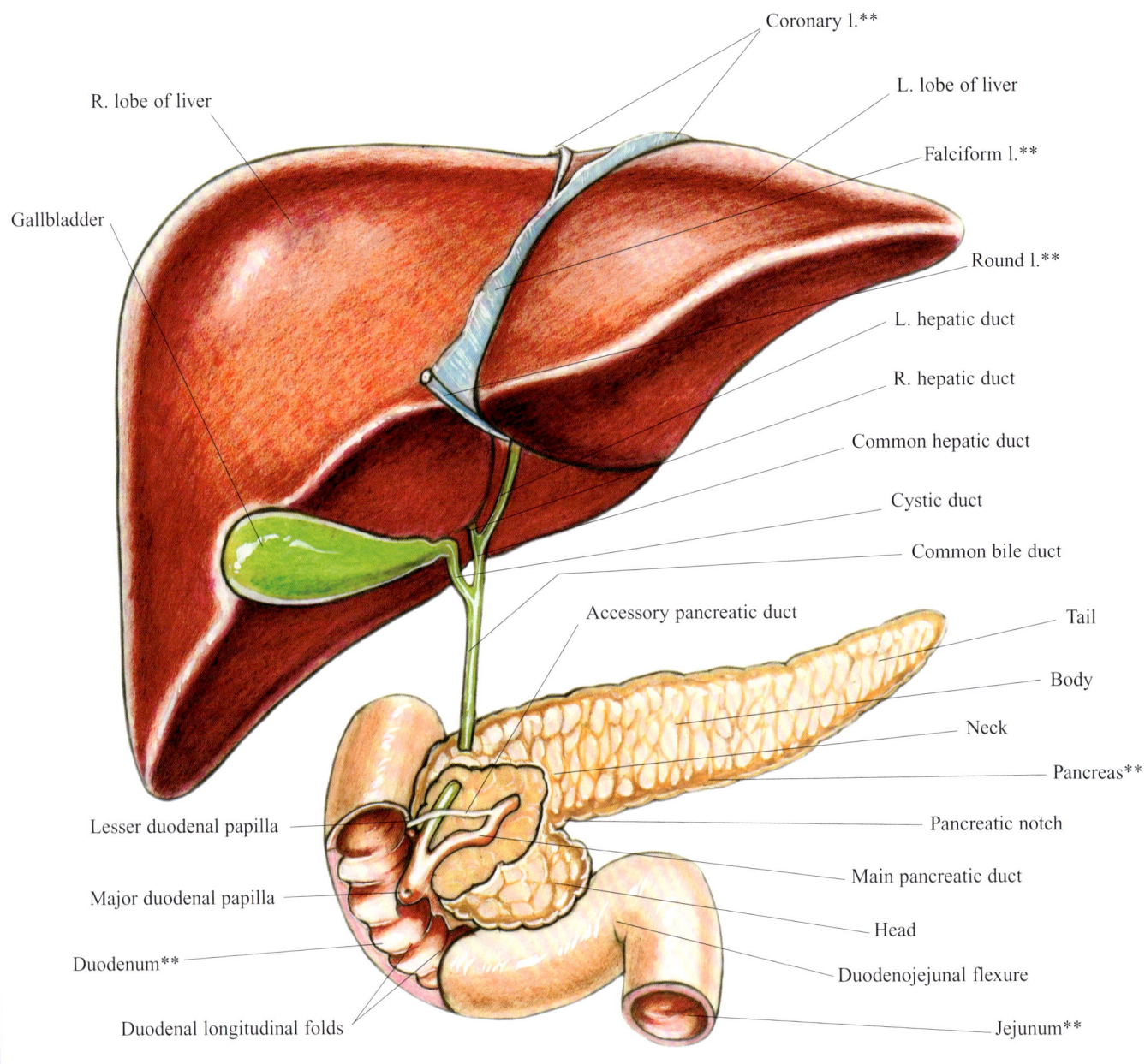

DIGESTIVE SYSTEM

SMALL INTESTINE
(SCHEMATIC)

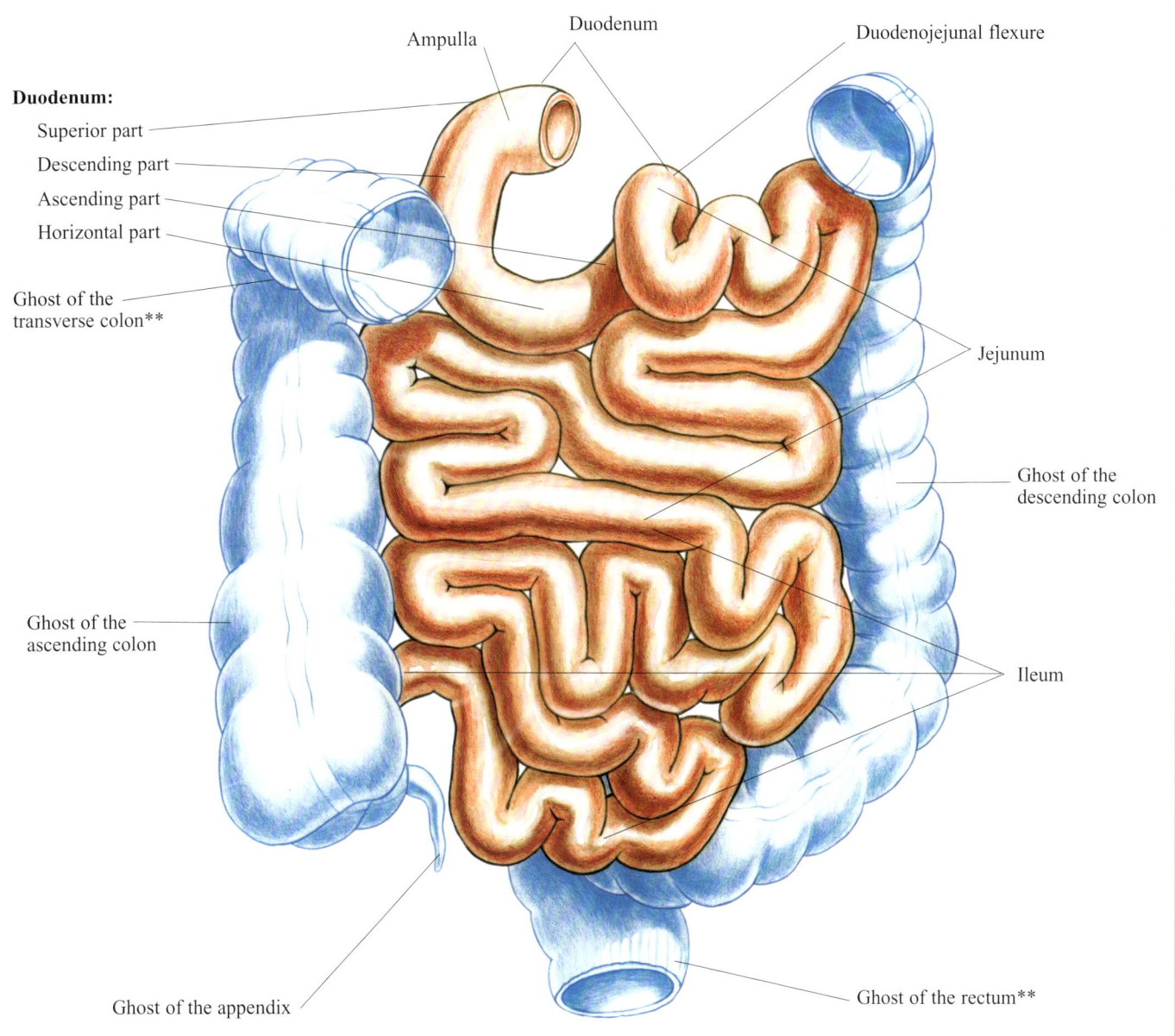

DIGESTIVE SYSTEM

LARGE INTESTINE

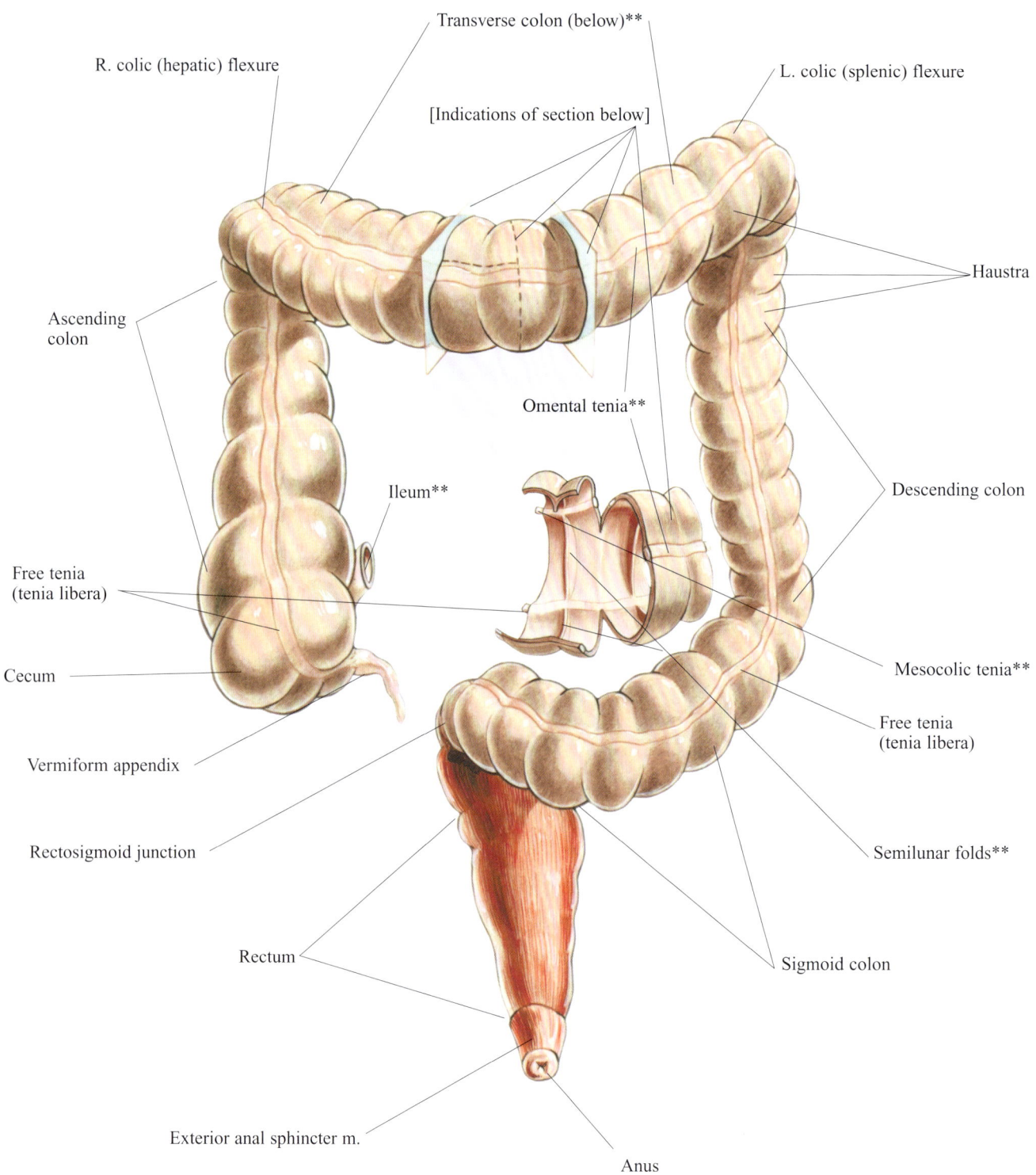

DIGESTIVE SYSTEM

ILEOCECAL SPHINCTER & APPENDIX

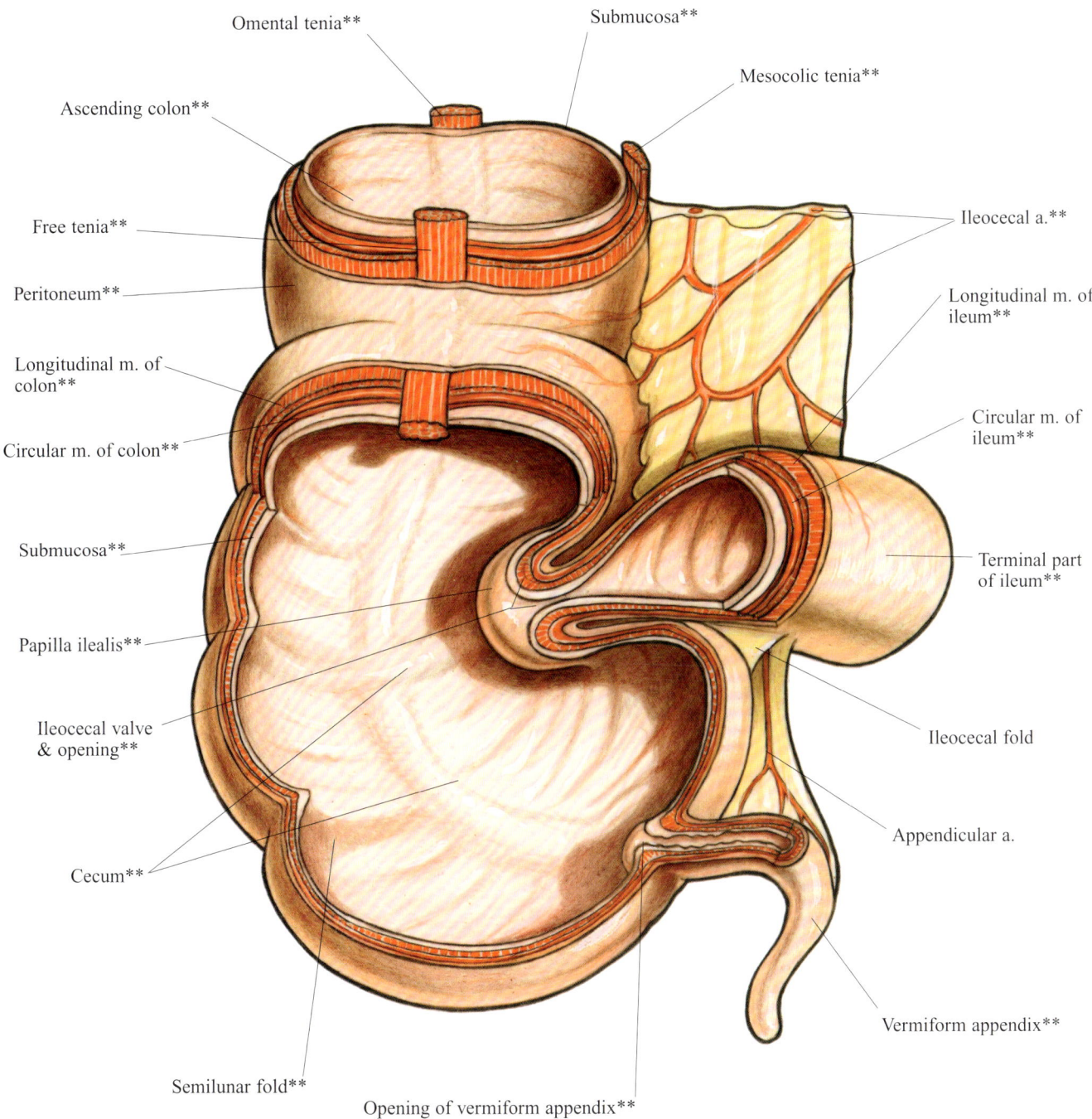

DIGESTIVE SYSTEM

RECTUM

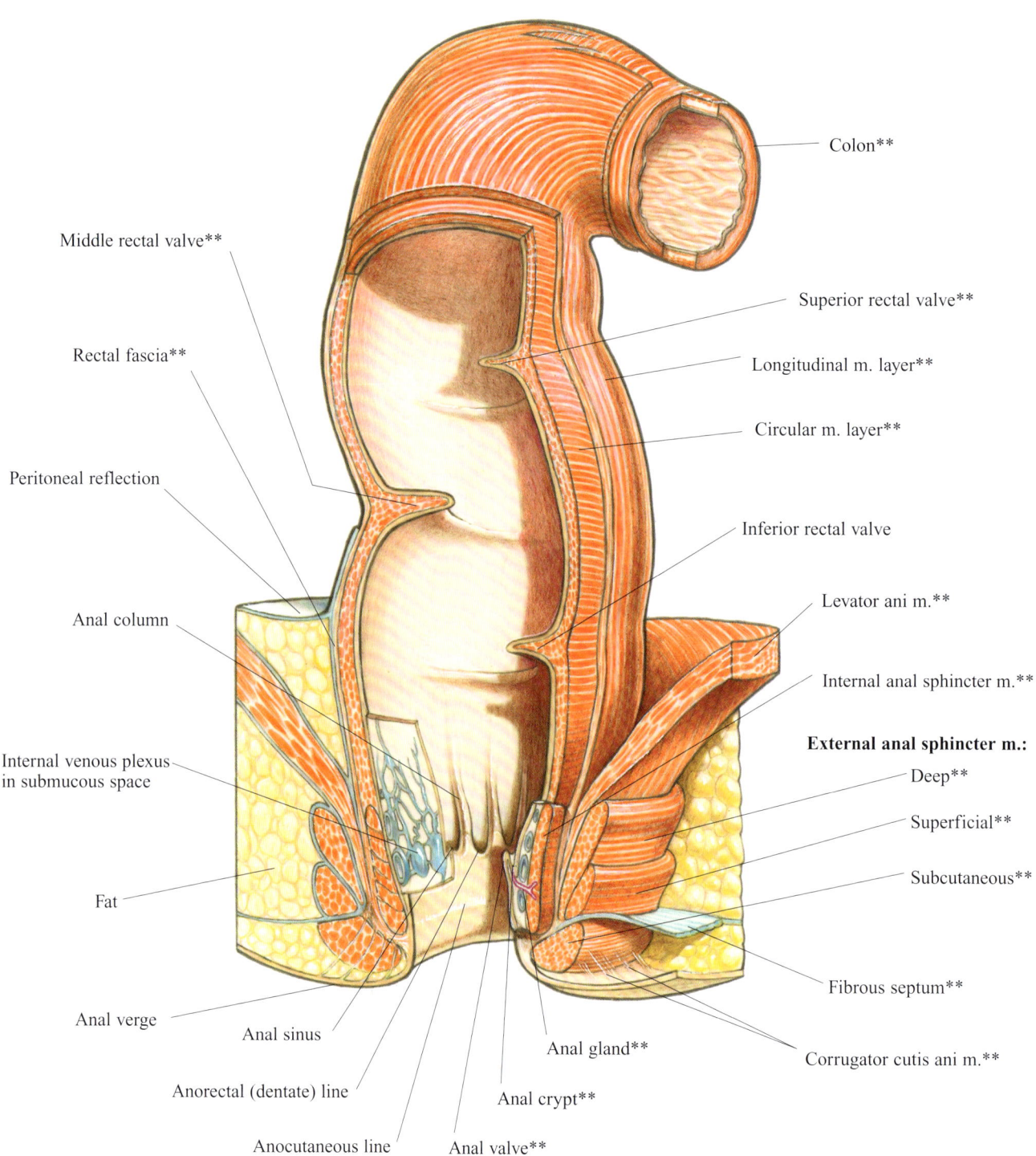

DIGESTIVE SYSTEM

NOTES

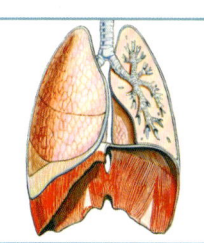

10

RESPIRATORY SYSTEM

RESPIRATORY SYSTEM

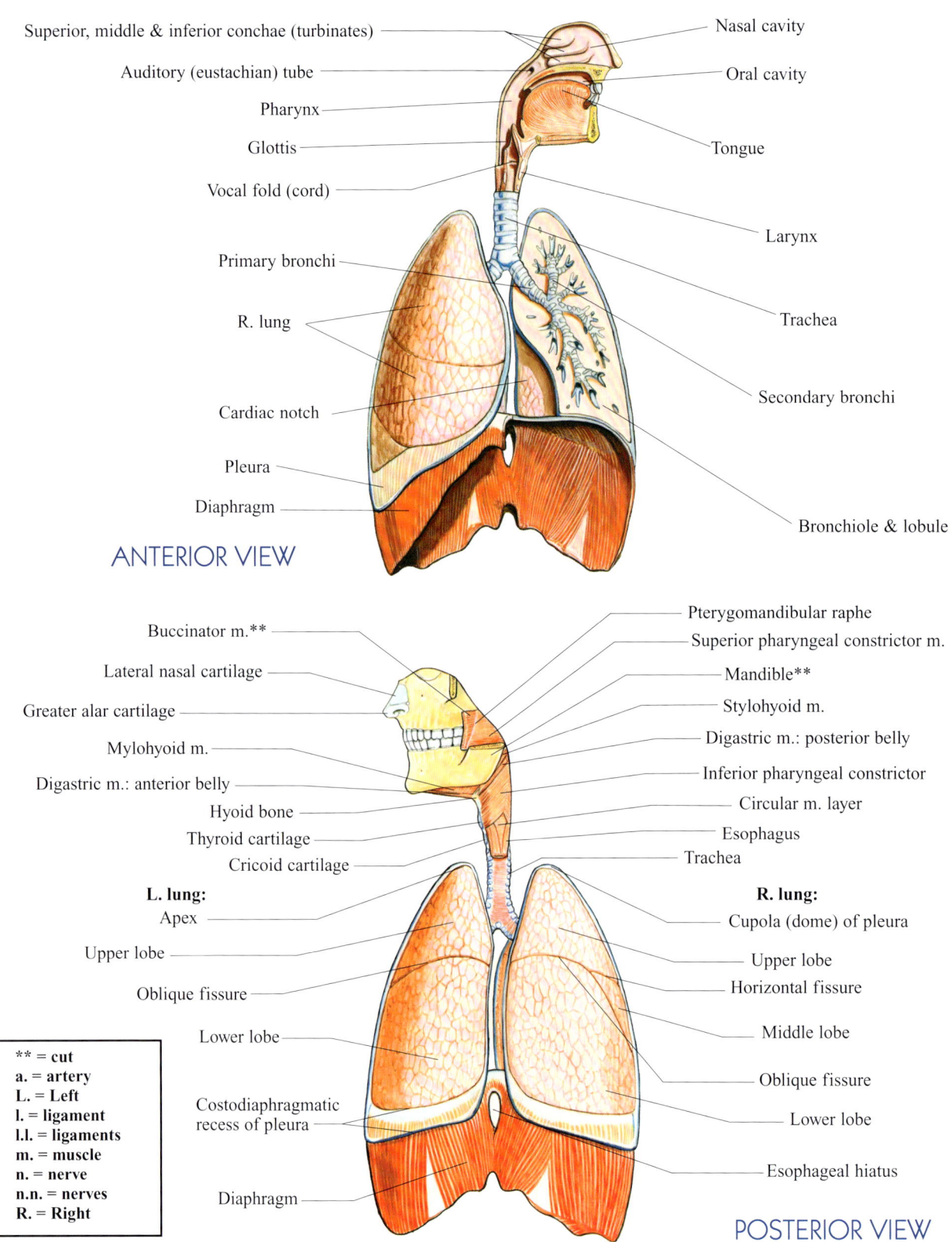

NASAL & ORAL CAVITY

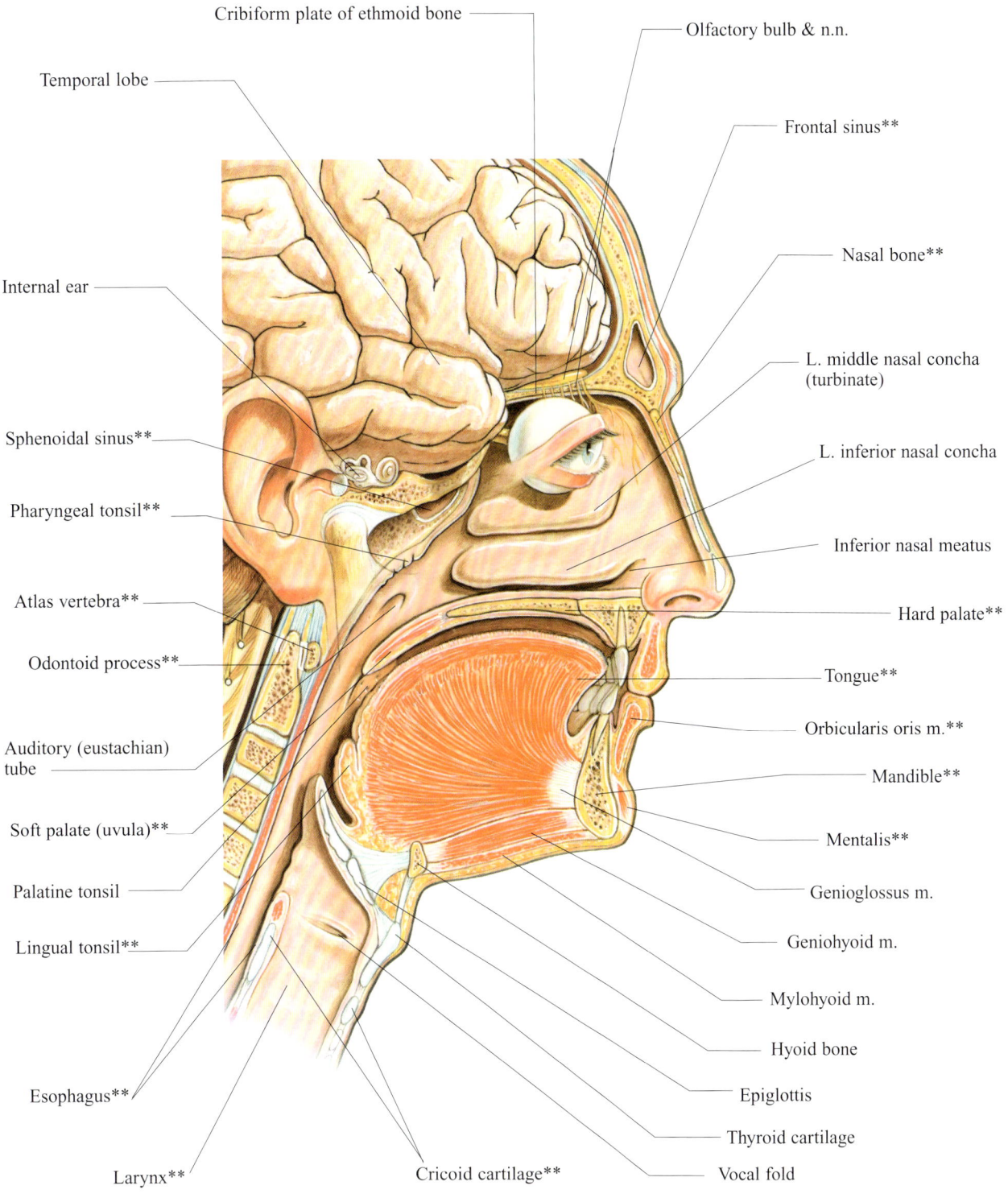

NASAL SEPTUM

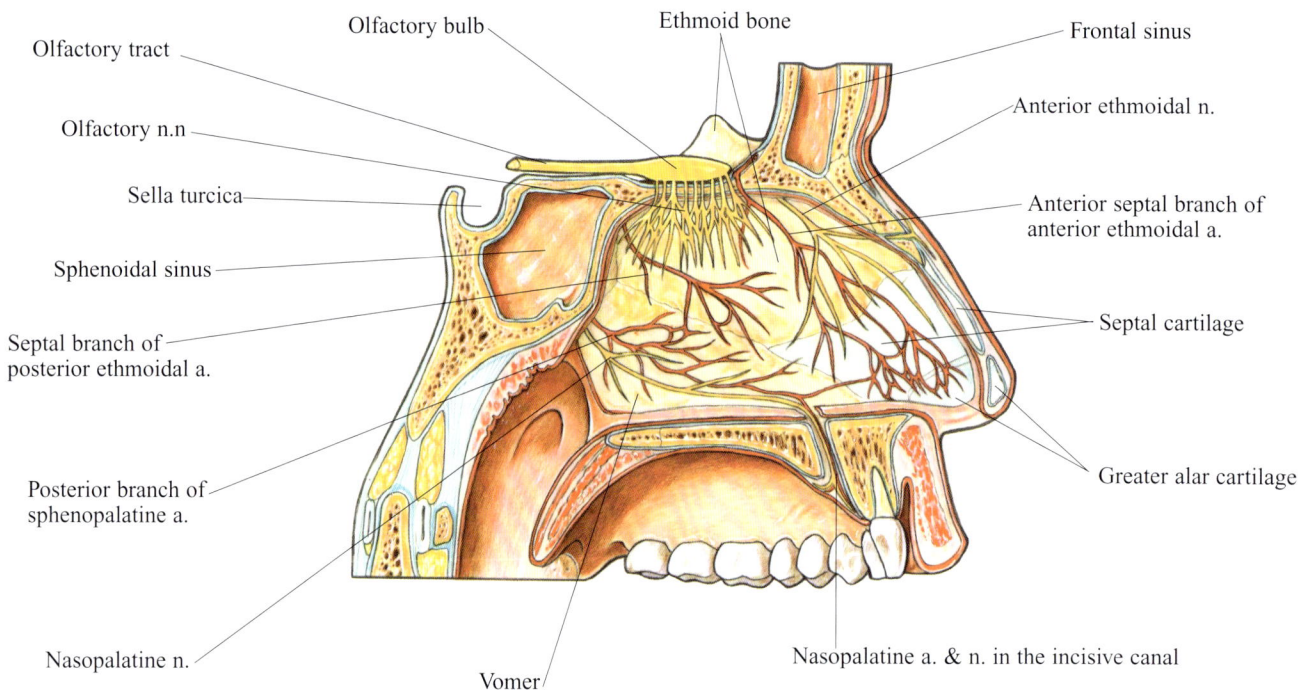

PARANASAL SINUSES

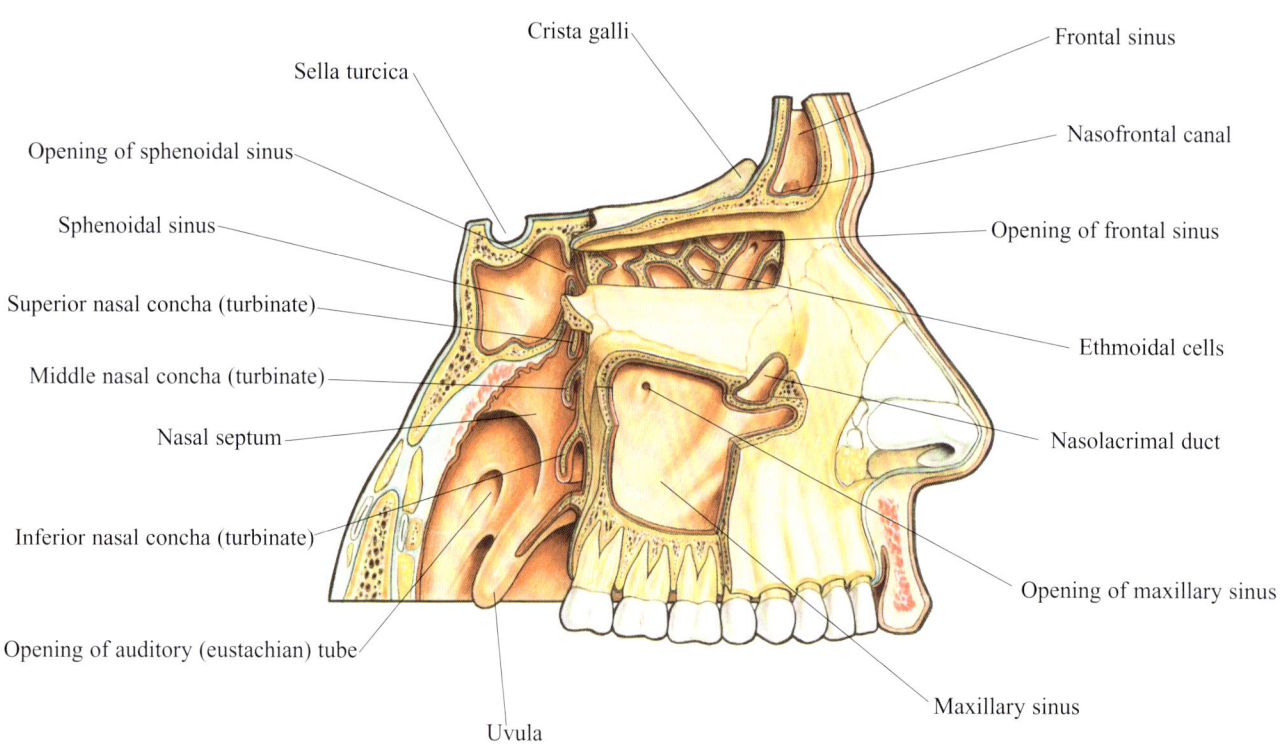

LARYNX

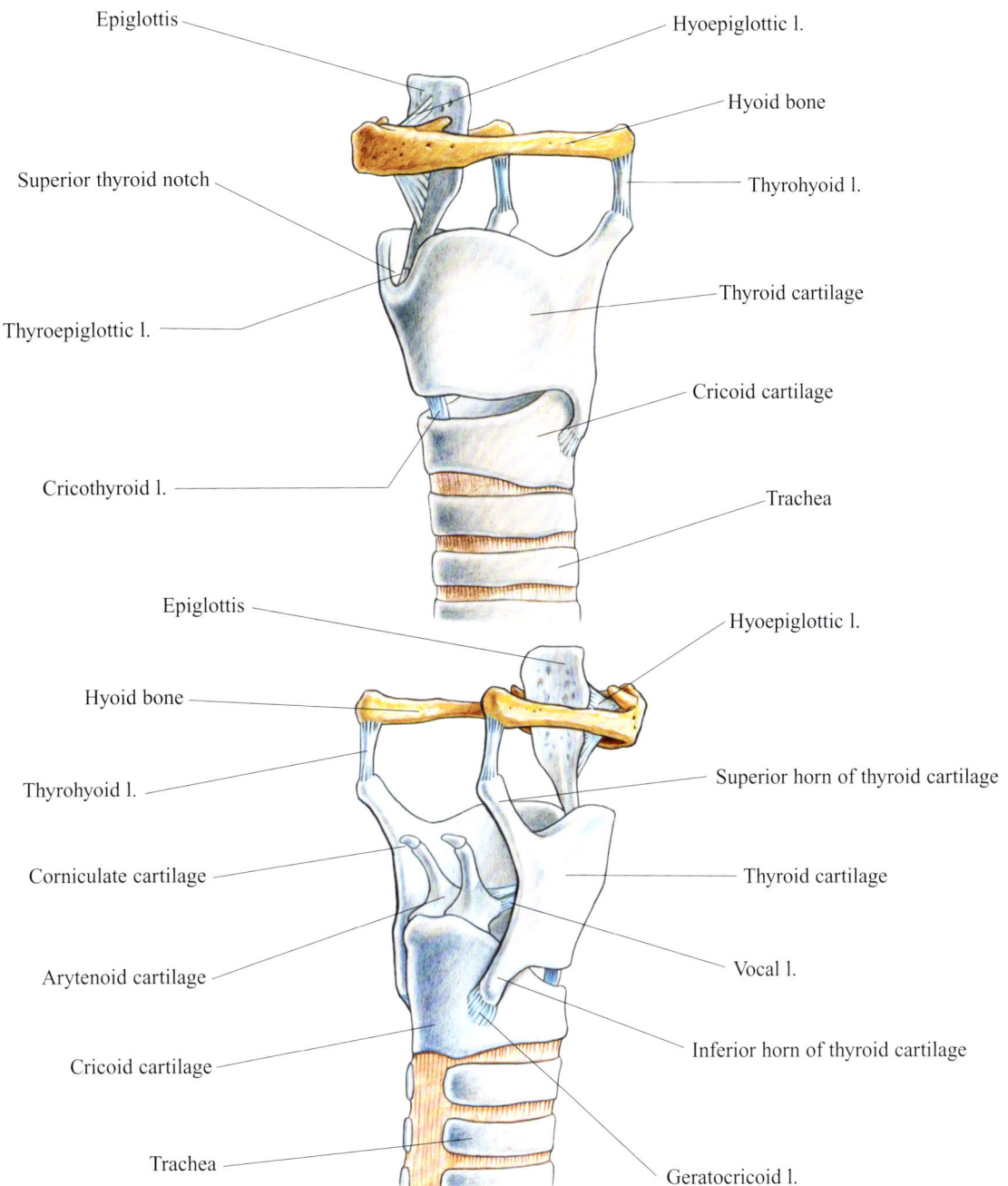

BRONCHIAL TREE

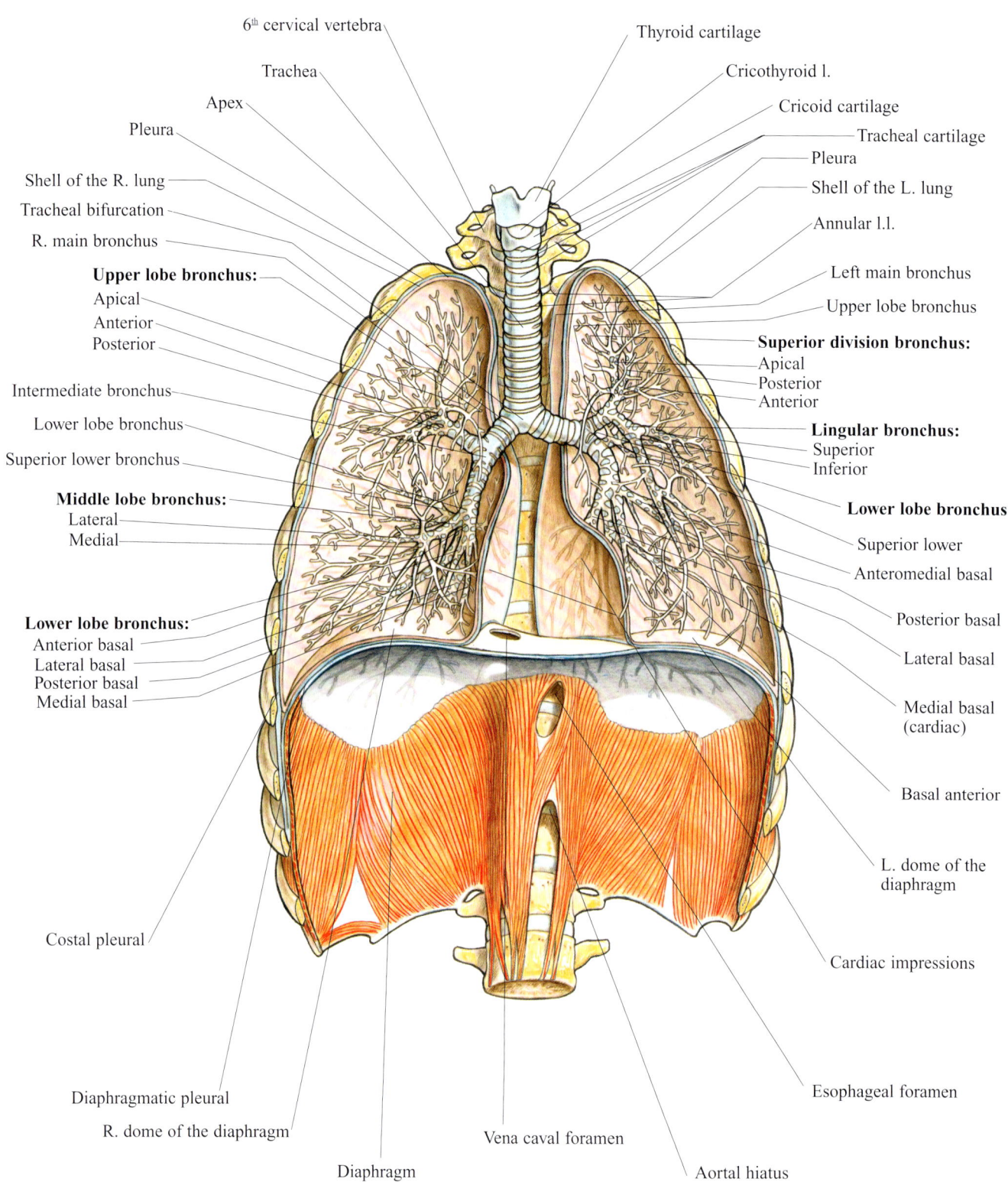

MUSCLES OF RESPIRATION

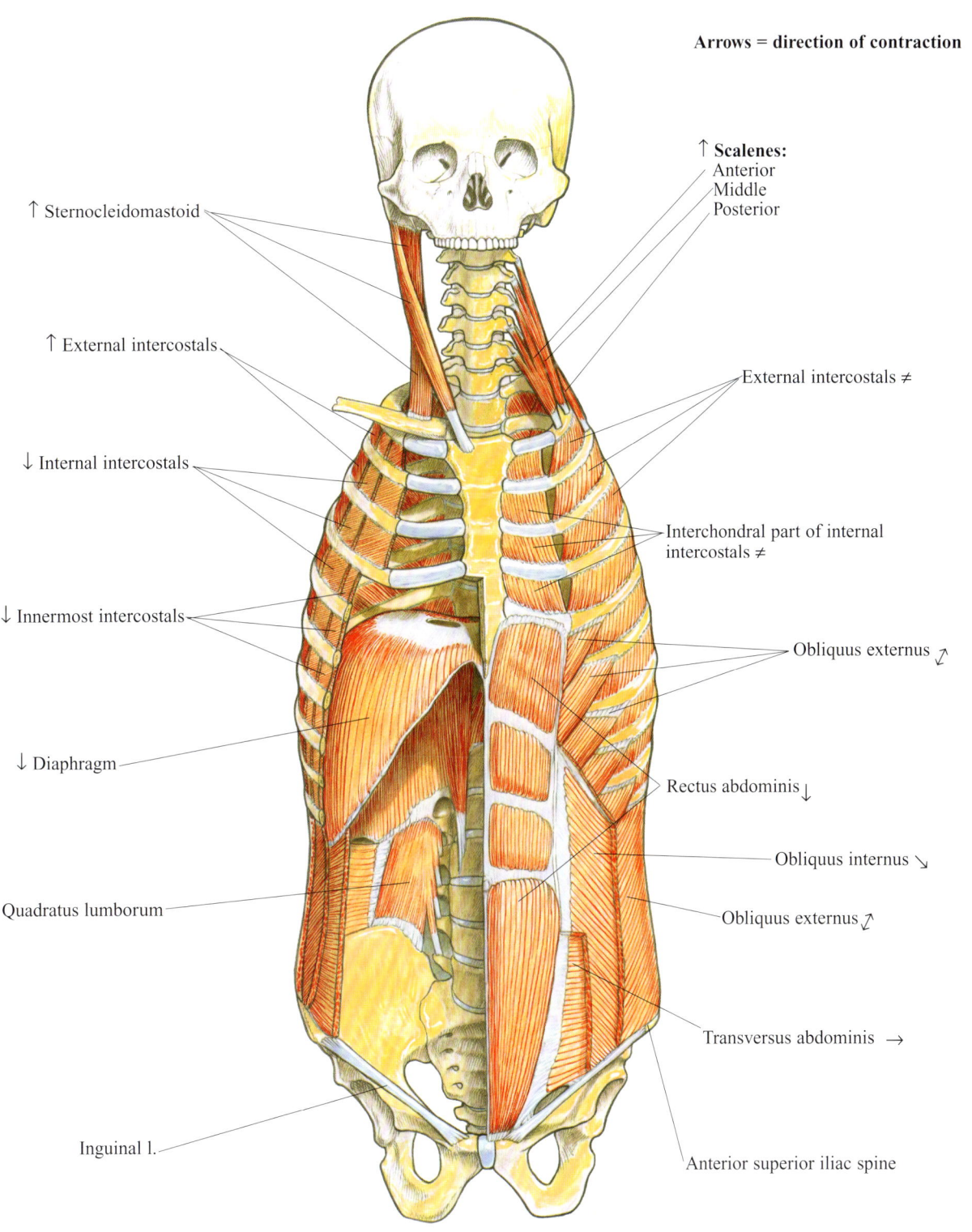

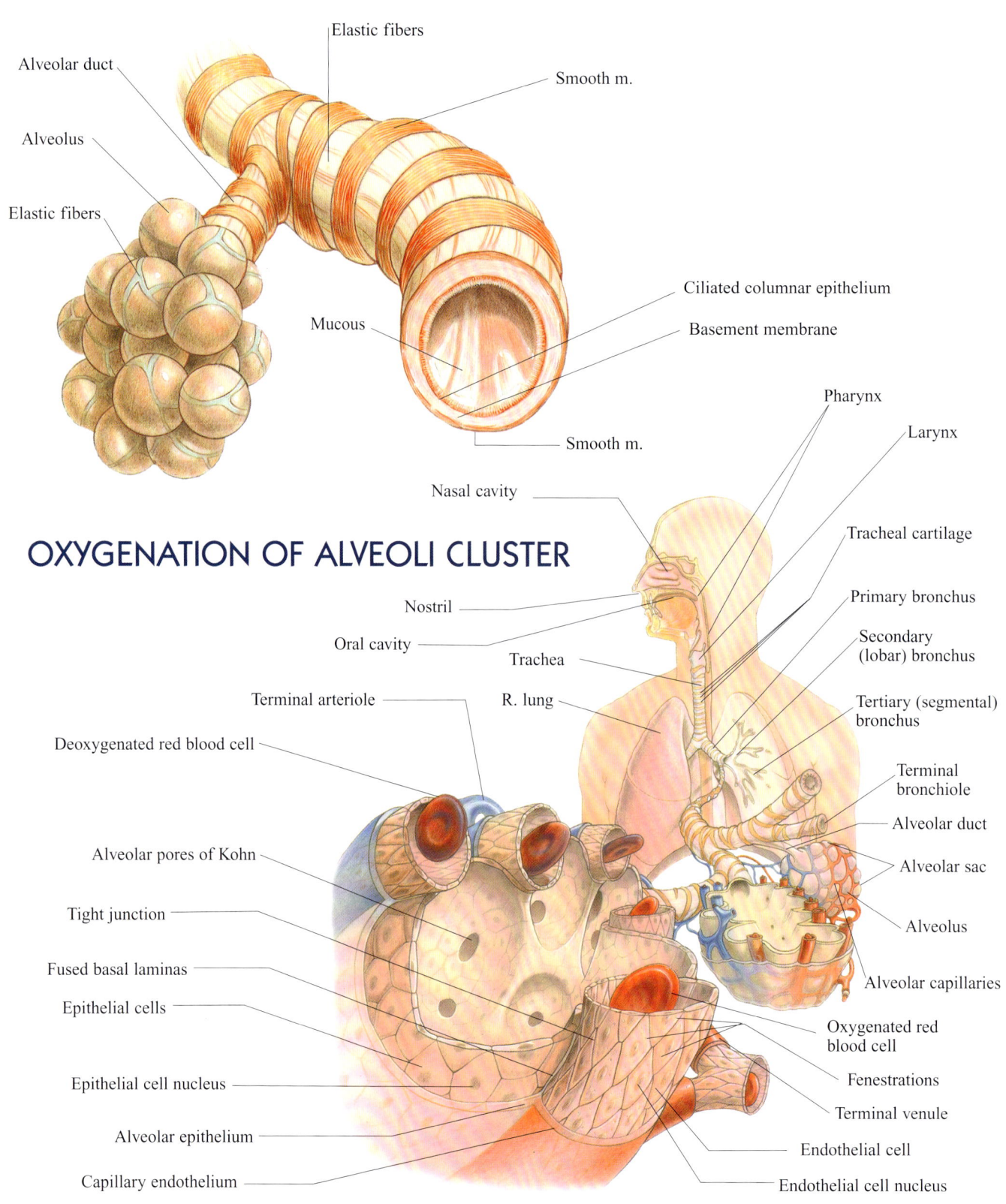

RESPIRATORY SYSTEM

NOTES

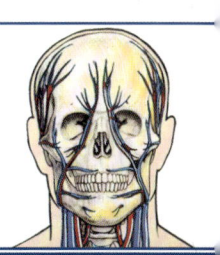

11
CIRCULATORY SYSTEM

CIRCULATORY SYSTEM

CIRCULATORY SYSTEM

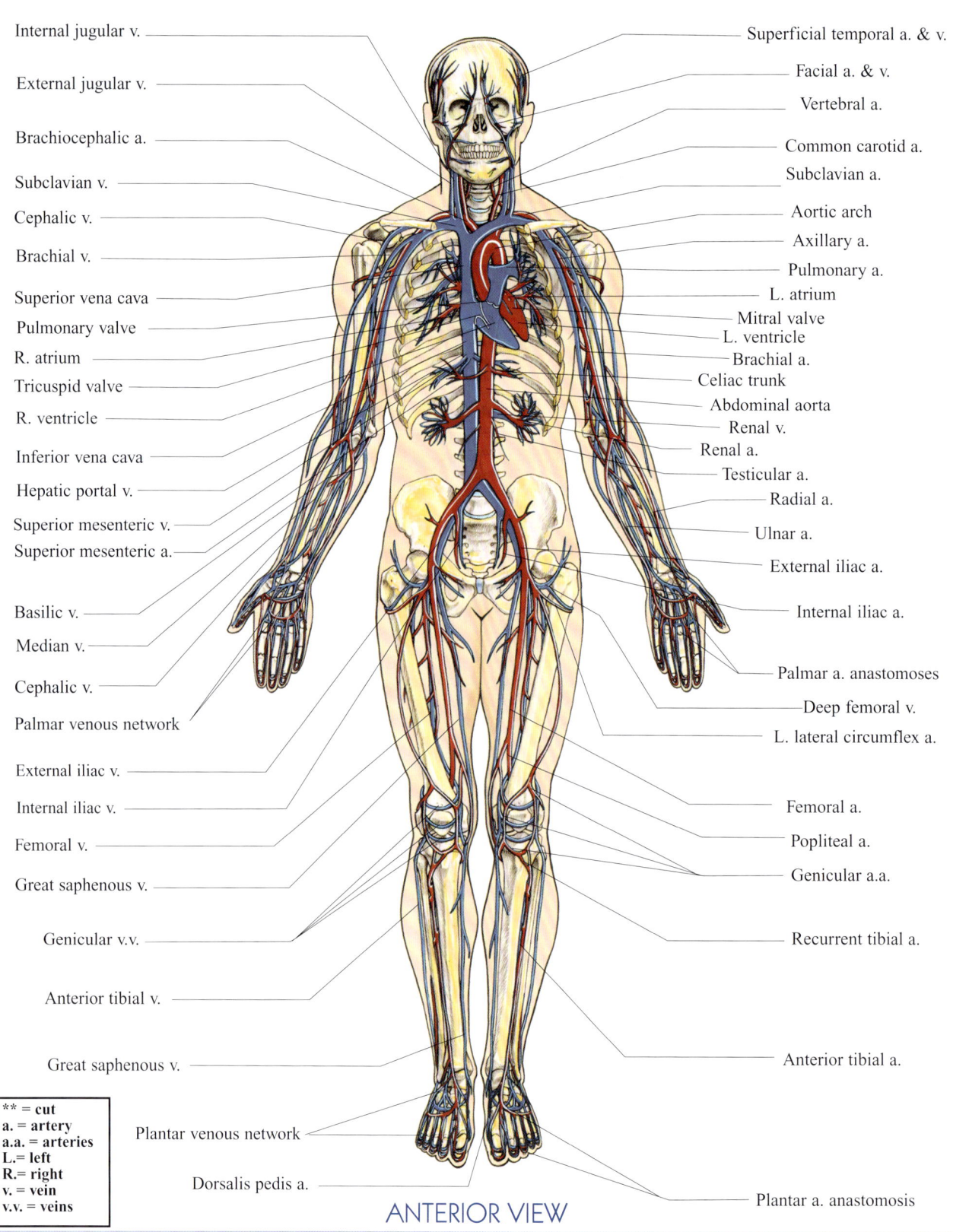

ANTERIOR VIEW

CIRCULATORY SYSTEM

CIRCULATORY SYSTEM

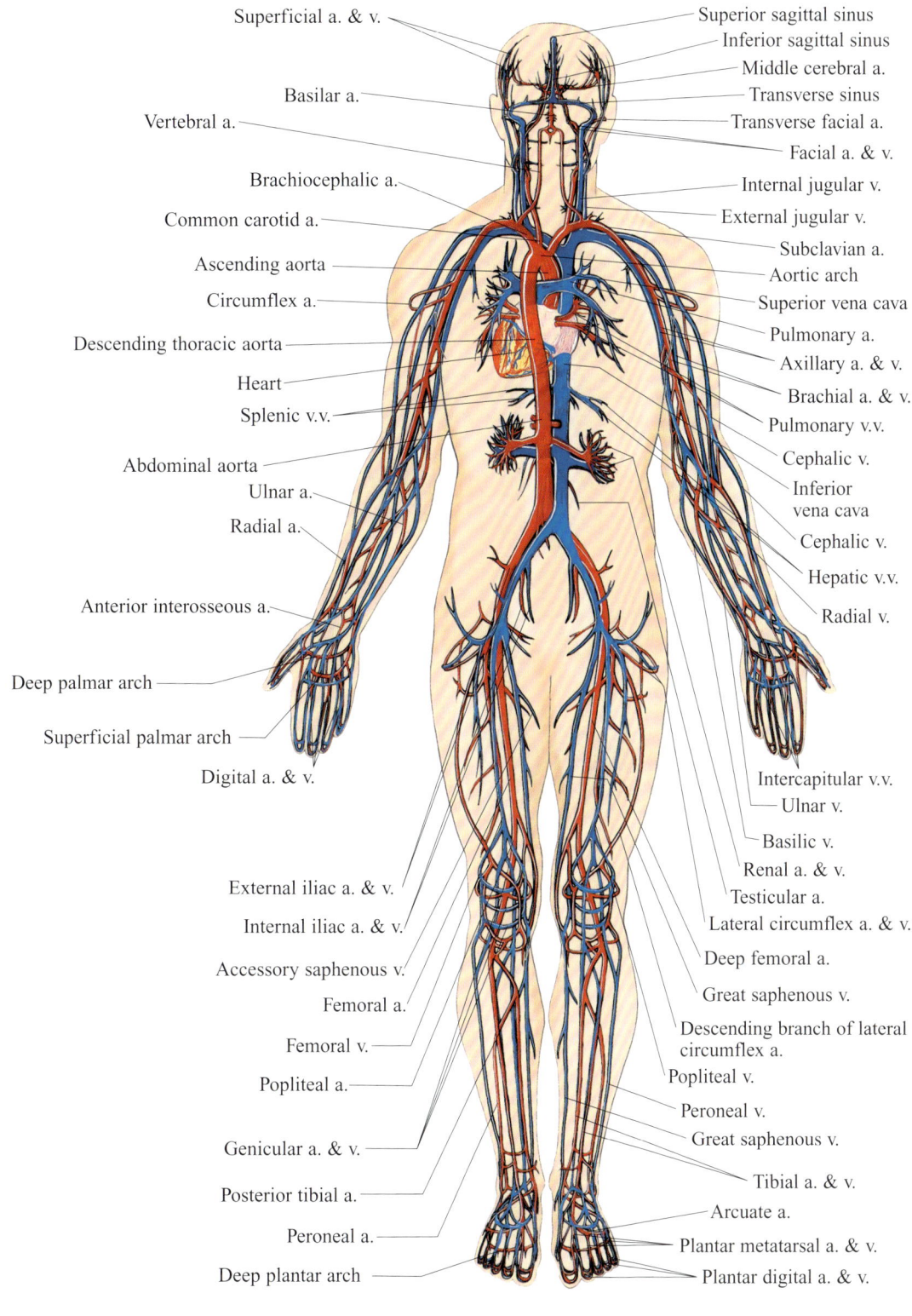

POSTERIOR VIEW

CIRCULATORY SYSTEM

VENOUS SYSTEM

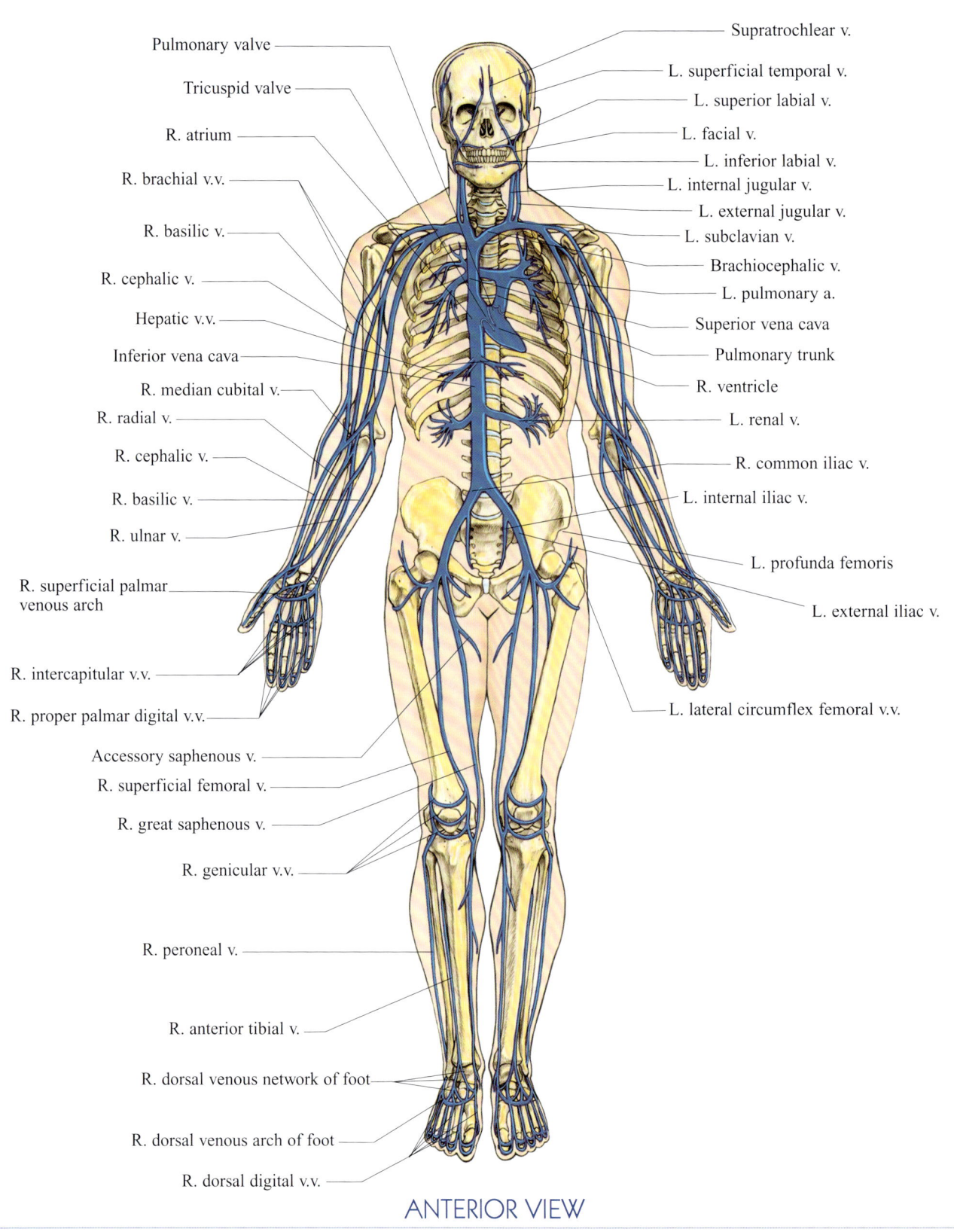

ANTERIOR VIEW

CIRCULATORY SYSTEM

ARTERIAL SYSTEM

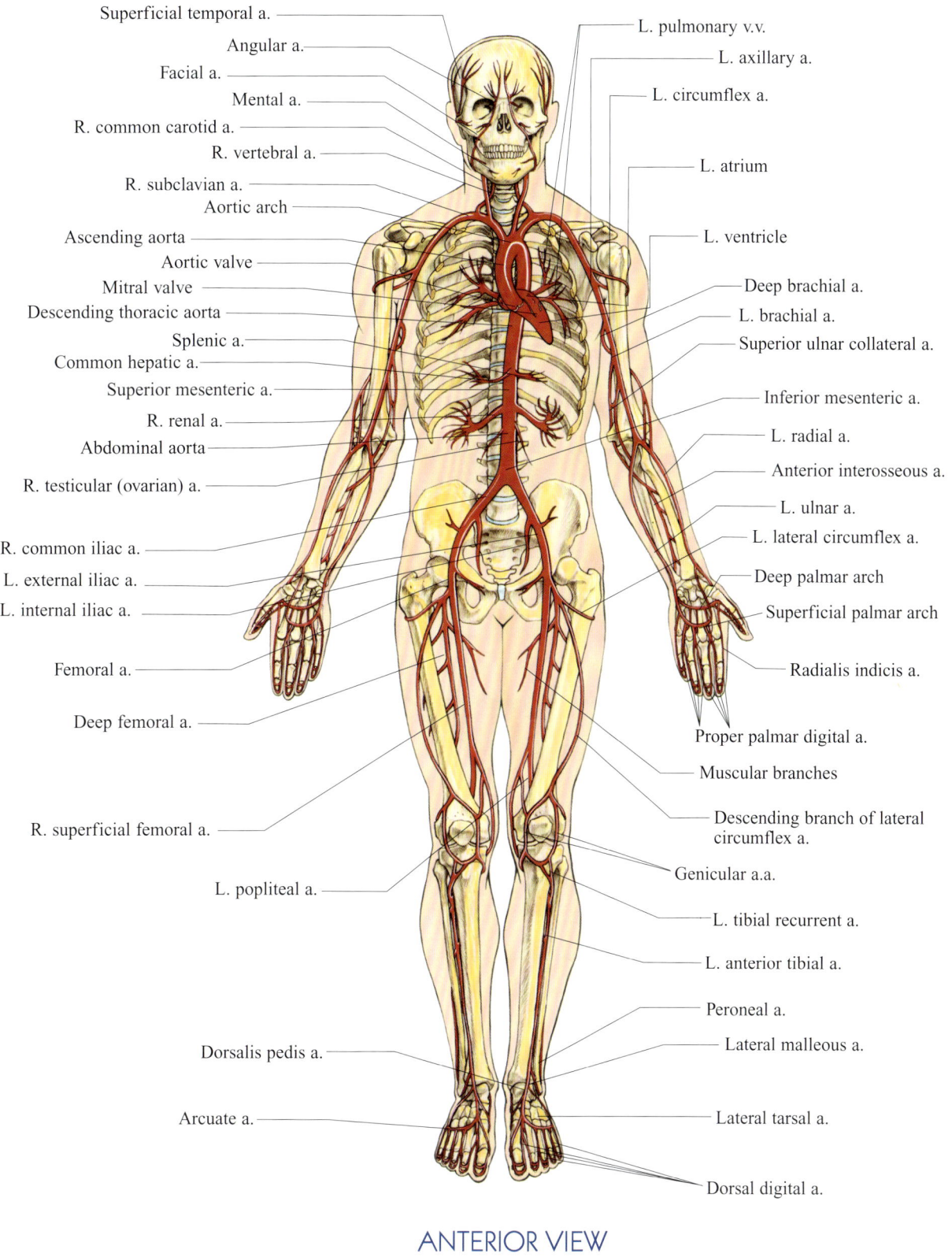

ANTERIOR VIEW

CIRCULATORY SYSTEM

HEAD & NECK
(SCHEMATIC)

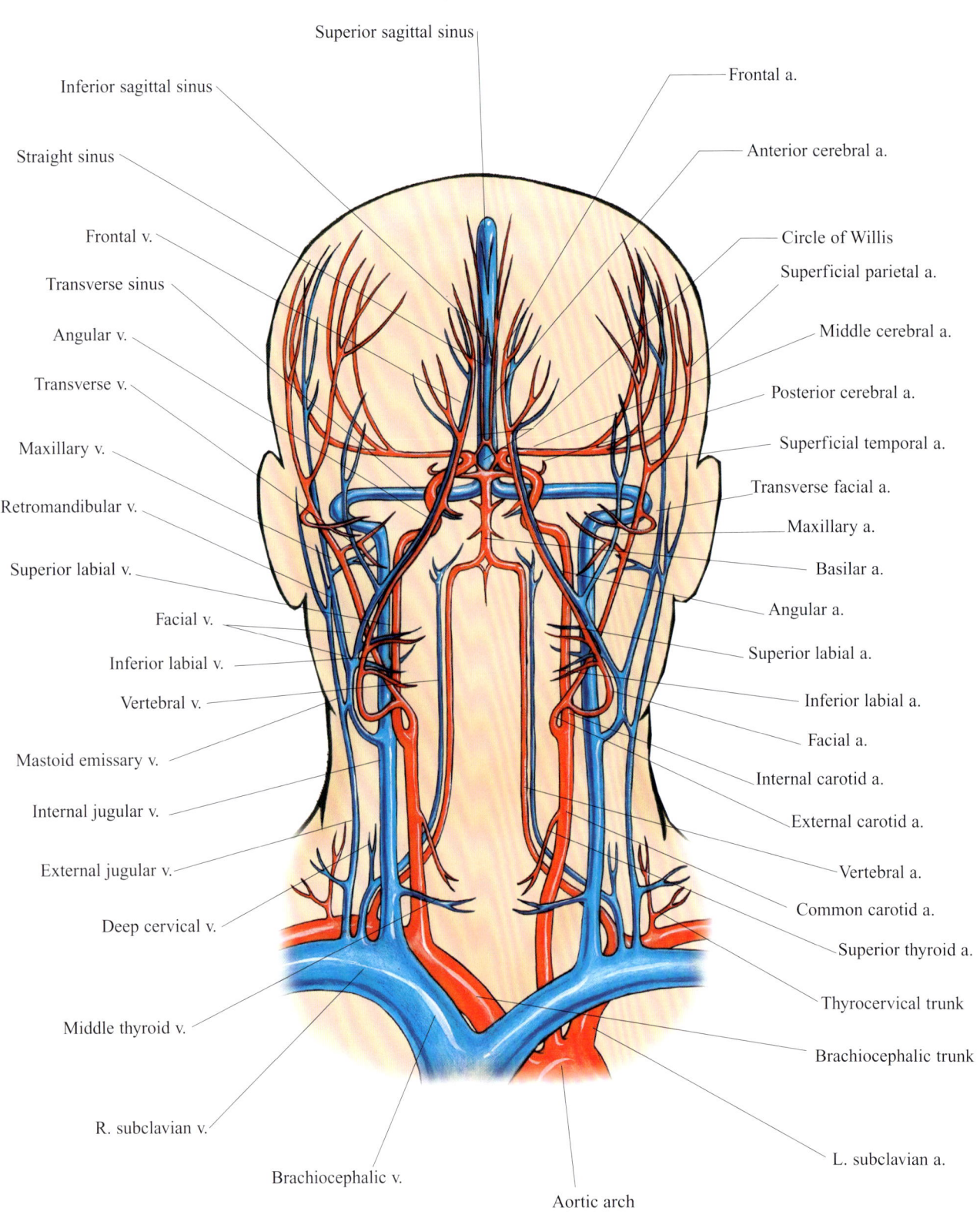

CIRCULATORY SYSTEM

SKULL & ARTERIES

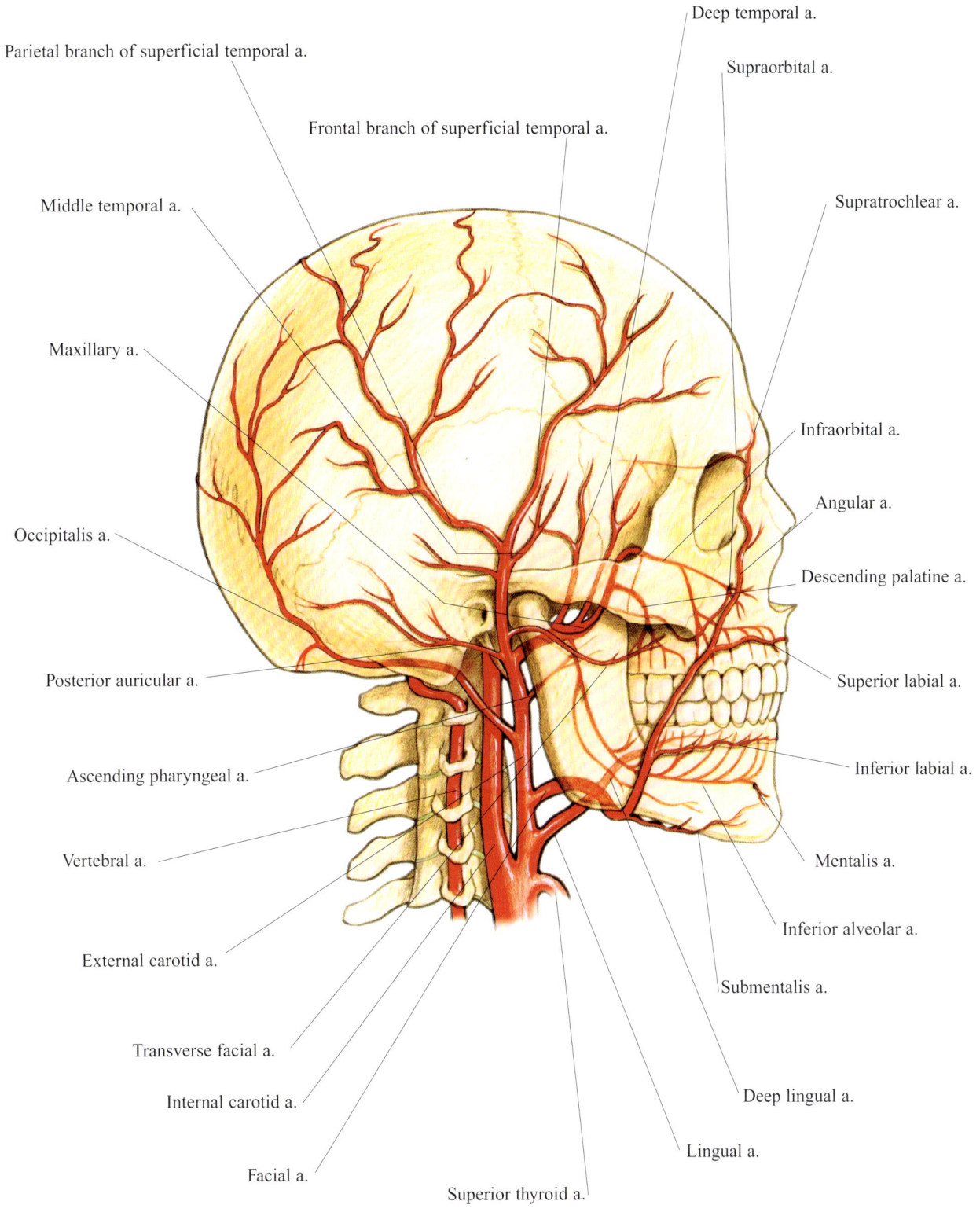

161

CIRCULATORY SYSTEM

ARTERIES OF BRAIN & CIRCLE OF WILLIS

INFERIOR VIEW

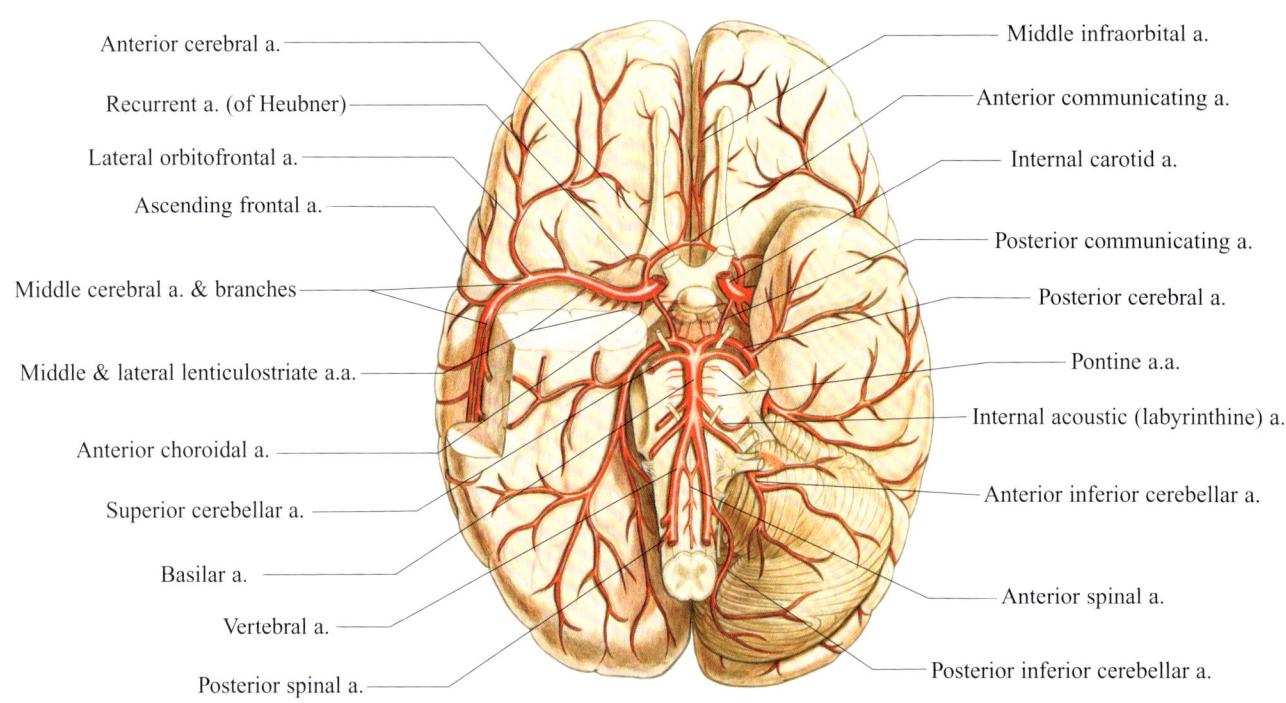

CIRCLE OF WILLIS

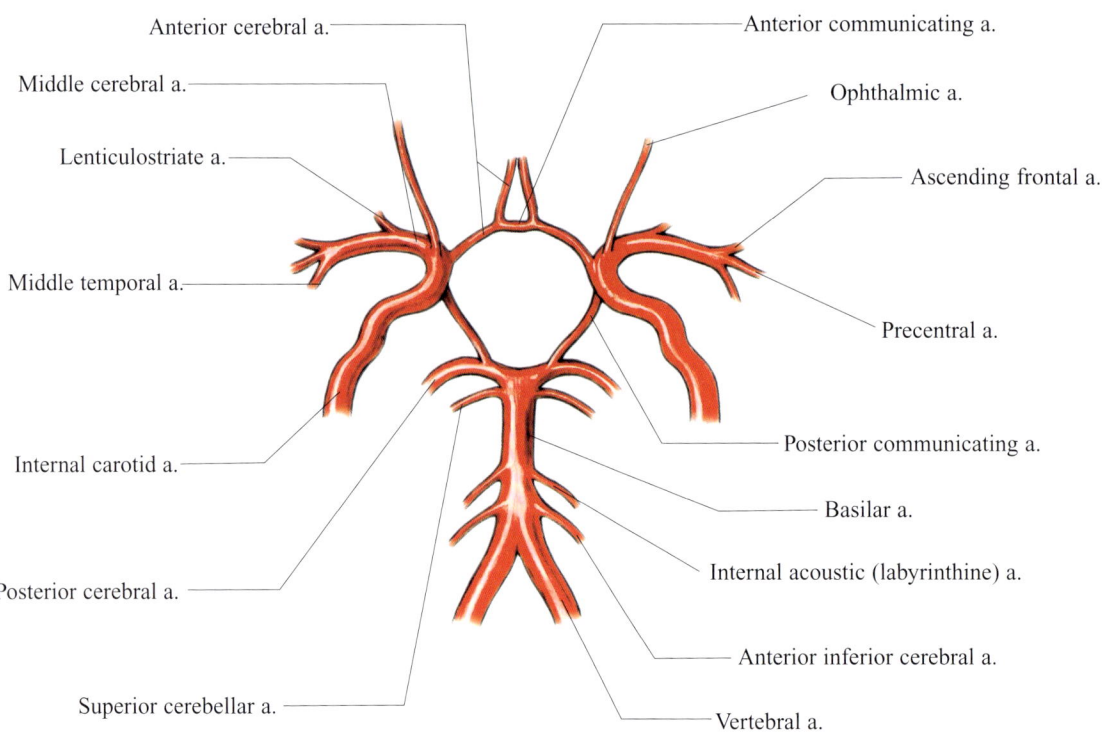

CIRCULATORY SYSTEM

BRAIN & NECK

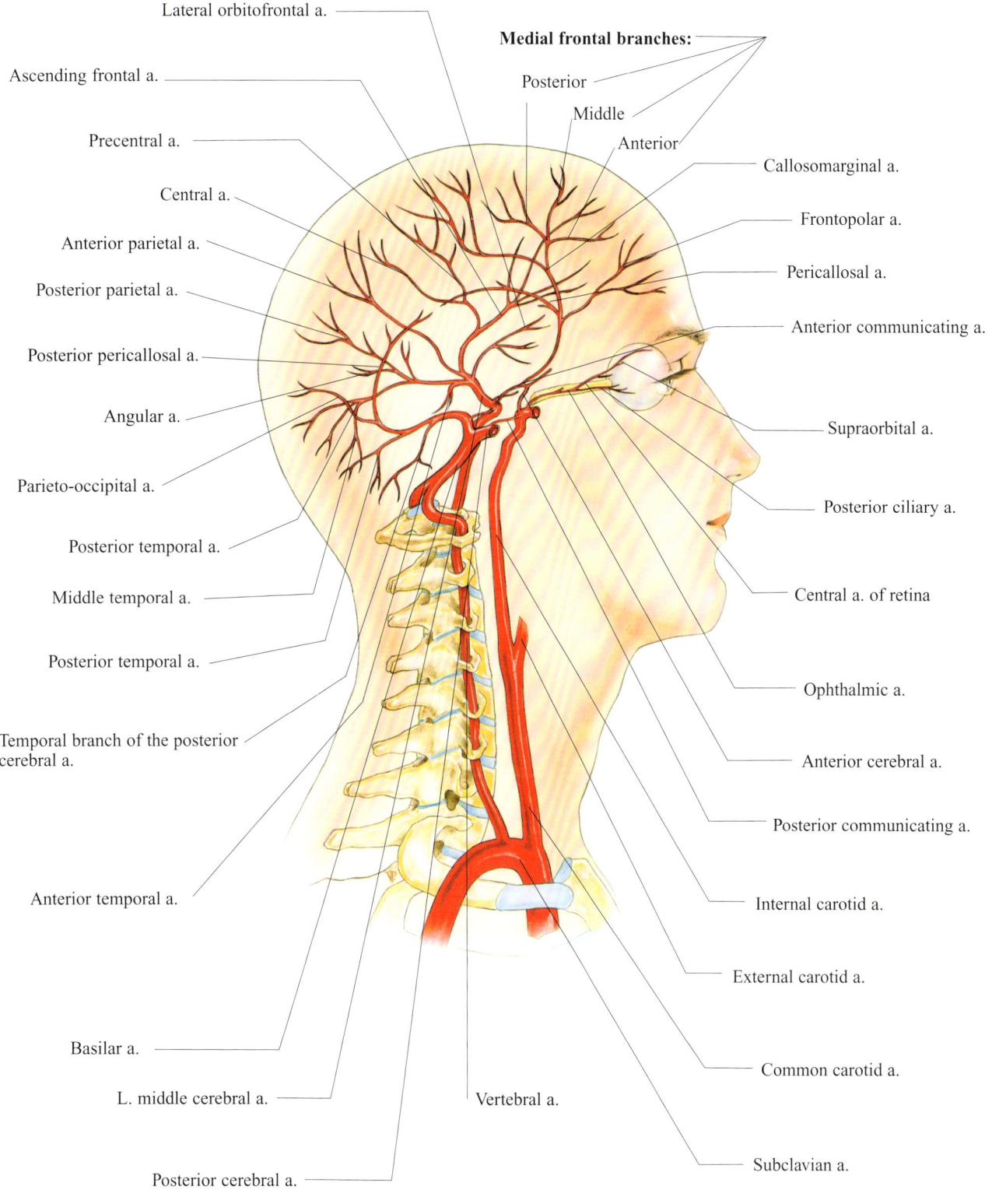

163

CIRCULATORY SYSTEM

BLOOD CIRCUITS

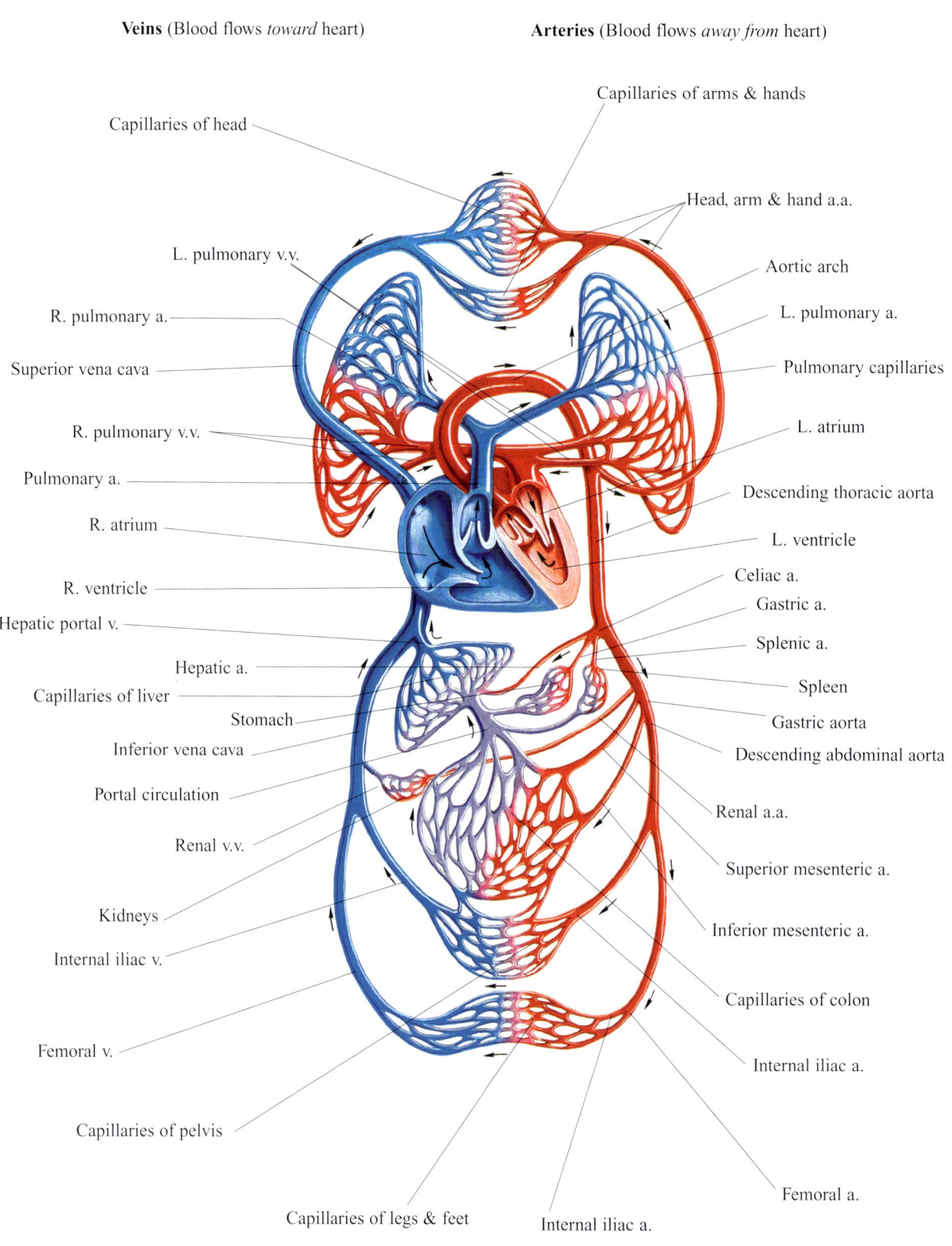

CIRCULATORY SYSTEM

HEPATIC PORTAL VEINS

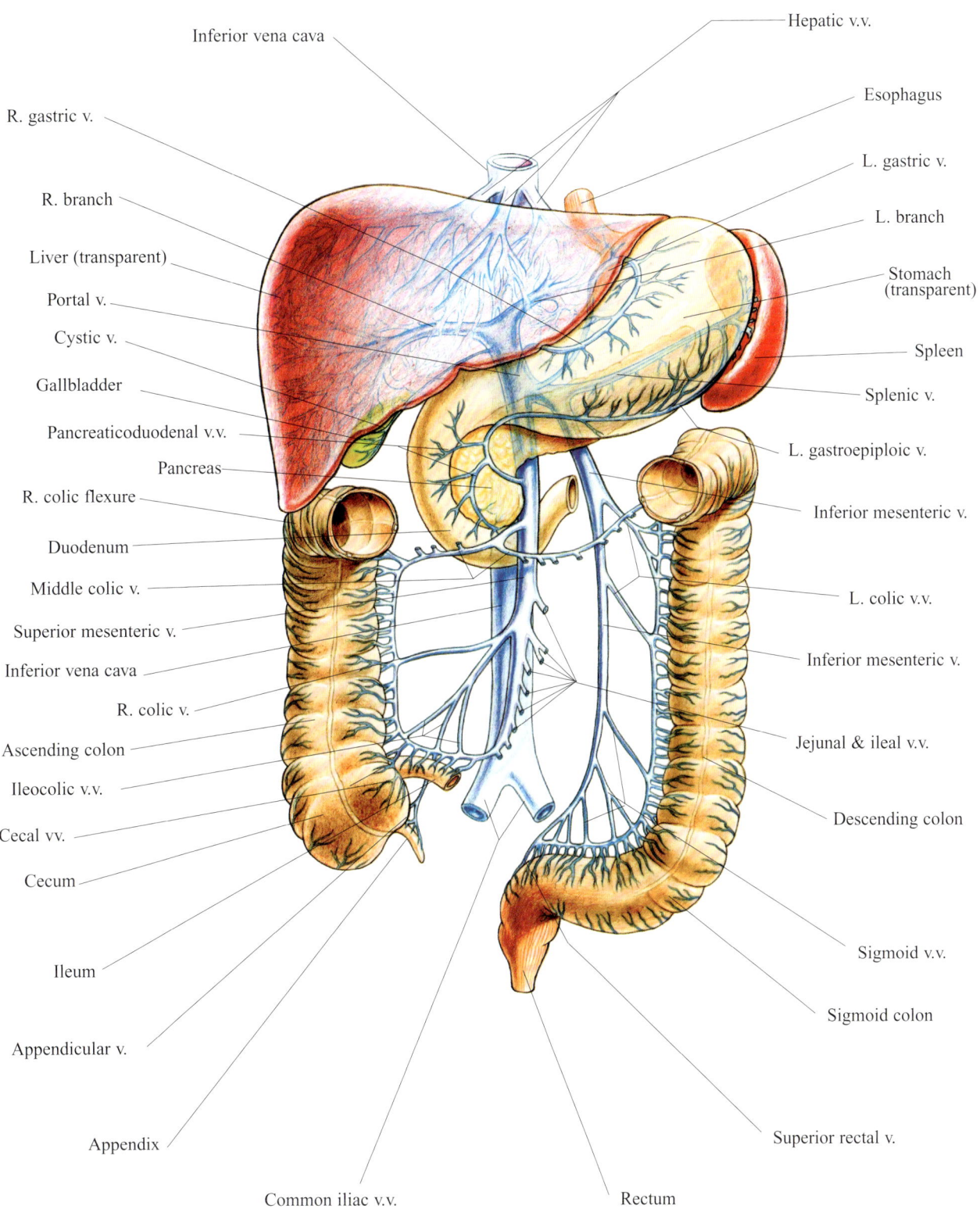

CIRCULATORY SYSTEM

BLOOD VESSELS

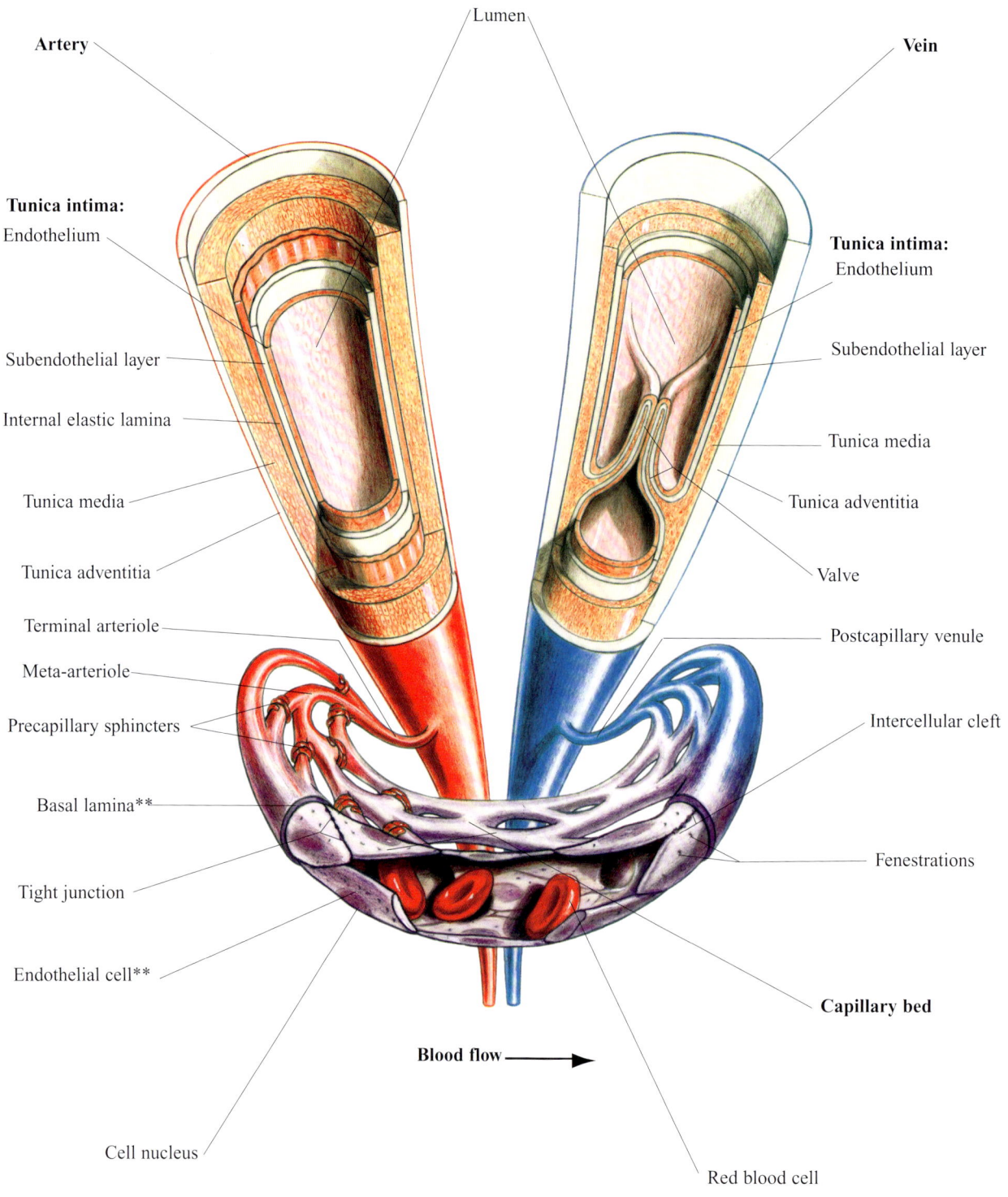

CIRCULATORY SYSTEM

NOTES

12
THE HEART

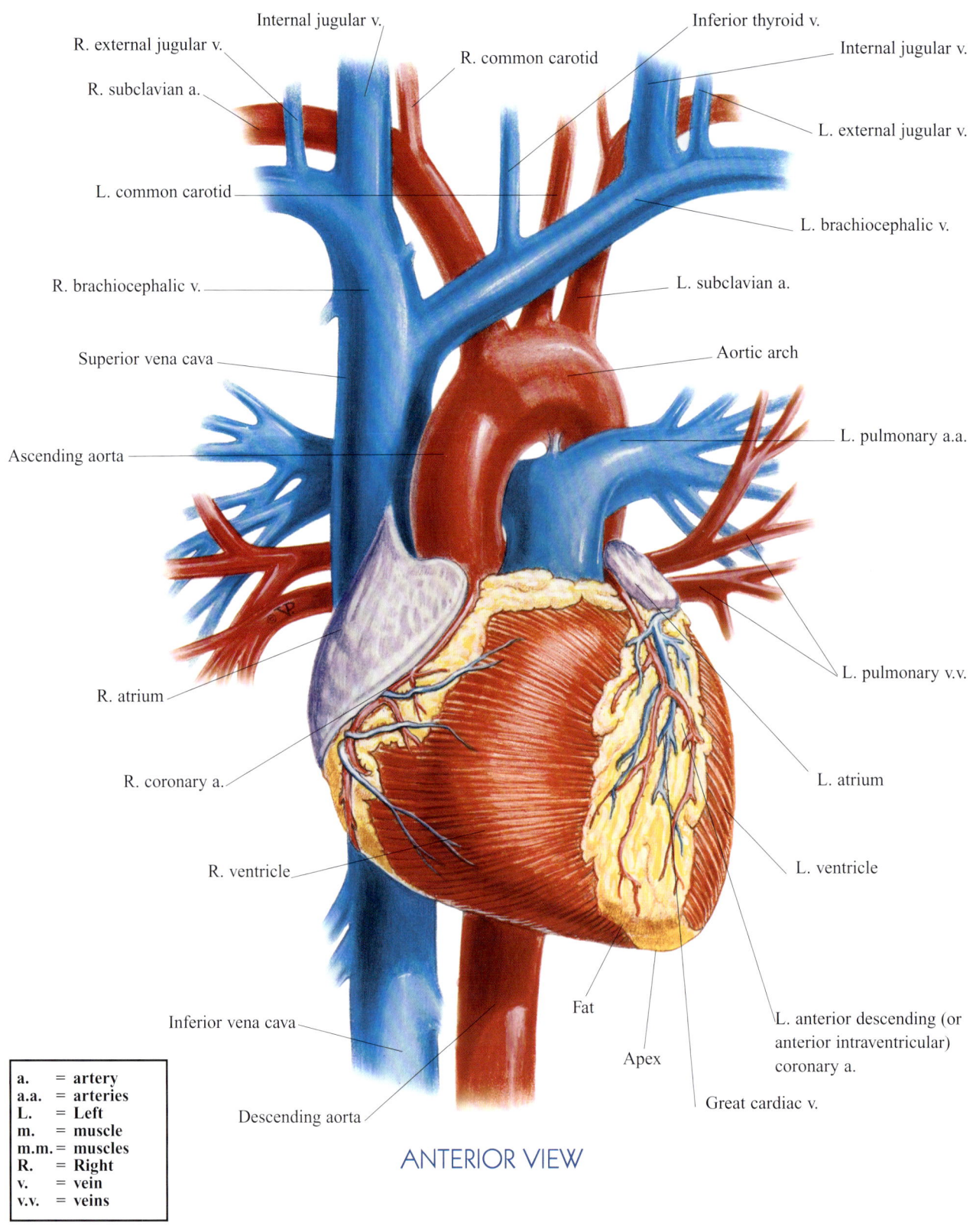

HEART

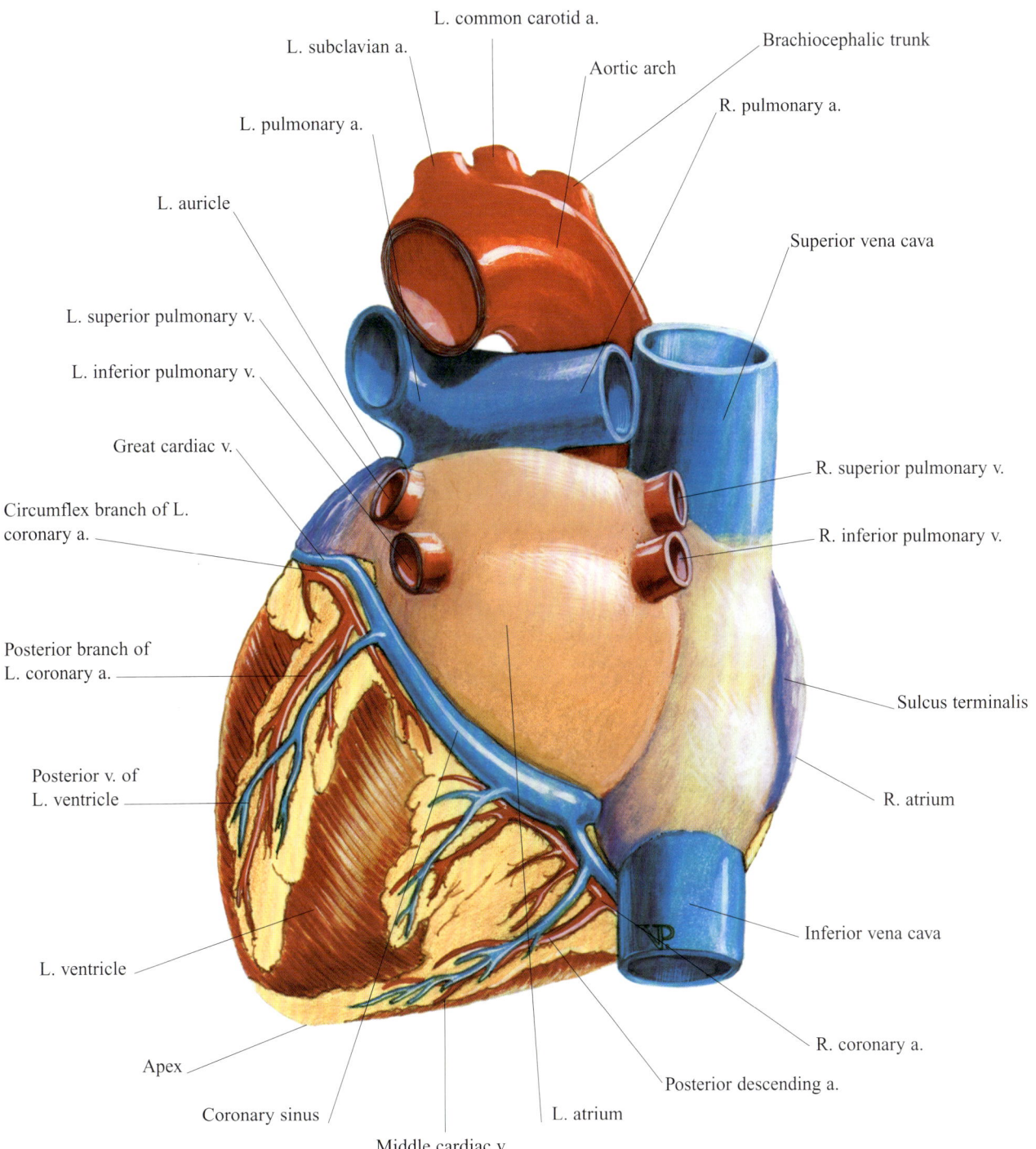

POSTERIOR VIEW

THE HEART

CORONARY ARTERIES & CARDIAC VEINS

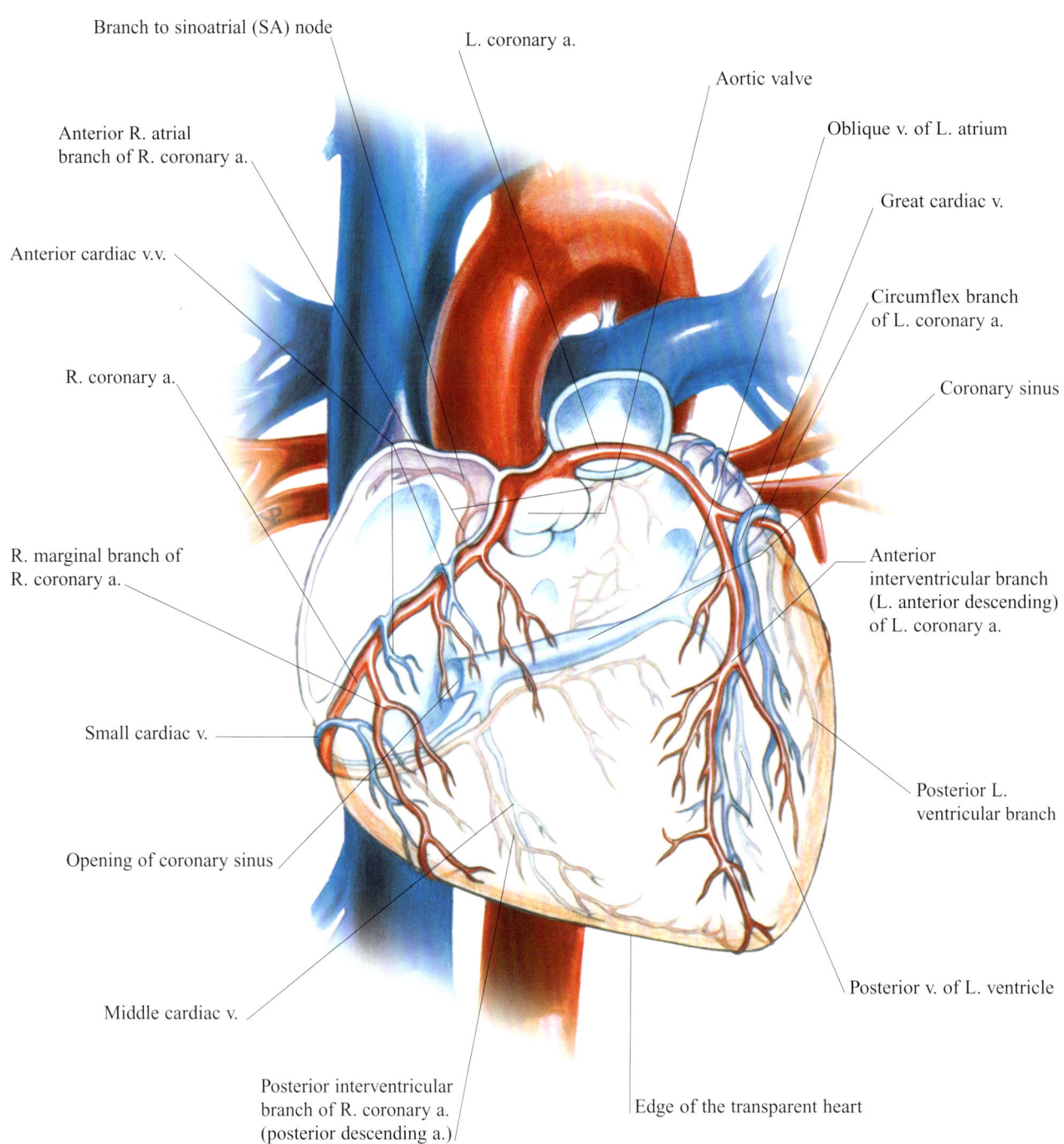

THE HEART

HEART

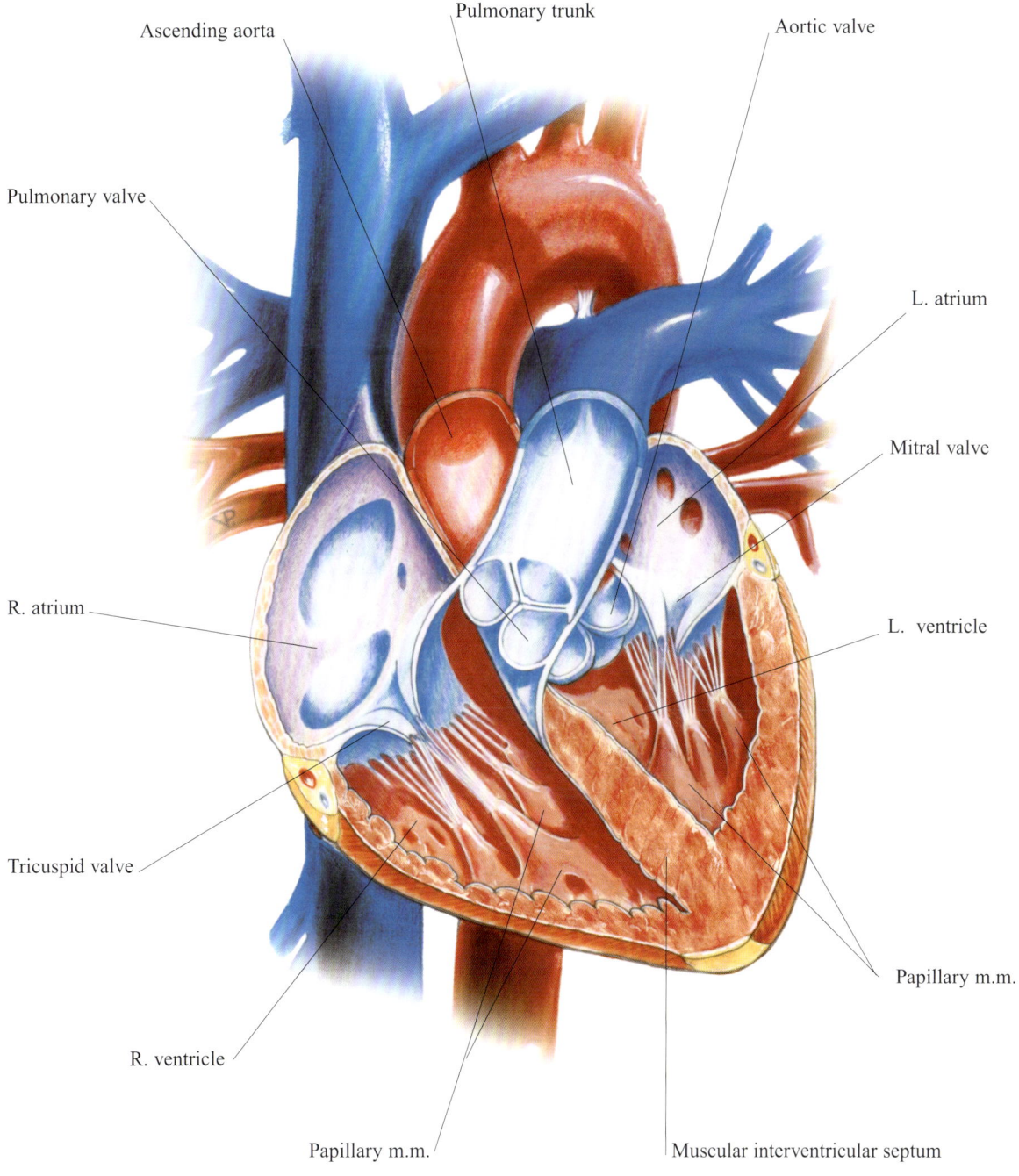

INTERIOR VIEW

THE HEART

CIRCULATION

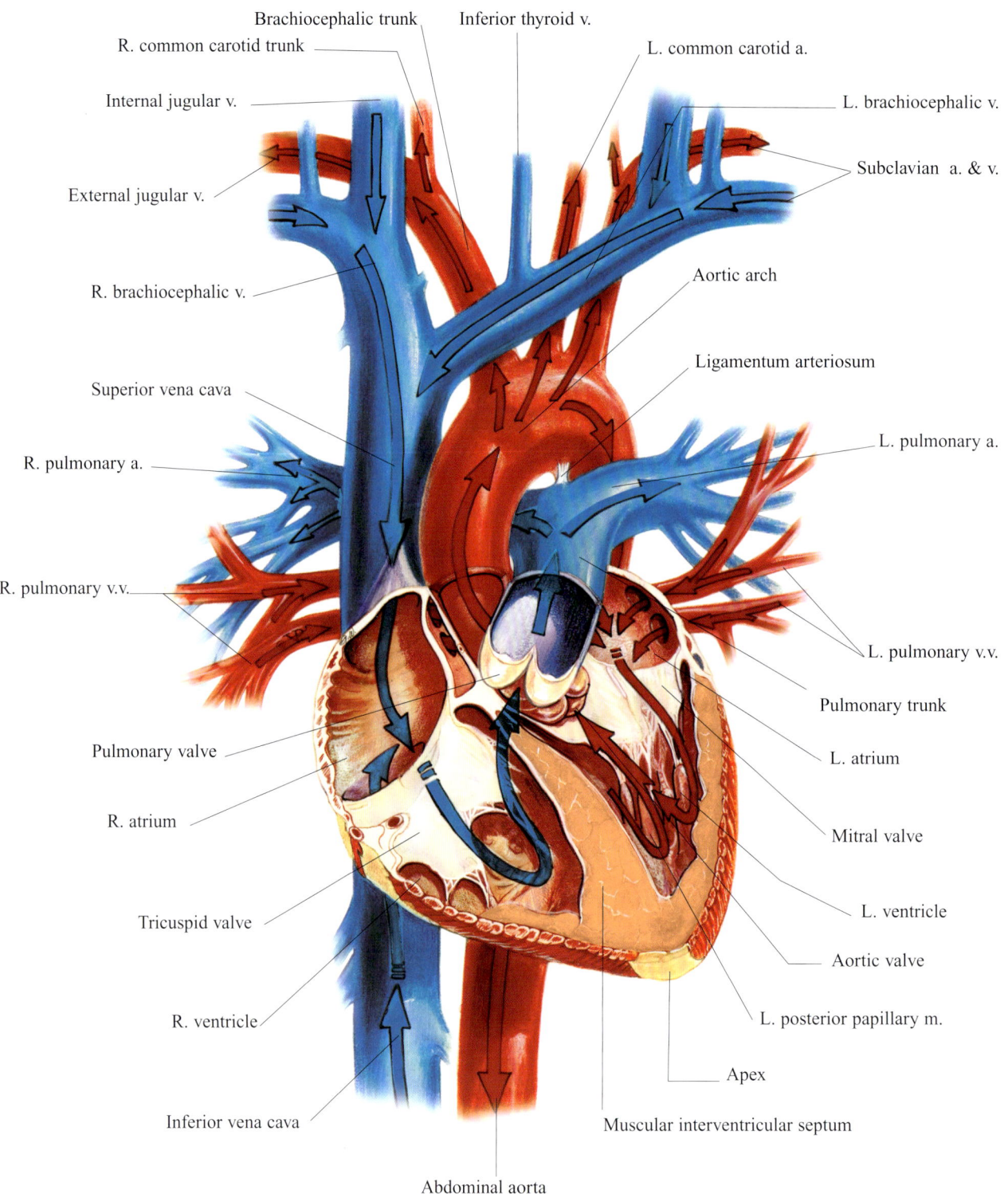

THE HEART

NERVES & ARTERIES

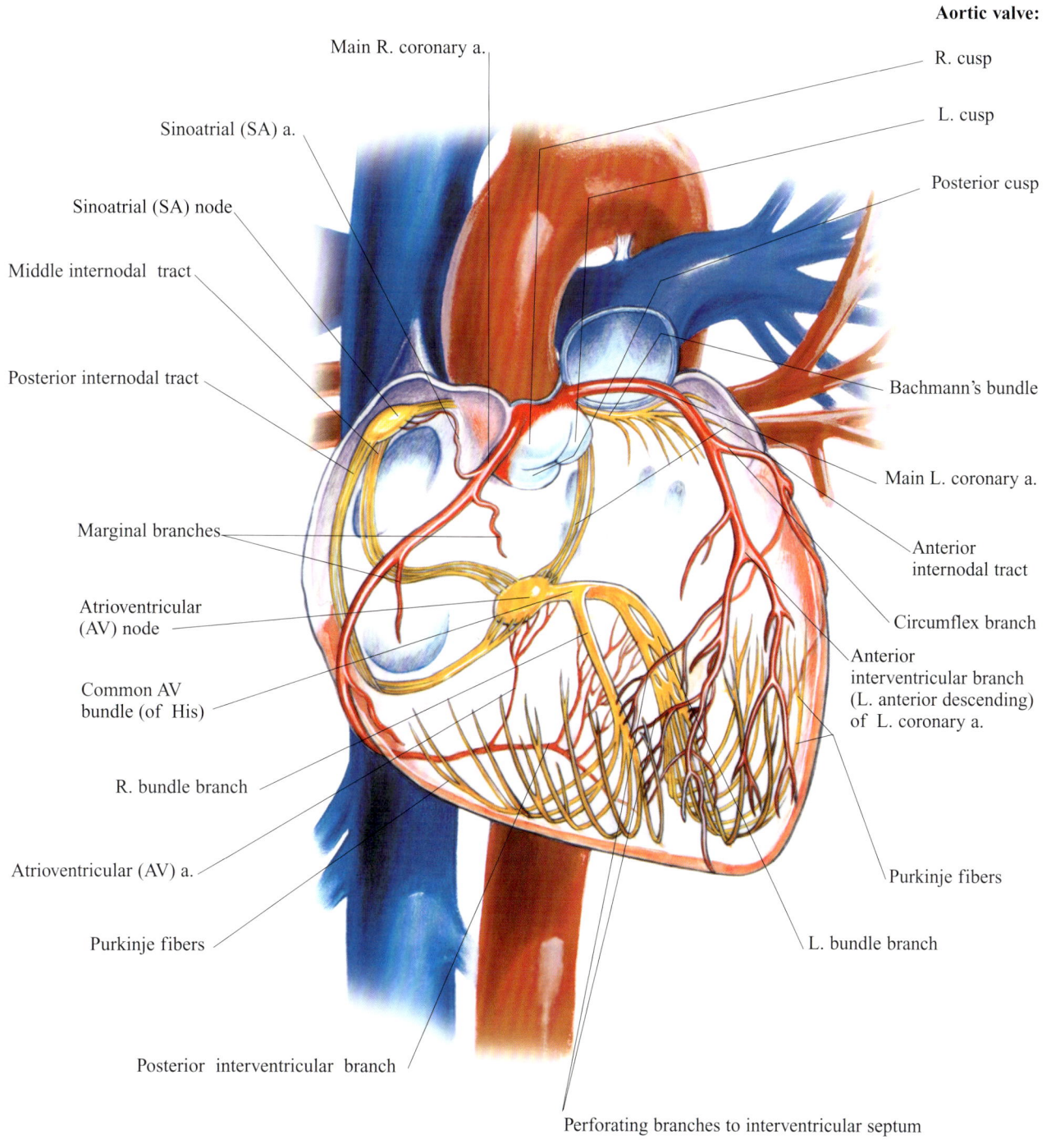

THE HEART

HEART IN DIASTOLE

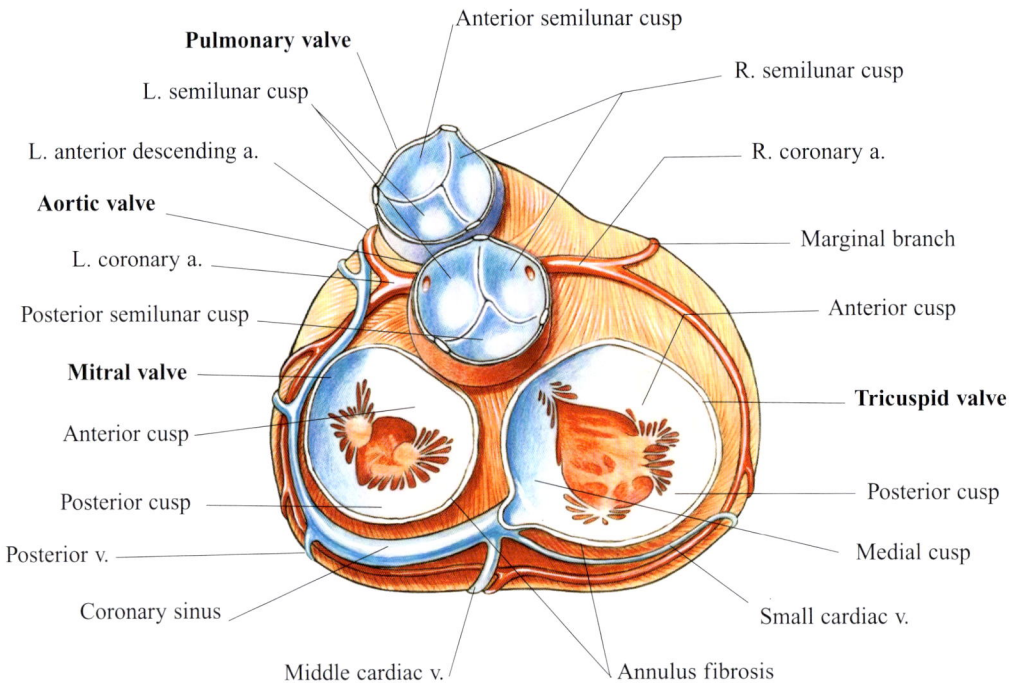

HEART IN SYSTOLE

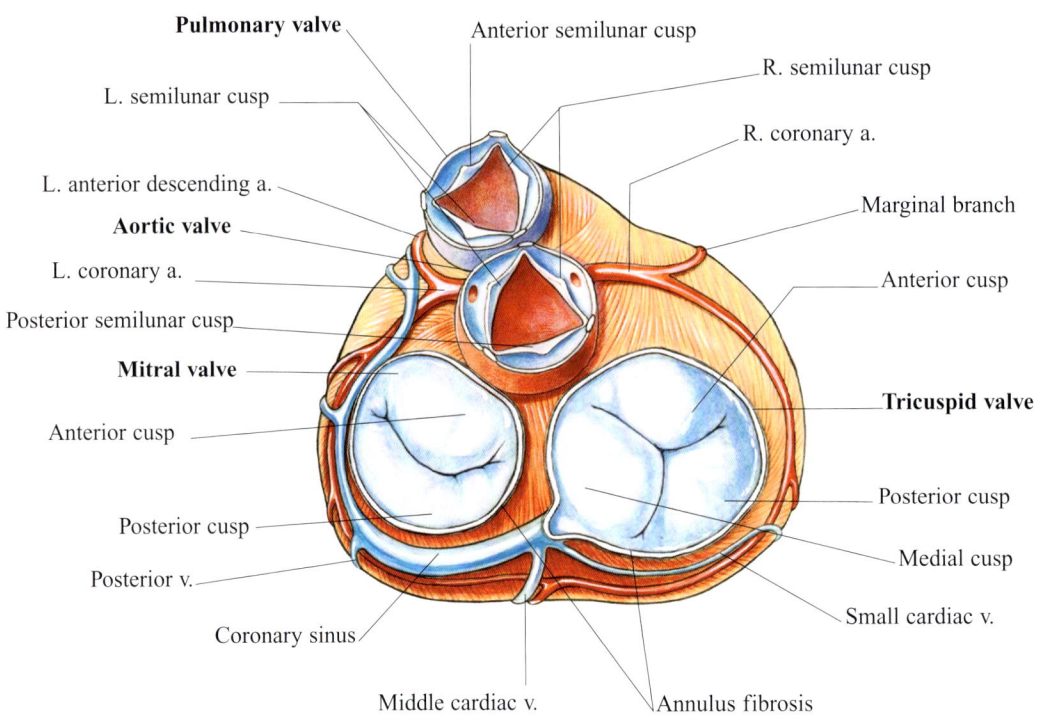

THE HEART

BEGINNING OF DIASTOLE

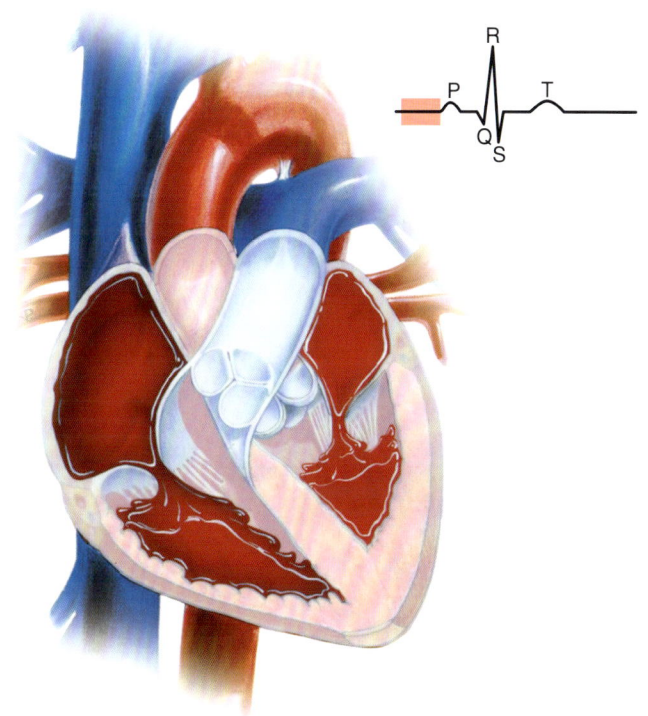

END OF DIASTOLE

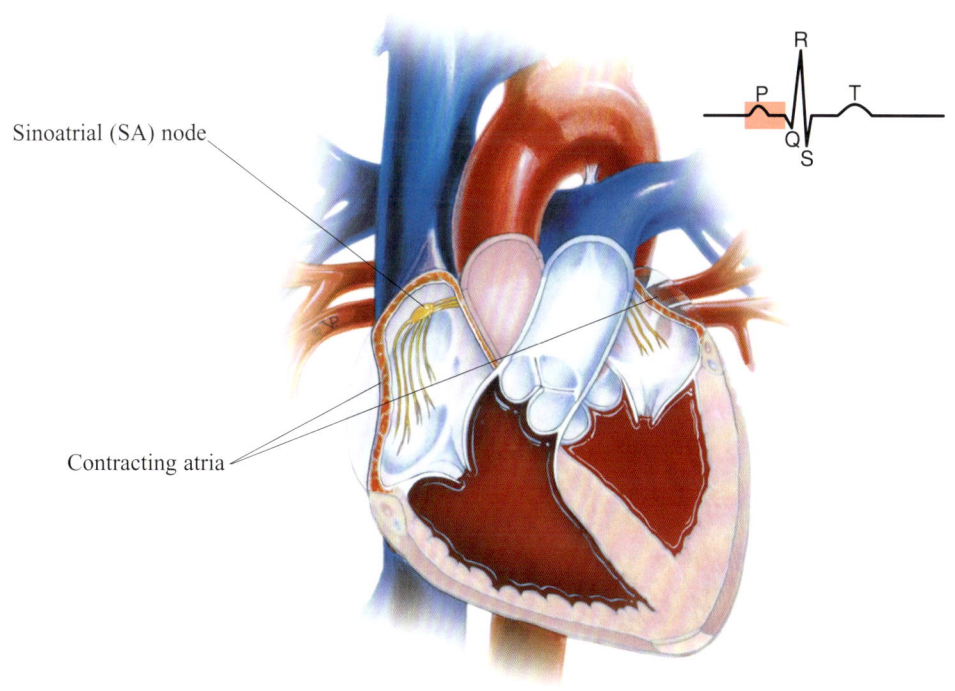

Sinoatrial (SA) node

Contracting atria

THE HEART

BEGINNING OF SYSTOLE

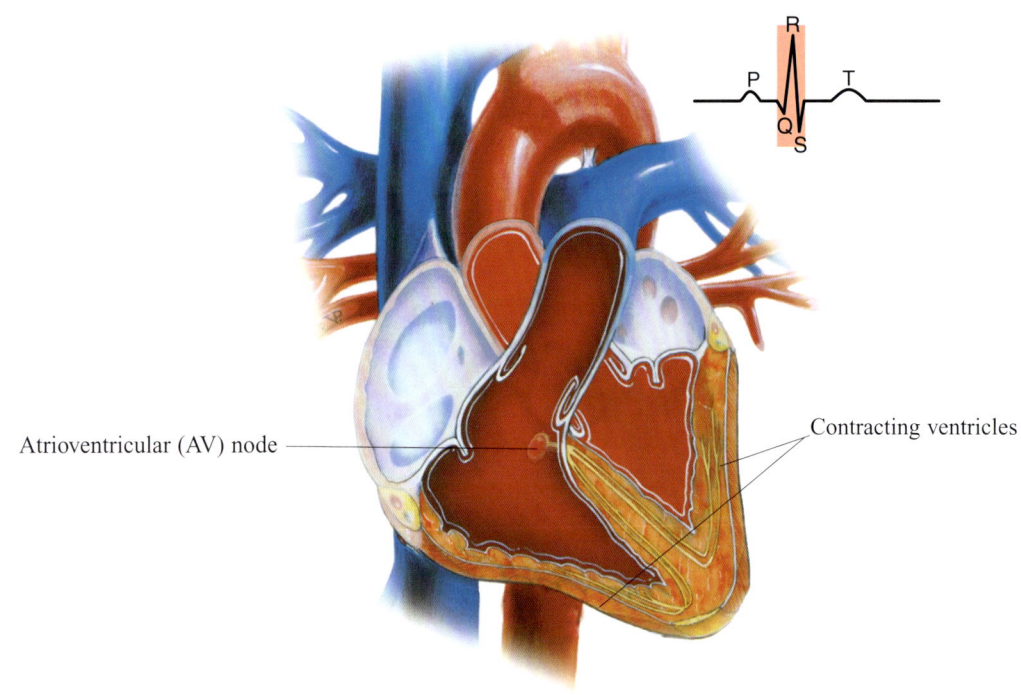

Atrioventricular (AV) node

Contracting ventricles

END OF SYSTOLE

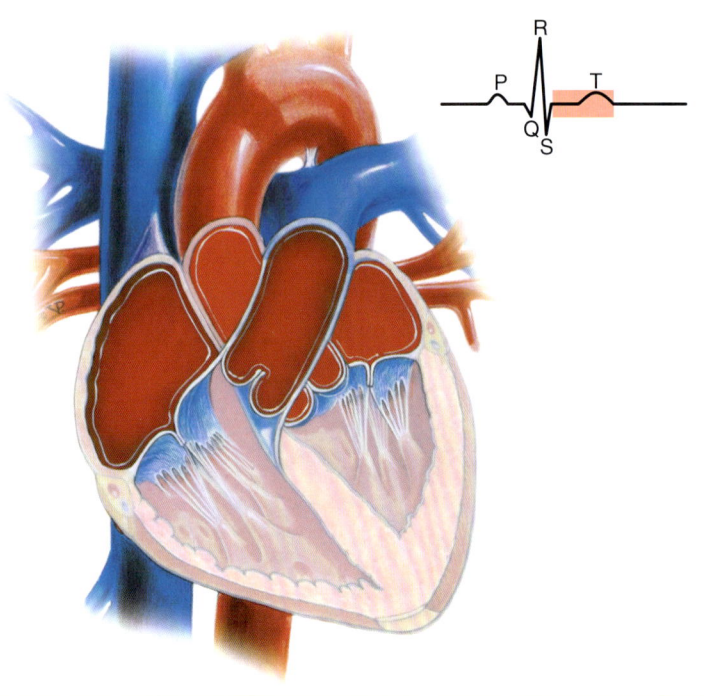

THE HEART

NOTES

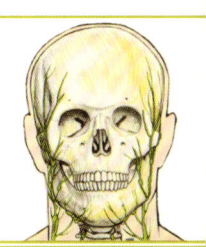

13
LYMPHATIC SYSTEM

LYMPHATIC SYSTEM

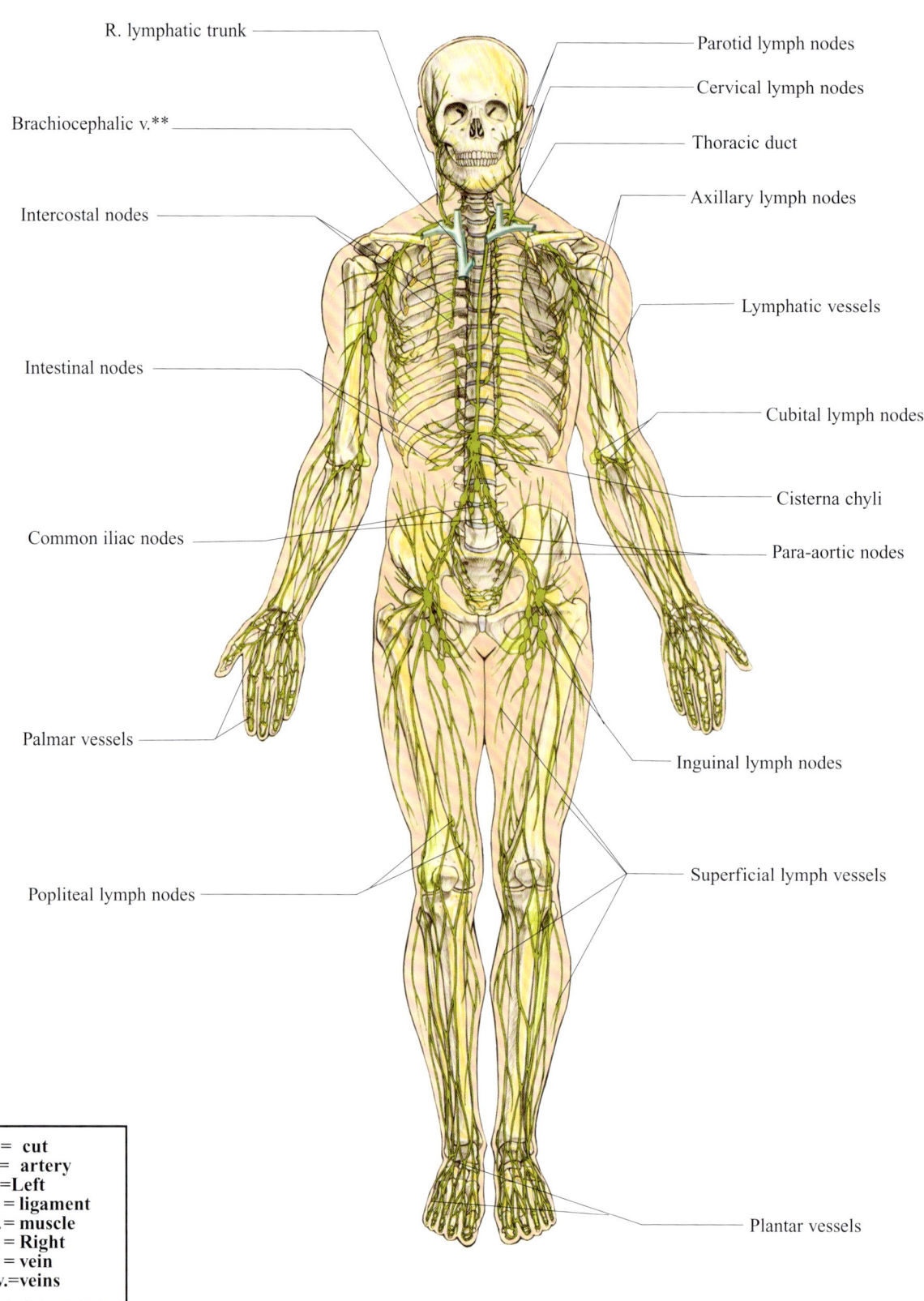

HEAD & NECK

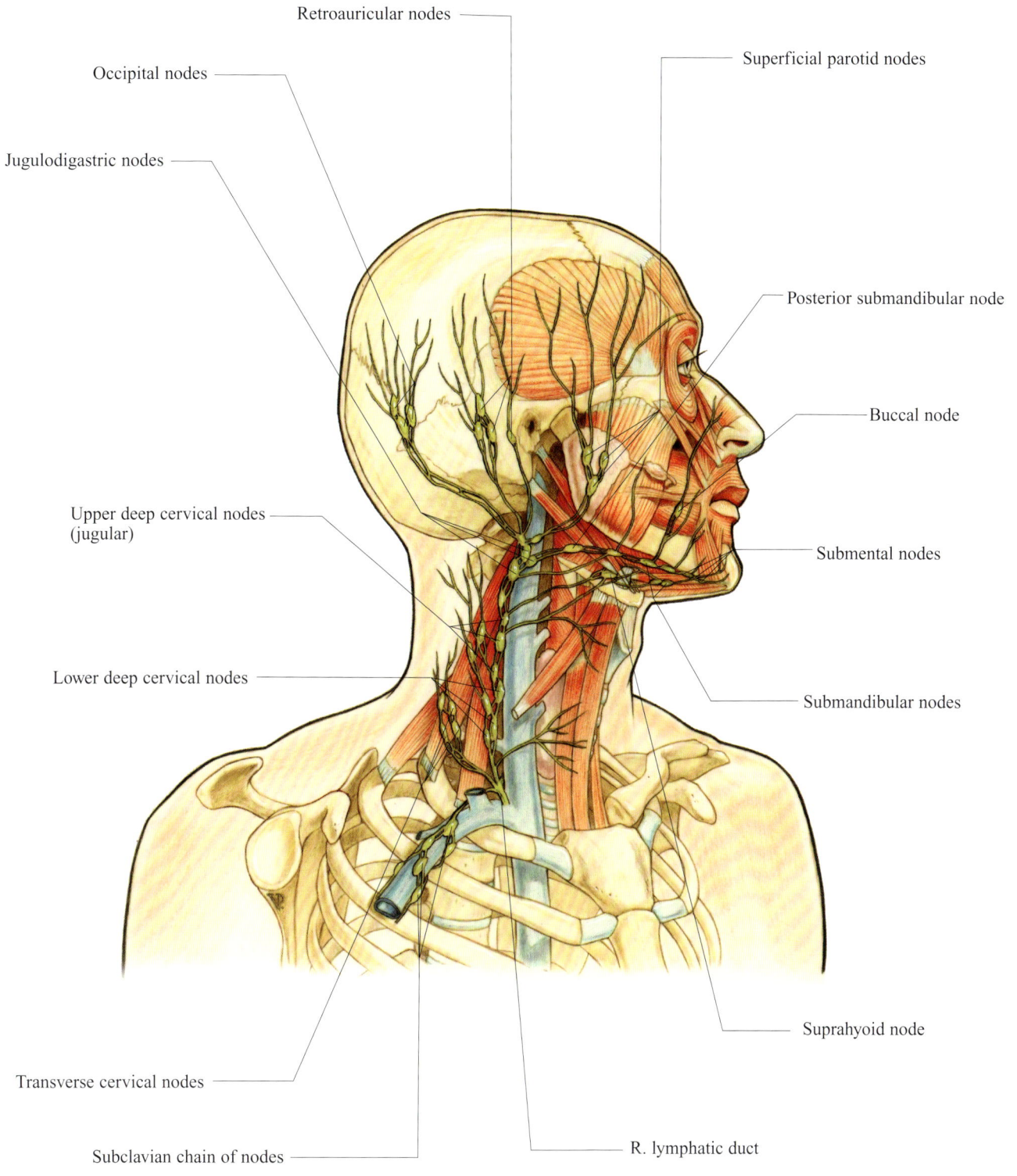

ARM AXILLA & THORAX

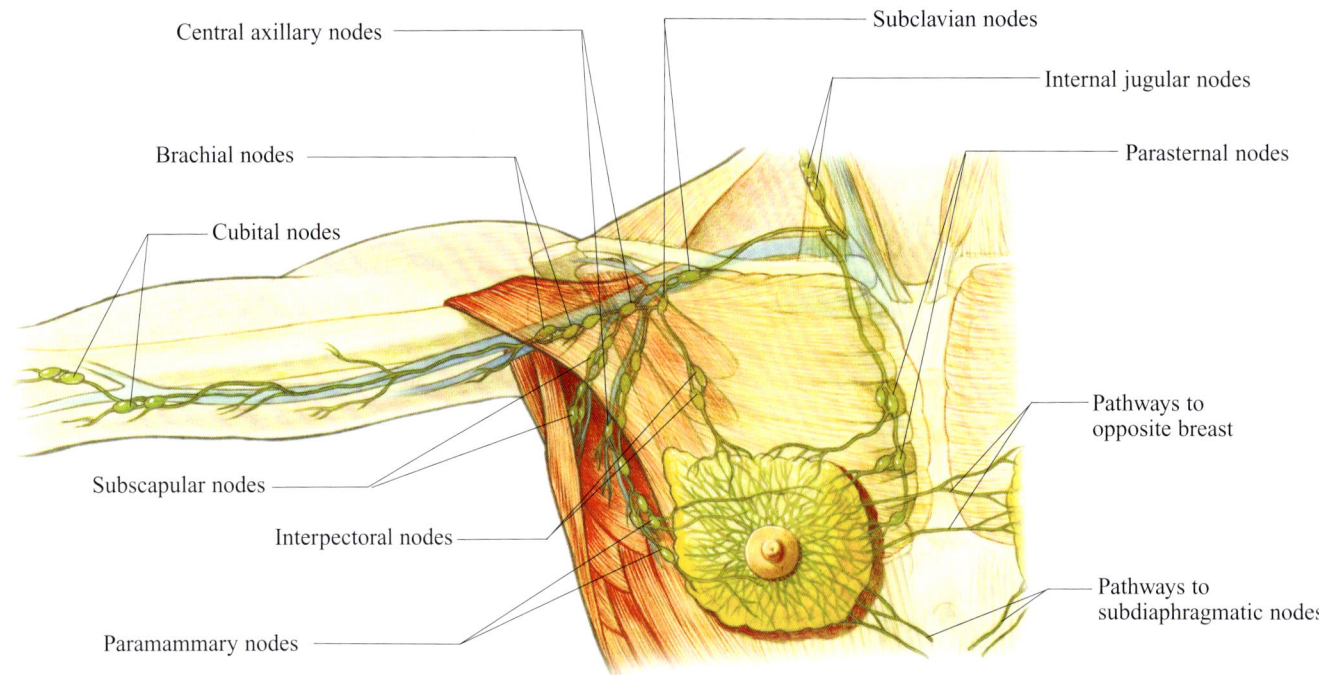

HEART & LUNGS

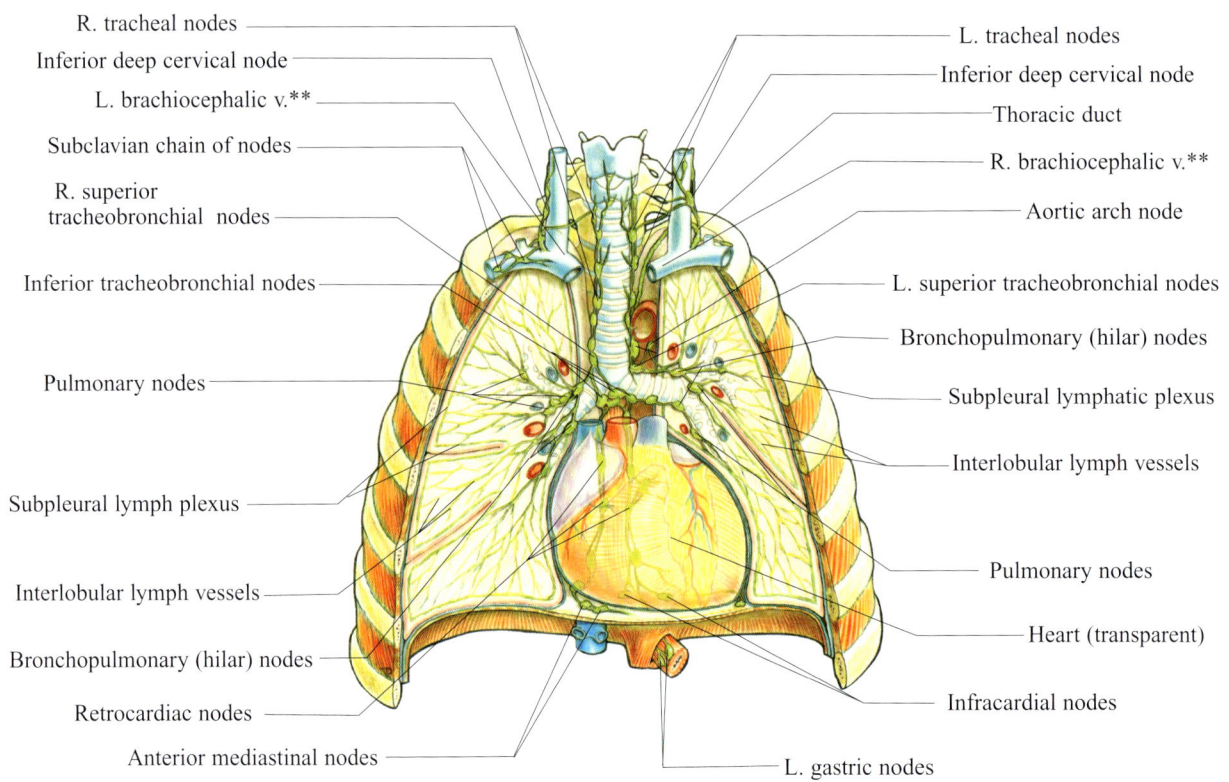

THORACIC DUCT

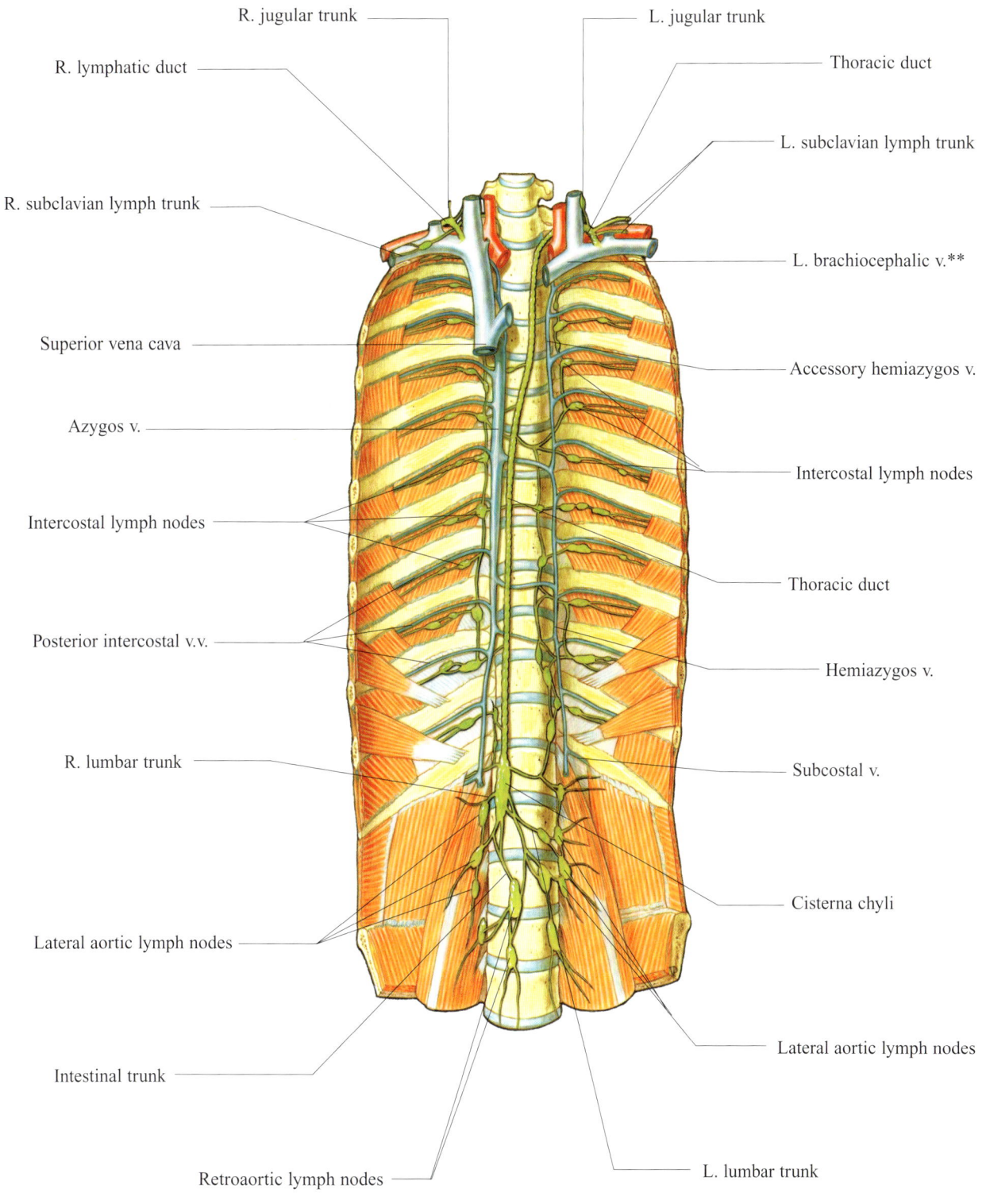

LYMPHATIC SYSTEM

DEEP ABDOMINAL & INGUINAL NODES

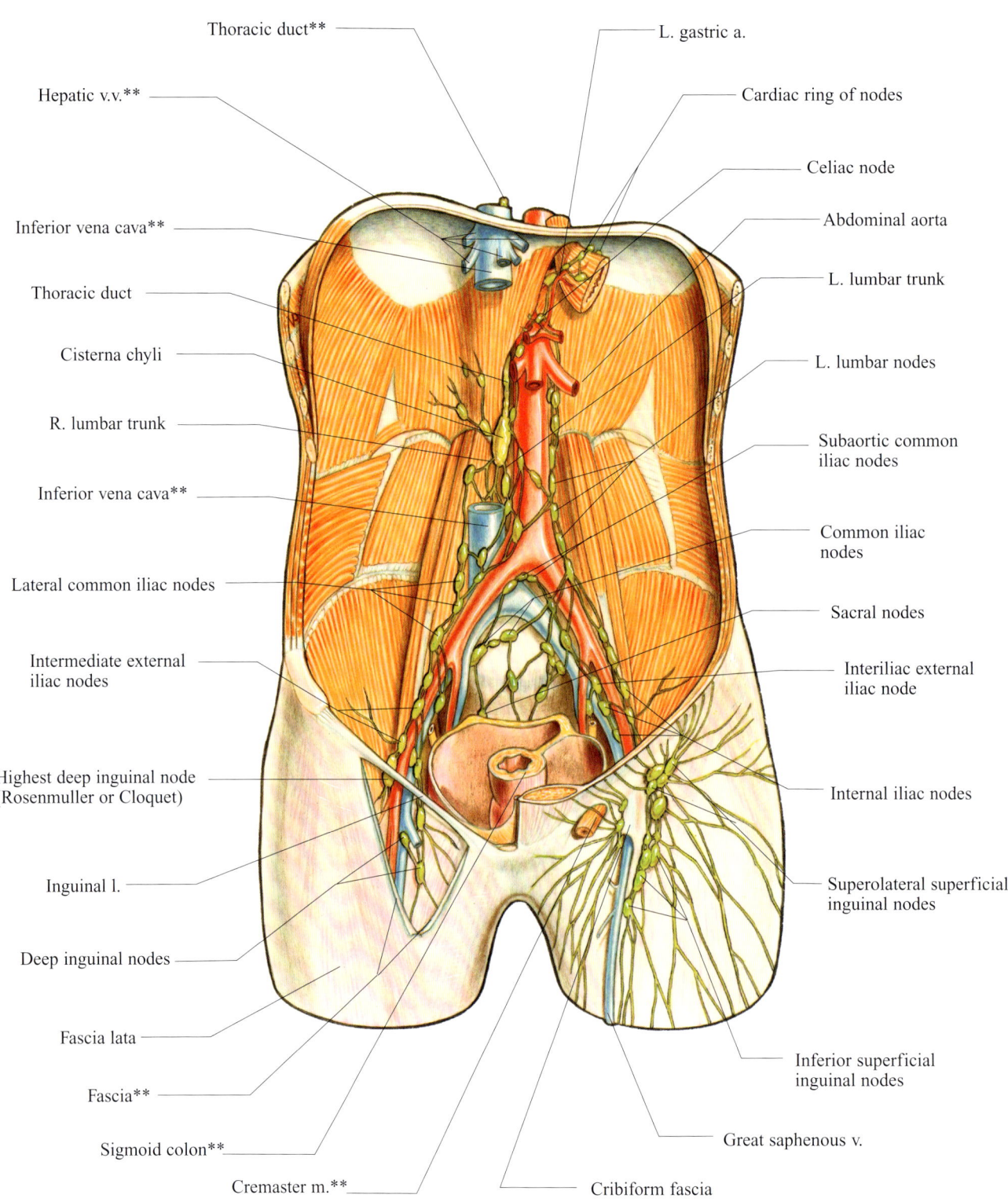

STOMACH & PANCREAS

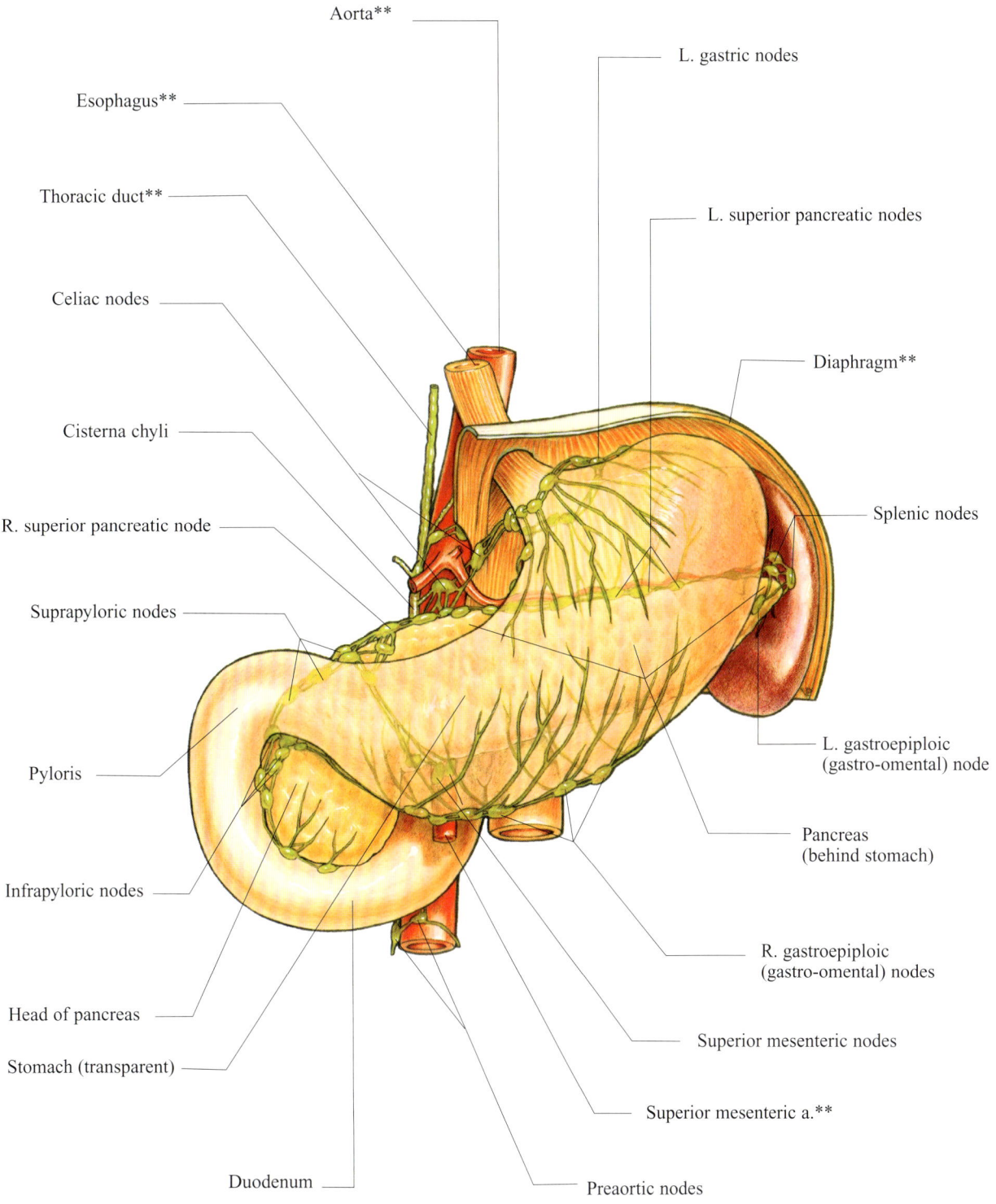

LARGE INTESTINE

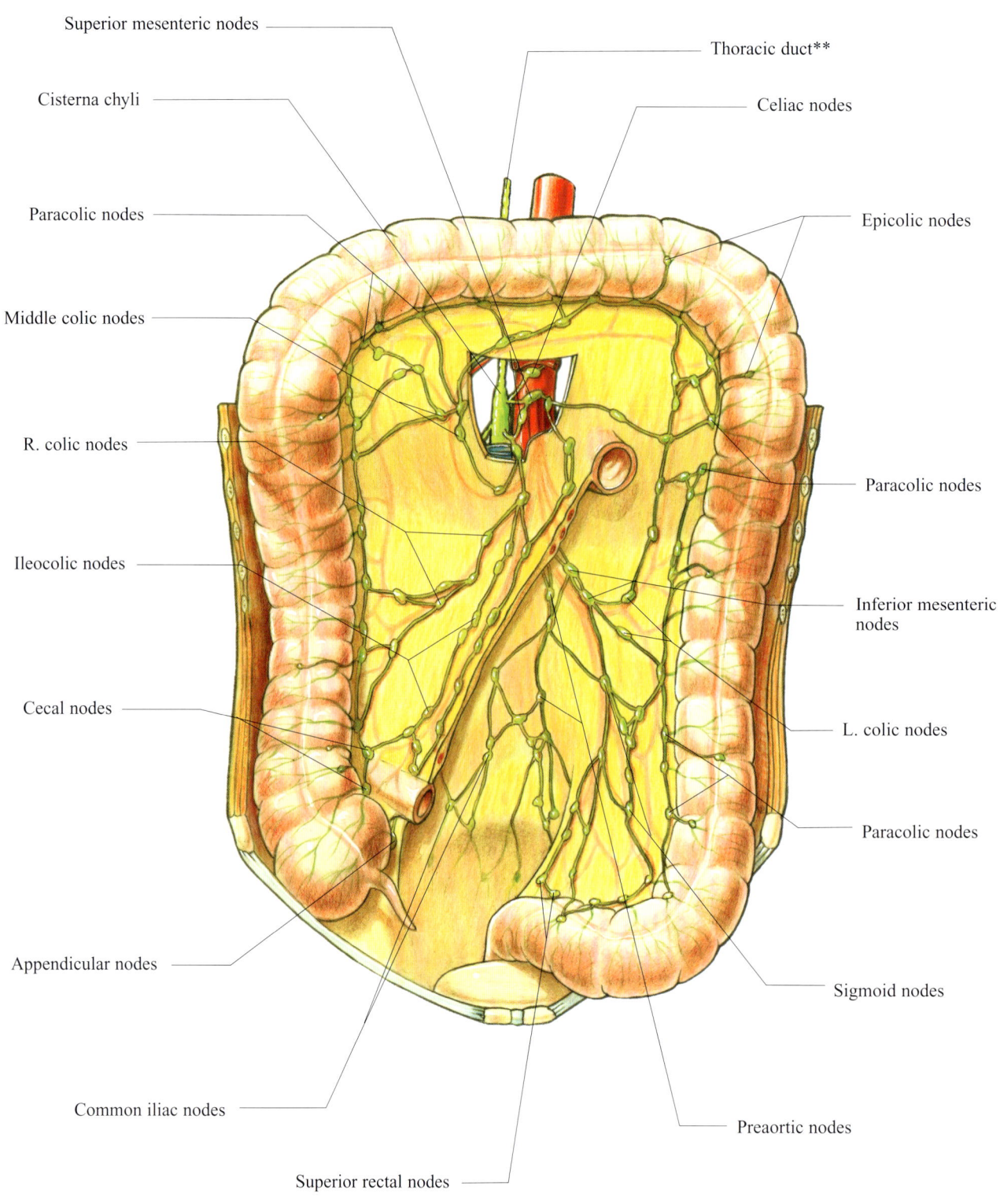

LYMPHATIC SYSTEM

NODES & VESSELS

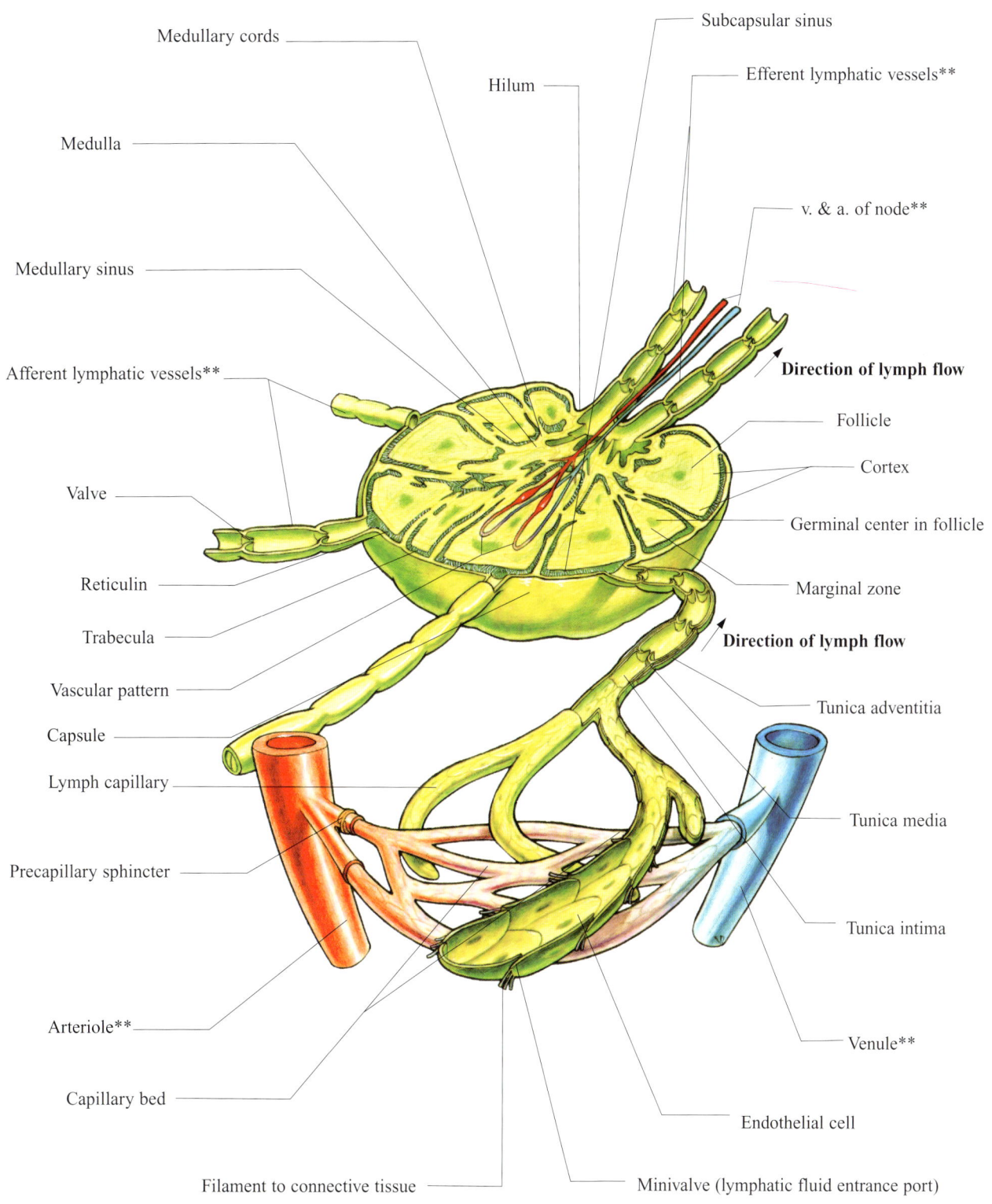

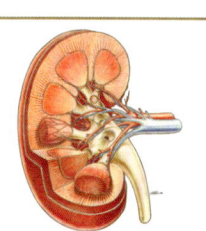

14
UROGENITAL SYSTEM

UROGENITAL SYSTEM

MALE UROGENITAL SYSTEM

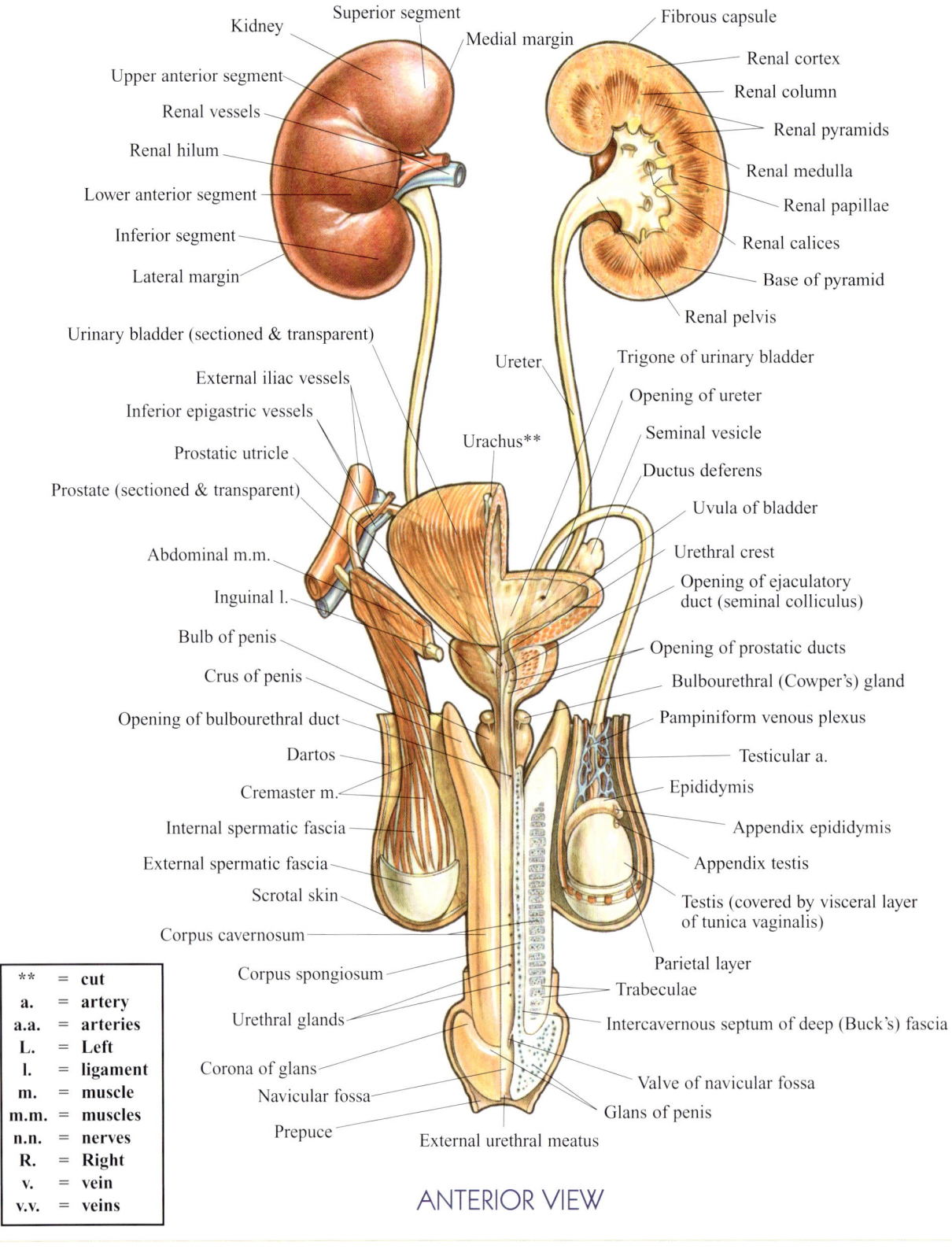

ANTERIOR VIEW

UROGENITAL SYSTEM

MALE URINARY SYSTEM

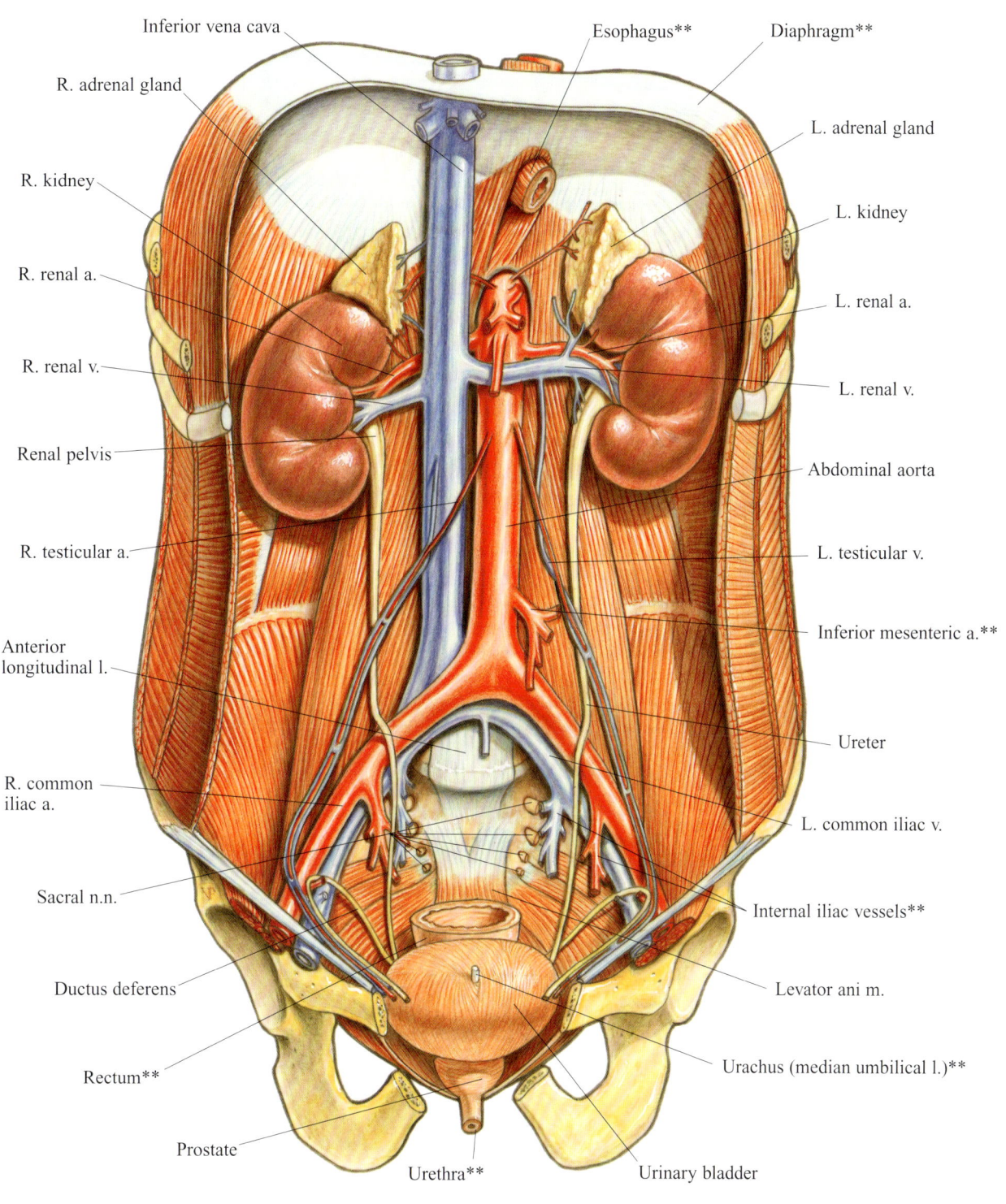

ANTERIOR VIEW

UROGENITAL SYSTEM

MALE UROGENITAL SYSTEM

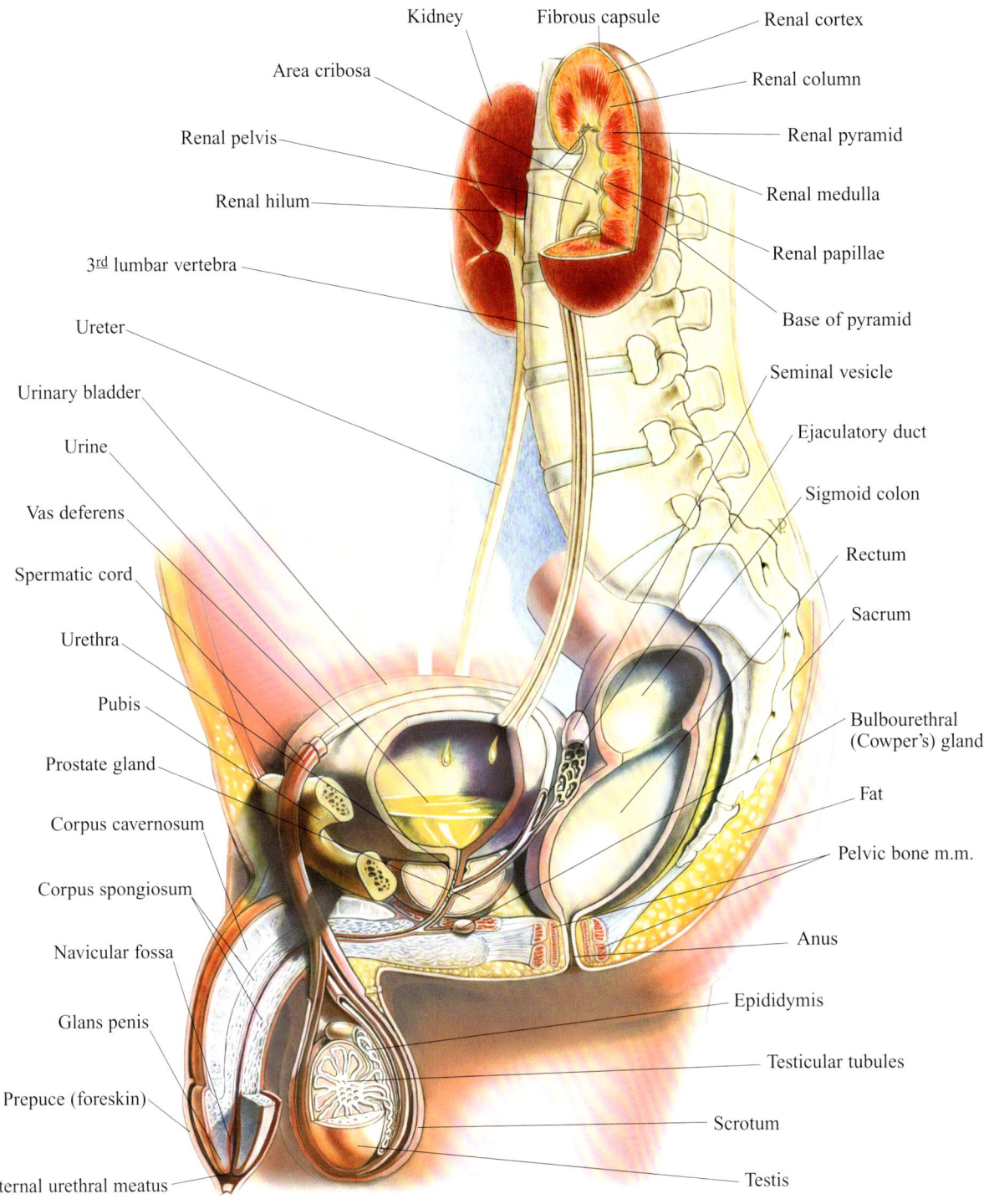

LATERAL VIEW

UROGENITAL SYSTEM

FEMALE UROGENITAL SYSTEM

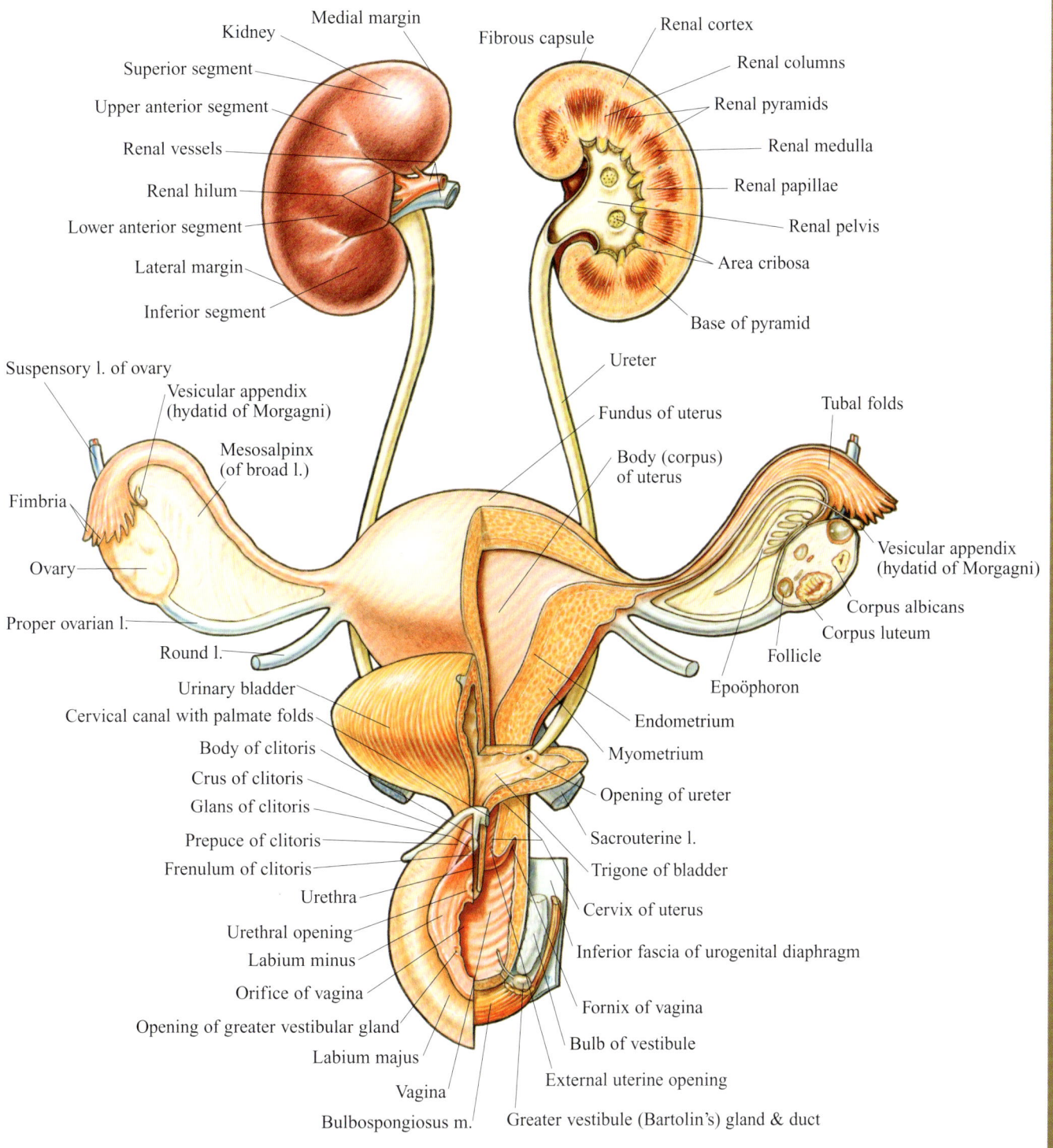

ANTERIOR VIEW

UROGENITAL SYSTEM

FEMALE UROGENITAL SYSTEM

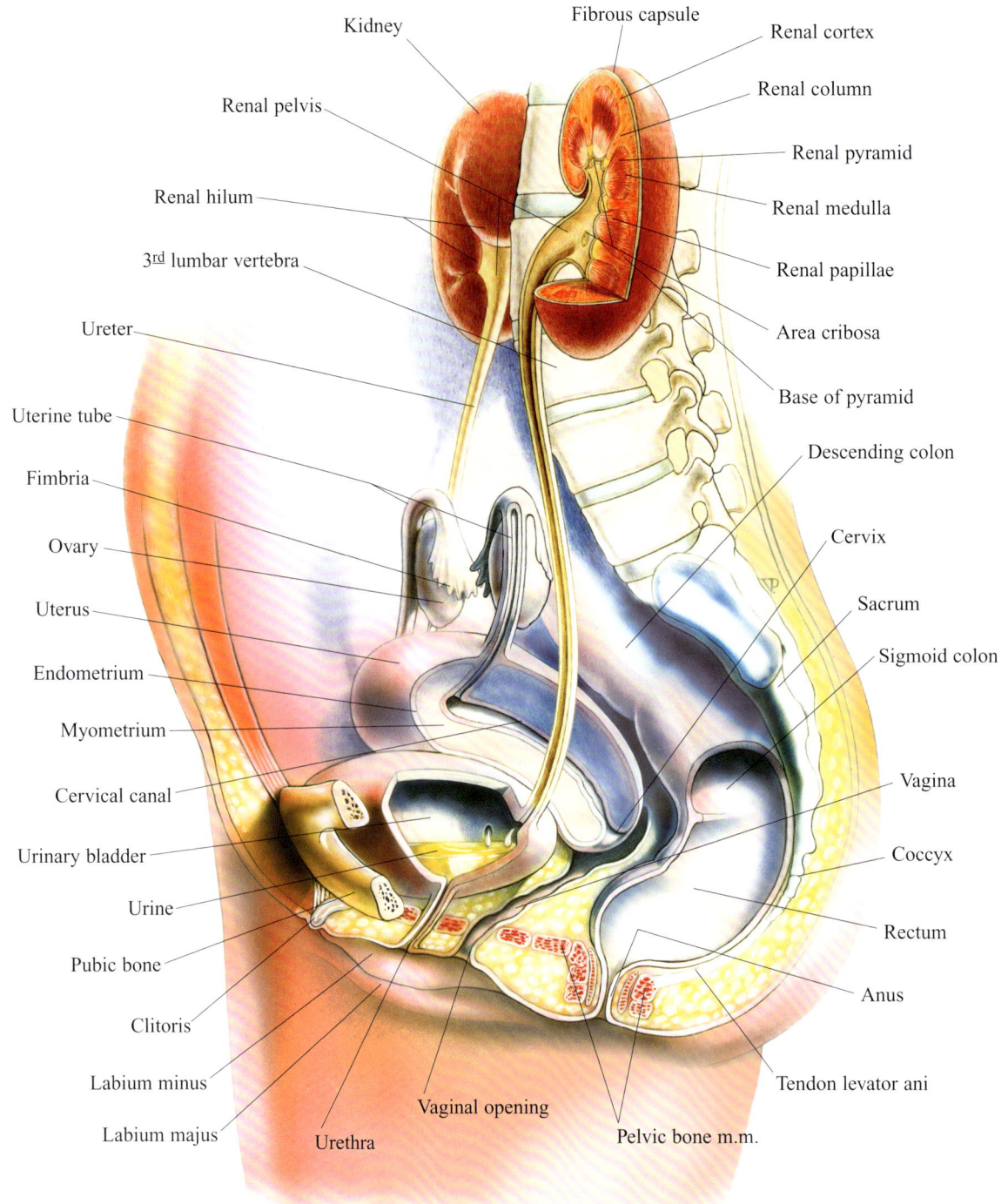

LATERAL VIEW

UROGENITAL SYSTEM

RIGHT KIDNEY

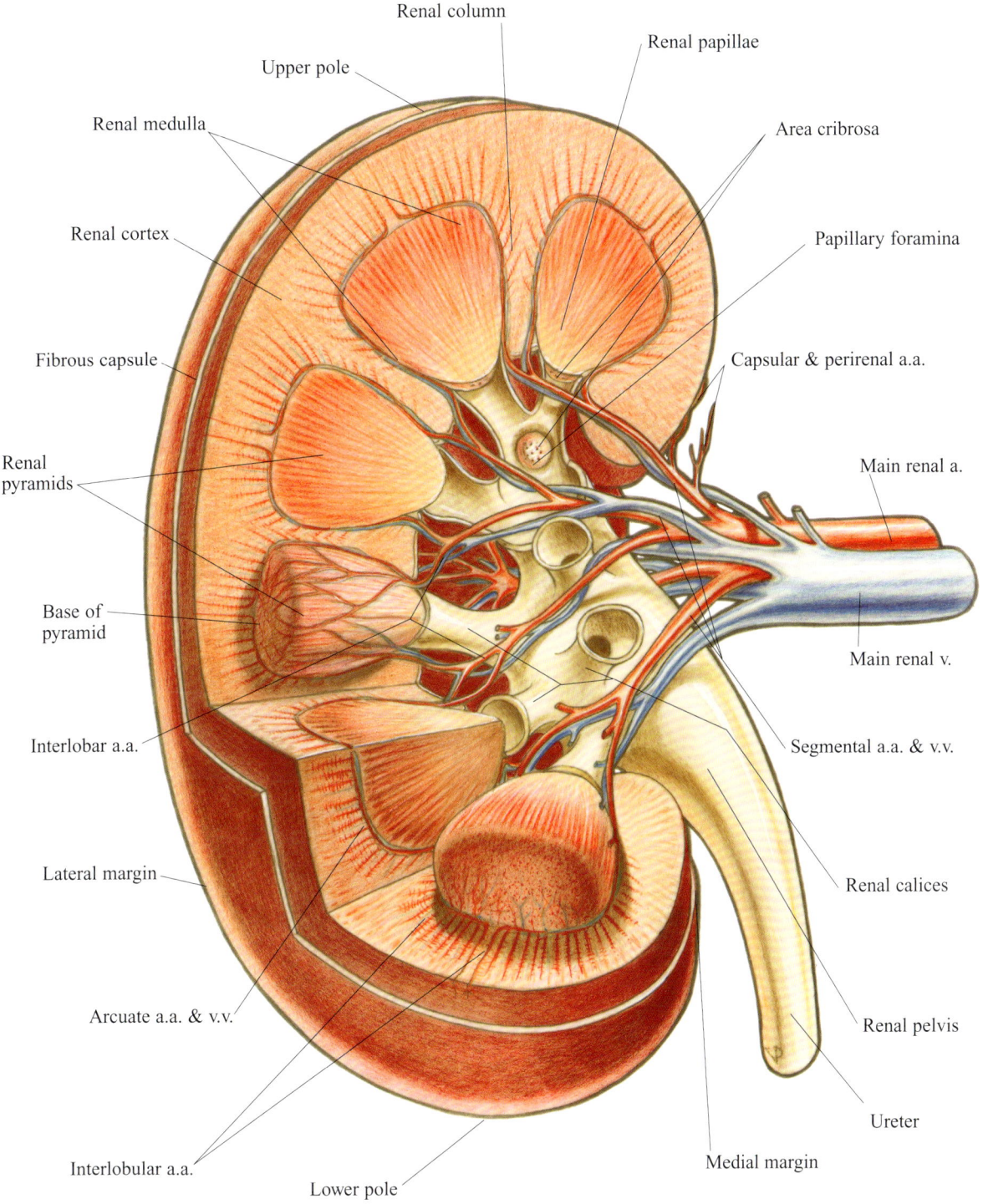

UROGENITAL SYSTEM

RENAL CORPUSCLE

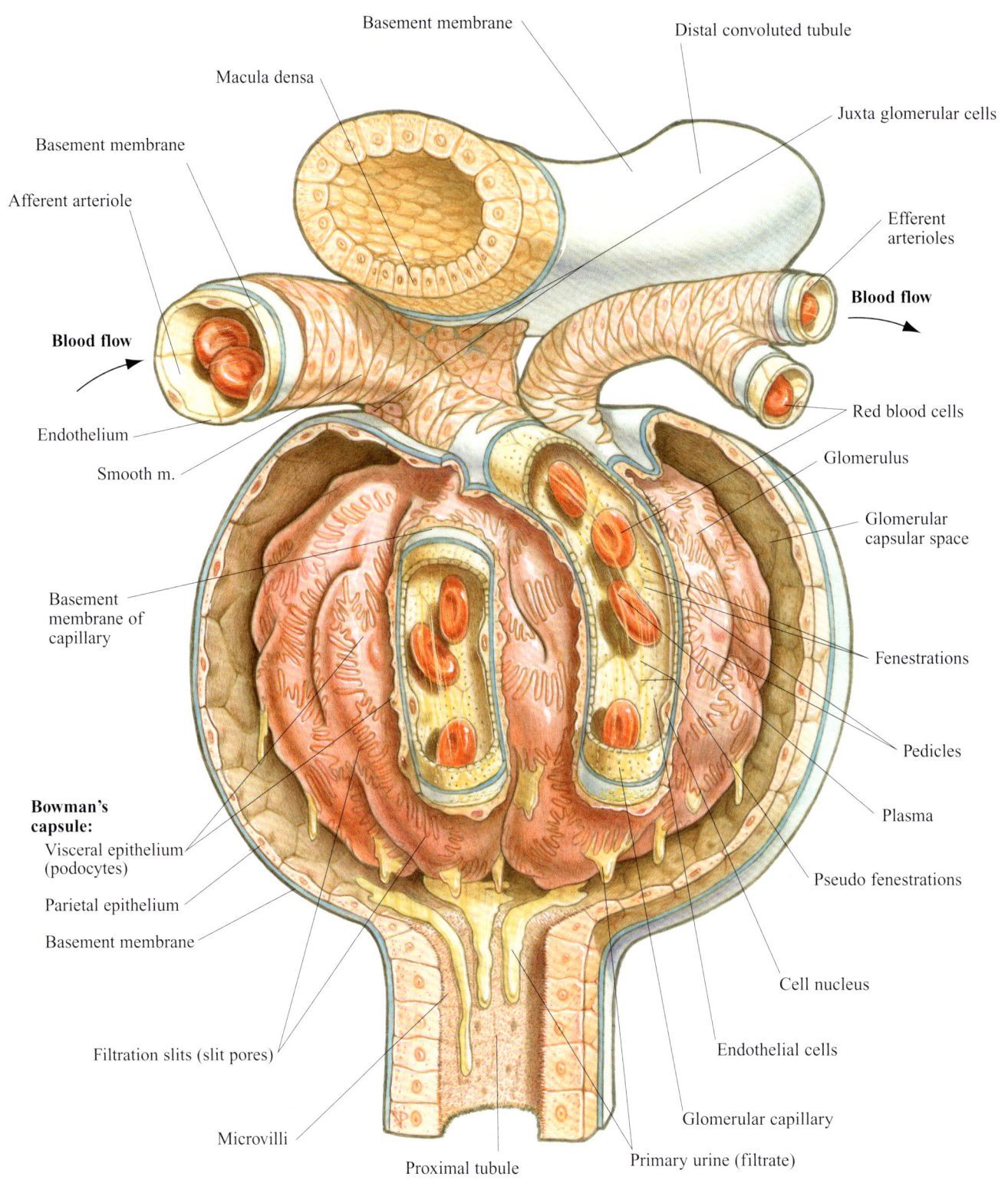

UROGENITAL SYSTEM

NEPHRON

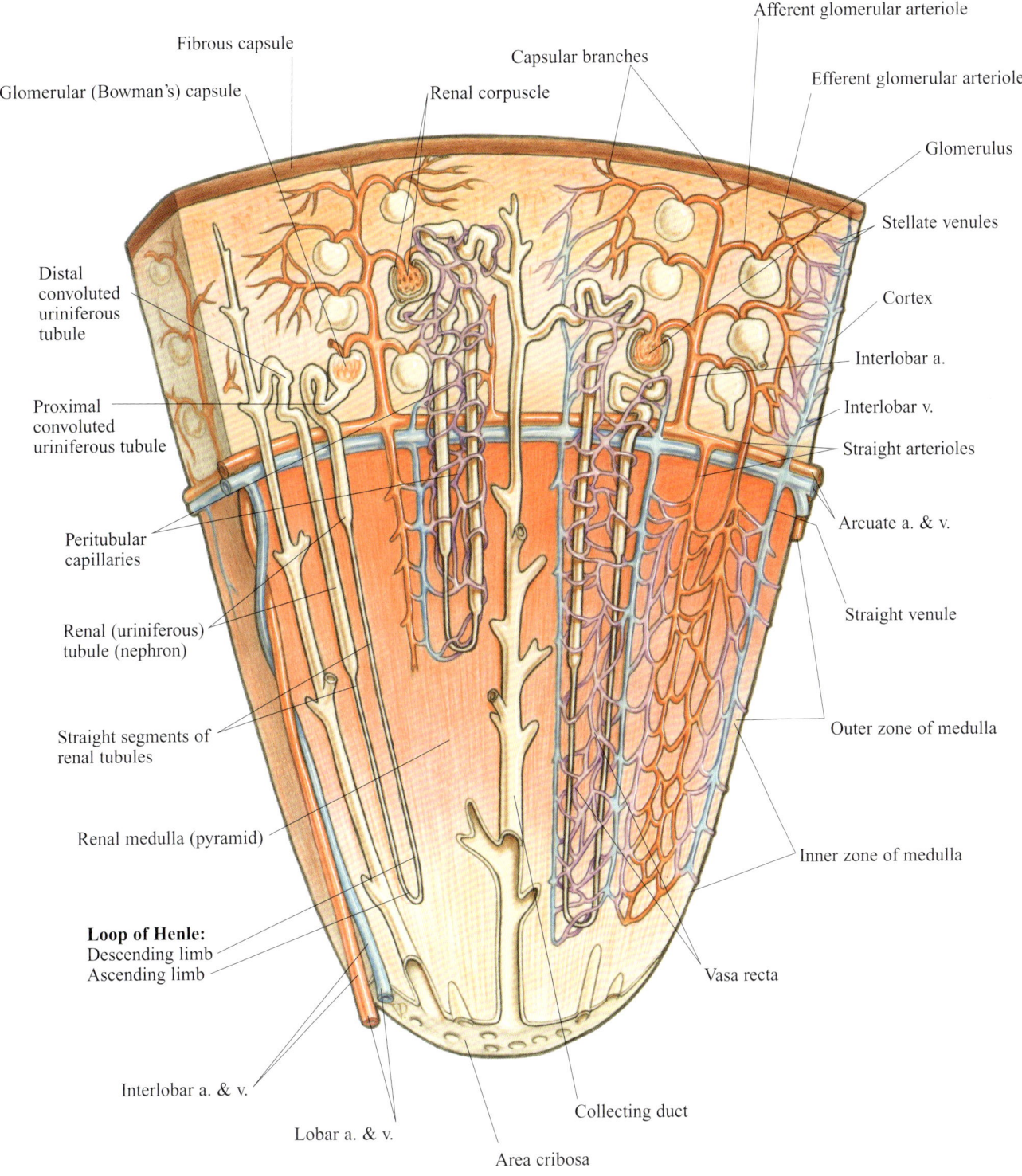

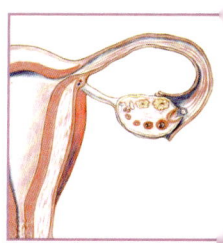

15
REPRODUCTIVE SYSTEM

MALE REPRODUCTIVE SYSTEM

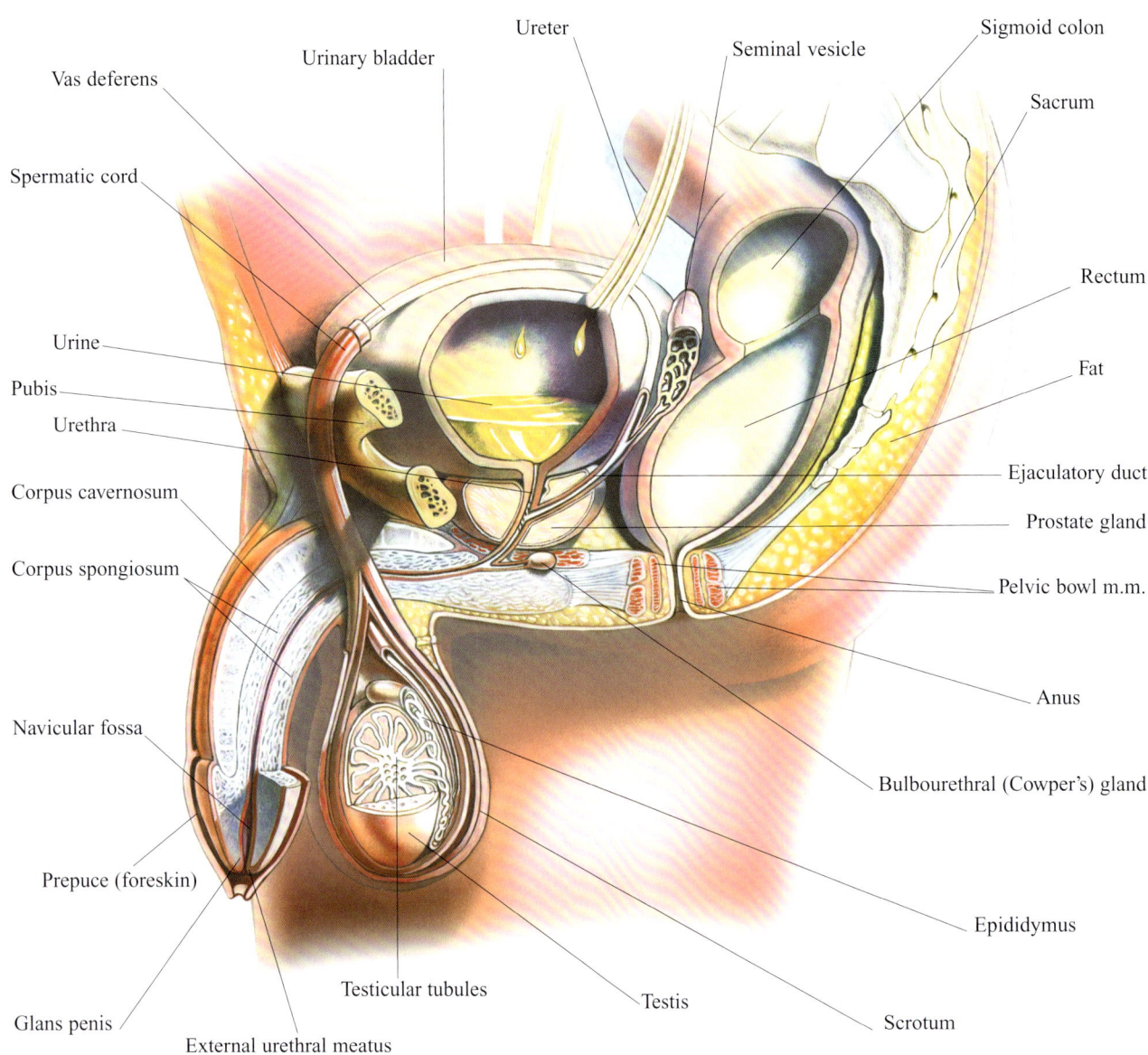

a.	= artery
a.a.	= arteries
L.	= Left
l.	= ligament
m.m.	= muscles
R.	= Right
v.	= vein
v.v.	= veins

FEMALE REPRODUCTIVE SYSTEM

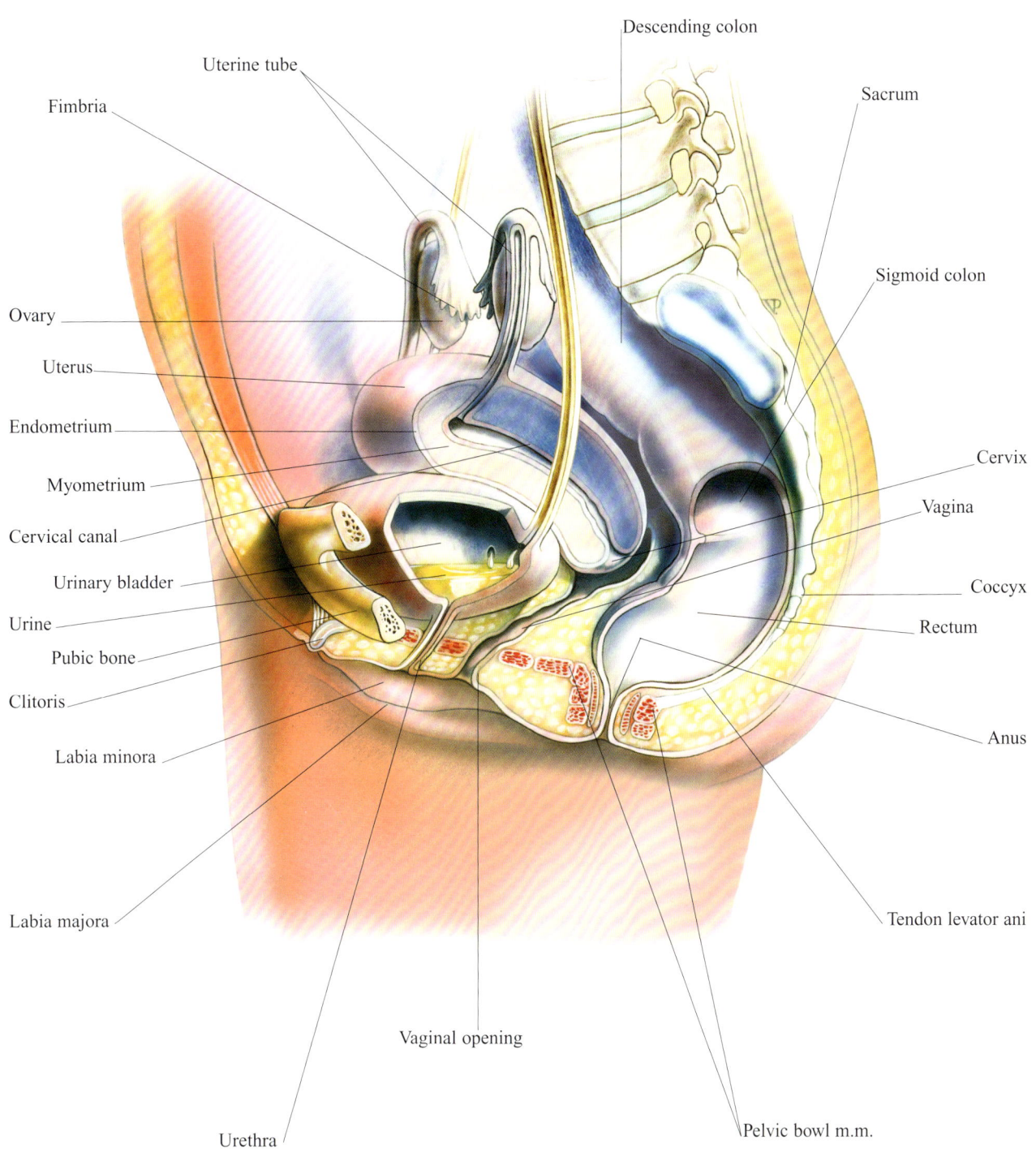

STAGES OF SPERM & OVUM

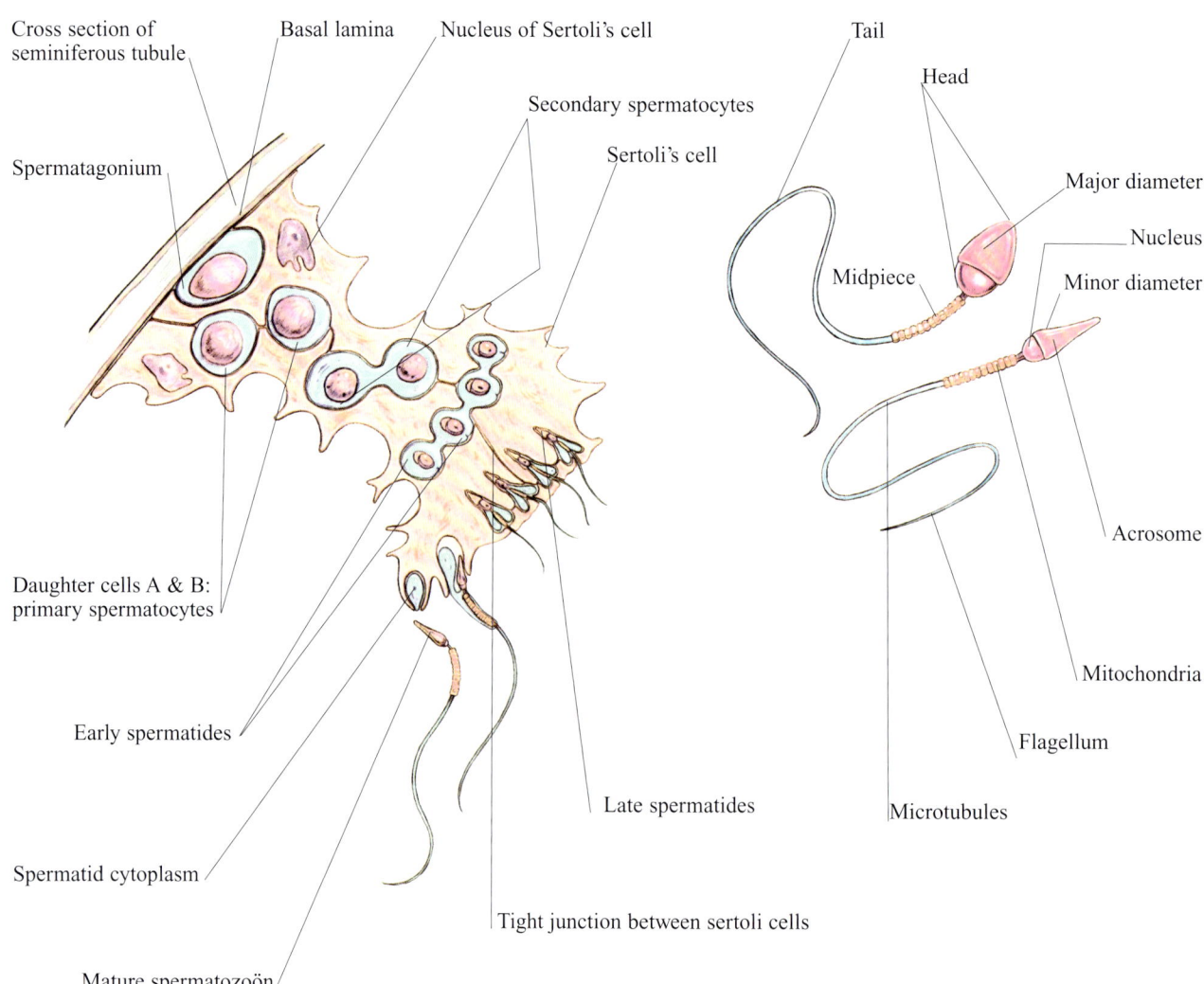

STAGES OF SPERM & OVUM

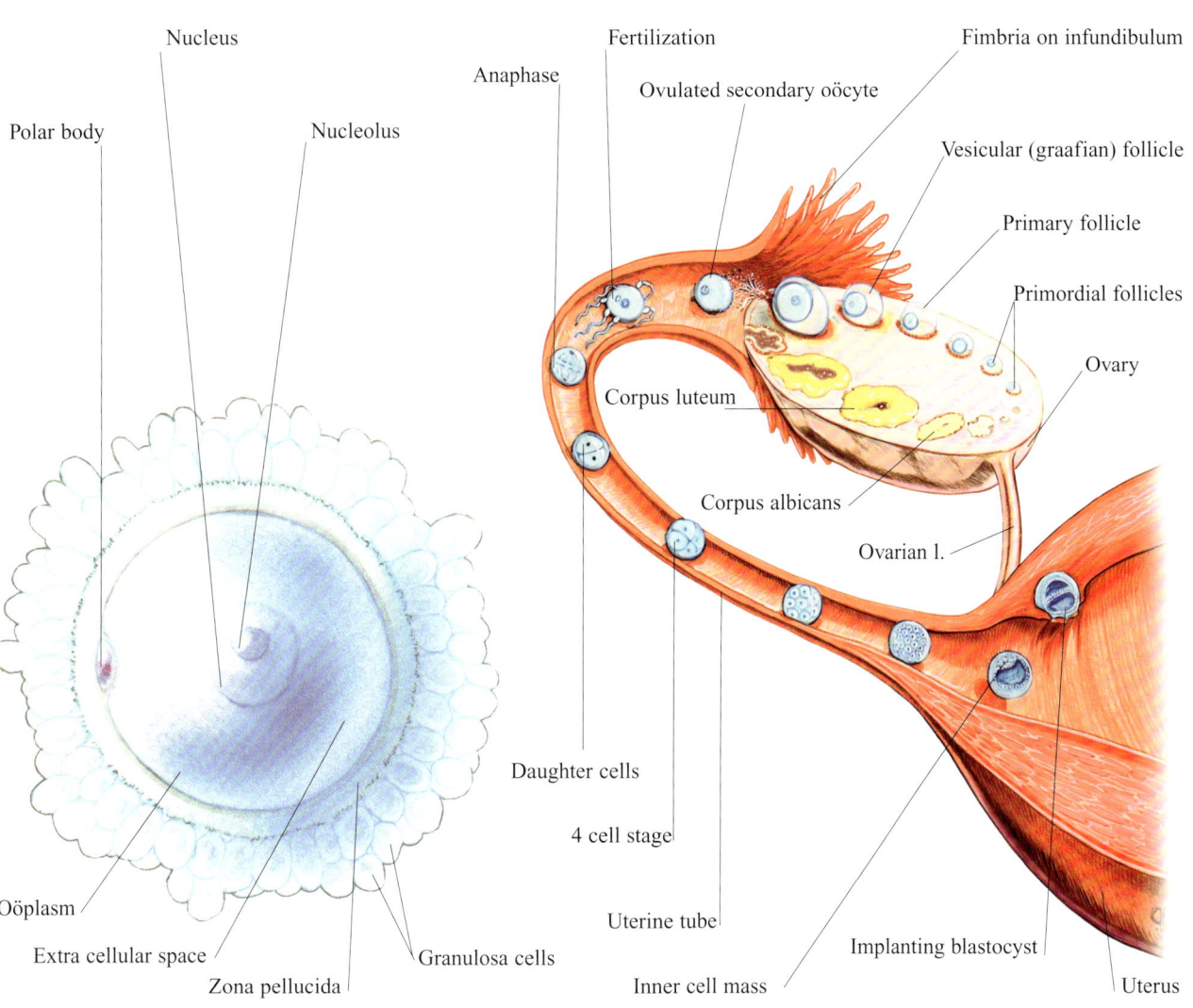

UTERINE CYCLE

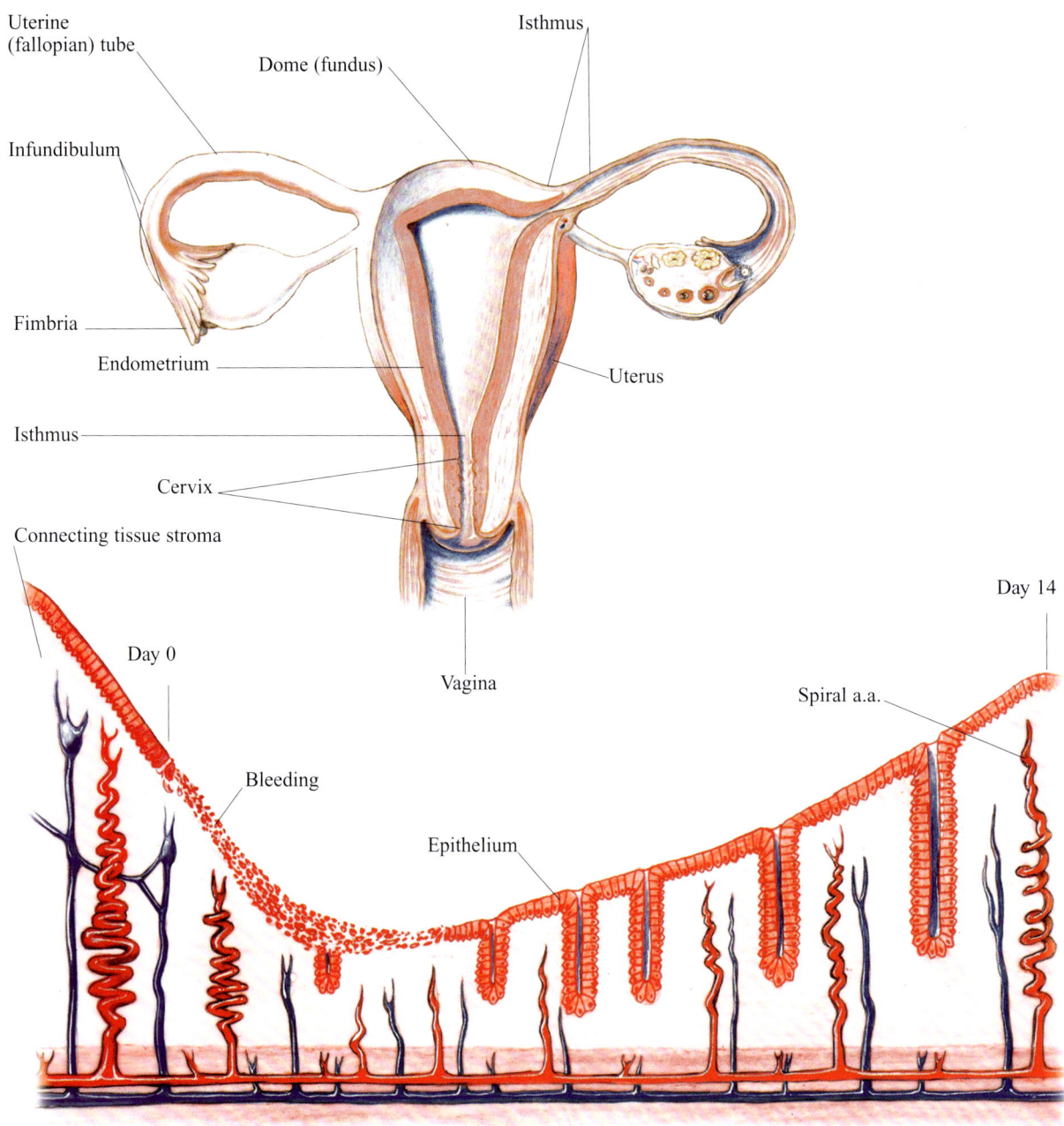

UTERINE CYCLE

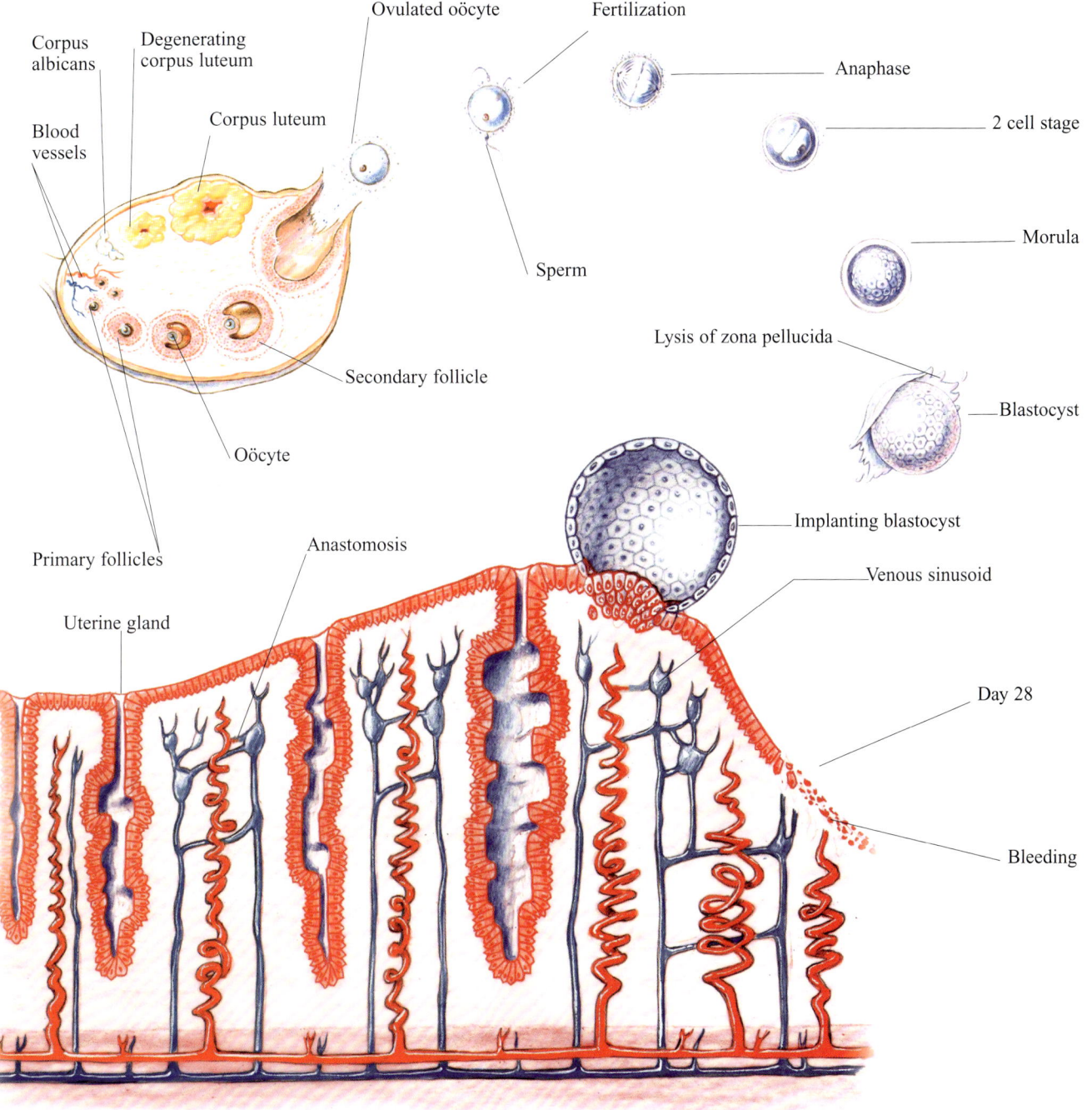

FETAL CIRCULATION

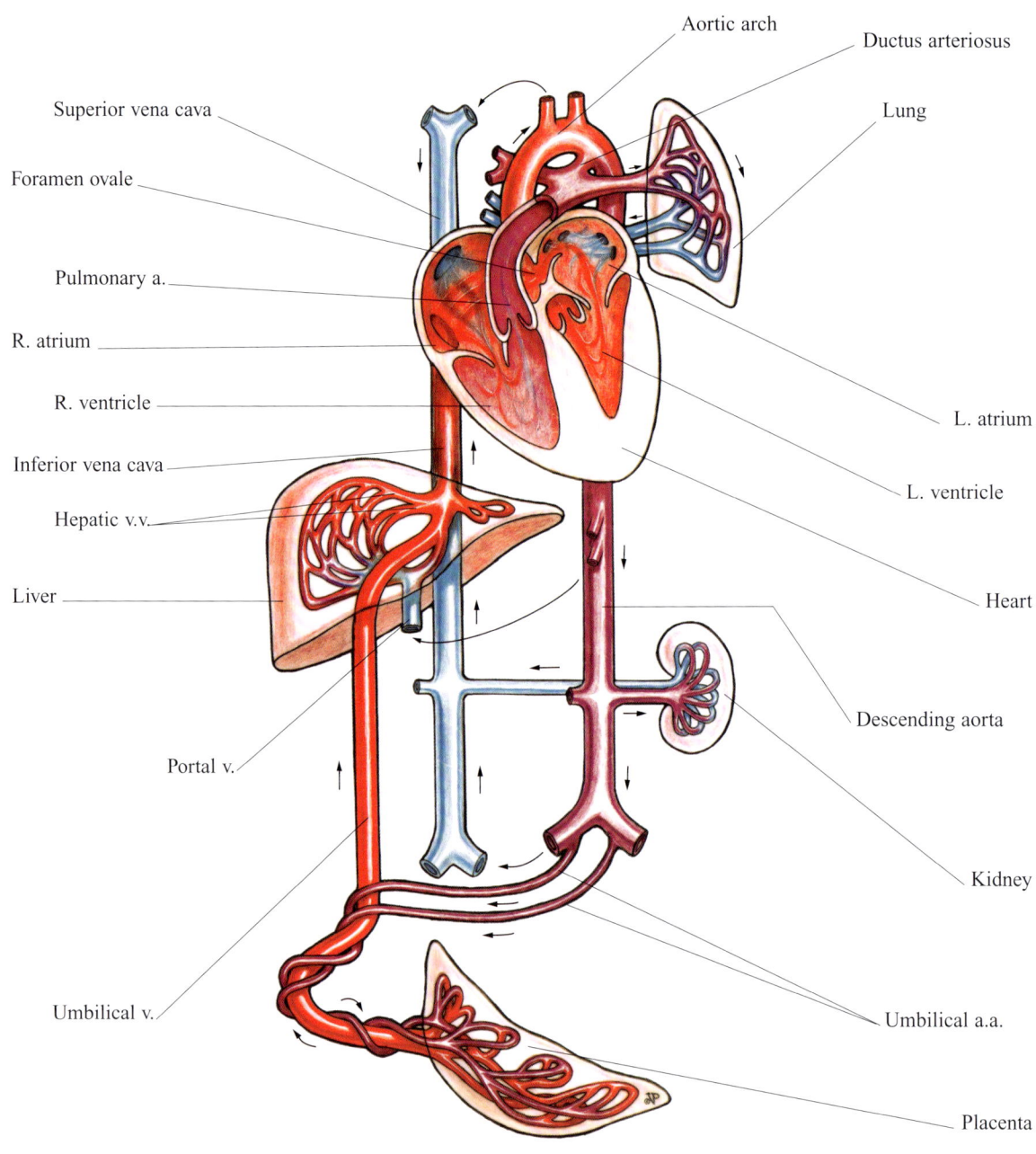

red = oxygenated blood
blue = unoxygenated blood
violet = mixed blood

FULL-TERM BABY

FULL-TERM BABY PRIOR TO DELIVERY

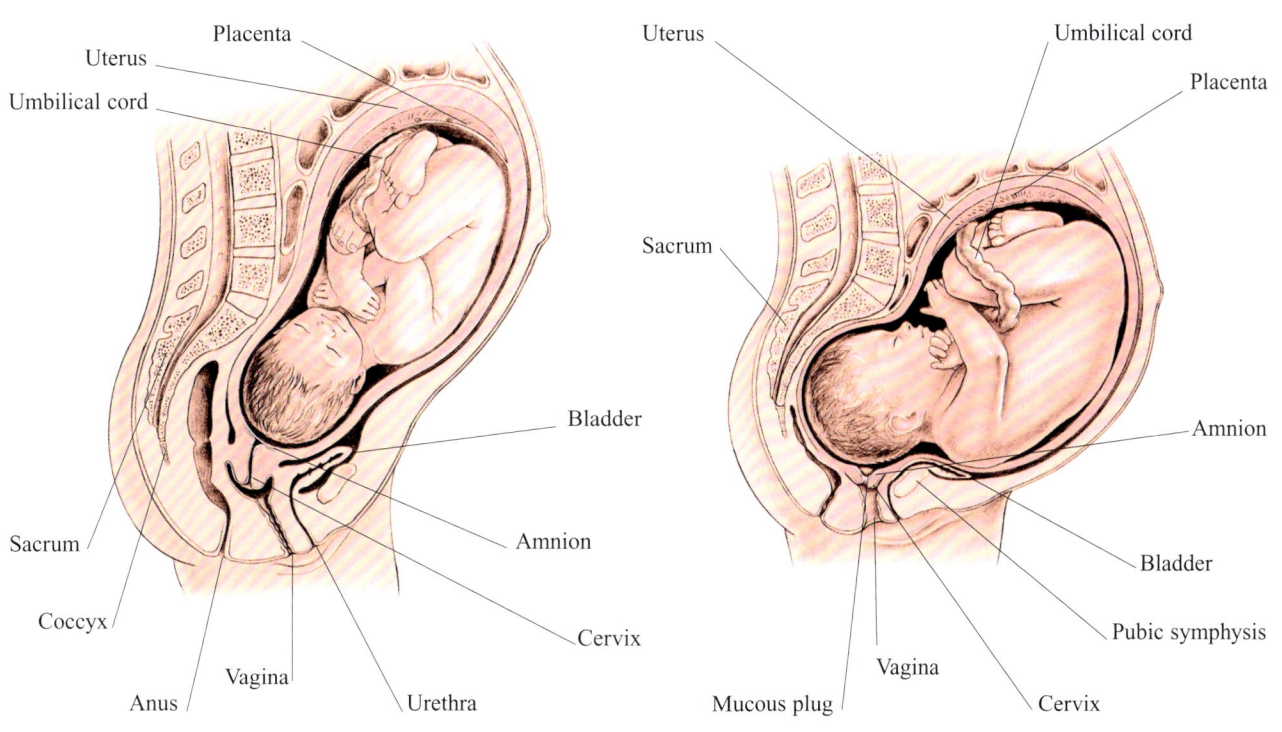

FULL-TERM BABY BEING DELIVERED

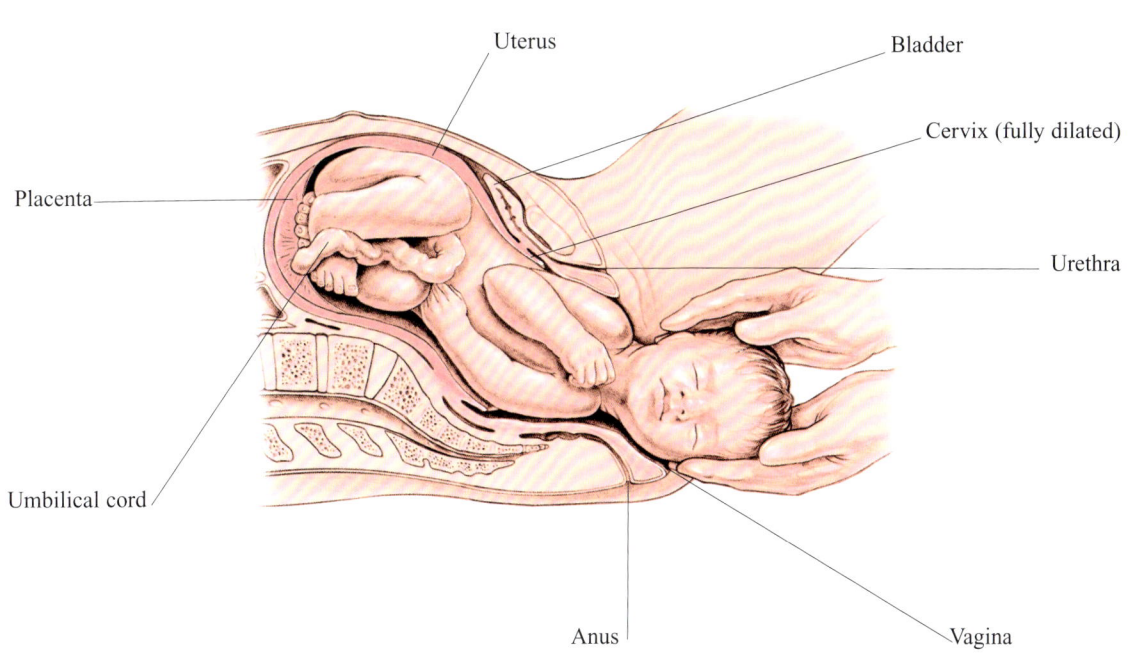

SEXUAL INTERCOURSE

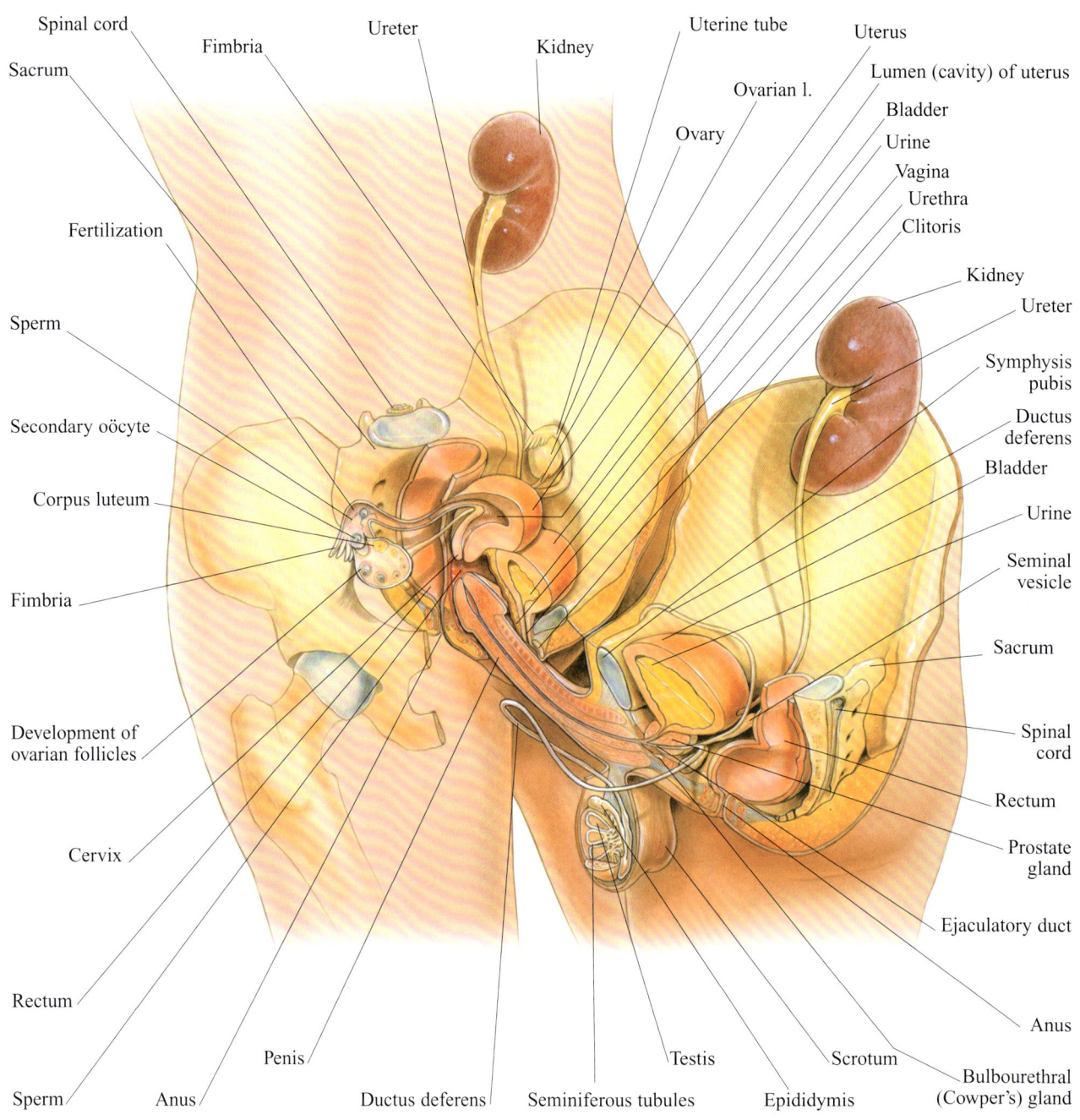

NOTES

INDEX

NOTE. Boldface entries indicate main headings; italics are bone features.

A

A band, muscle microstructure, 99
abdominal aorta, 156, 157, 159 174, 186, 193
abdominal crease, 9
abdominal esophagus, 137
abdominal muscles, 192
abducens nerve VI, 117, 119
abductor digiti minimi, 2, 3, 4, 7, 8, 10, 50, 51, 52, 53, 54, 55, 56, 67, 76, 77, 78, 79, 80, 81, 82, 83, 84, 85, 86, 89, 90, 91, 92, 94, 95, 96, 97, 98
abductor hallucis, 2, 4, 10, 54, 55, 56, 61, 63, 70, 88, 89, 90, 91, 92, 94, 95, 96, 97
abductor magnus muscle & fat, 4
abductor pollicis brevis, 8, 50, 52, 76, 77, 78, 79, 80, 81, 82, 86, 87
abductor pollicis longus, 4, 7, 8, 51, 52, 53, 64, 67, 77, 78, 80, 81, 82, 83, 84, 85, 87
abductor pollicis longus tendon, 8
abductor pollicis longus tendon in synovial sheath, 79
accessory hemiazygos vein, 185
accessory nerve XI, 117, 119
accessory pancreatic duct, 138
accessory process, 18, 20
accessory saphenous vein, 157, 158
acetabular cartilage, 41
acetabular labrum (lip), 41
acetabulum, 39, 109
acetabulum (socket), 14, 15
acoustic (or external auditory) meatus, 21
acromial articulation, 24
acromial end, 24
acromioclavicular joint, 25
acromioclavicular ligament, 36
acromion, 14, 15, 17, 25
acromion (scapula), 3, 4, 36, 68, 76
acromion process, 8
acrosome, 204
actin molecule, 99
adductor brevis, 48, 54, 60, 61,62, 64, 88
adductor group, 9
adductor hallucis, 54, 61, 89
adductor hallucis (oblique head), 55, 96, 97
adductor hallucis (transverse head), 55, 94, 95, 96, 97
adductor longus, 10, 48, 54, 60, 61, 64, 70, 88
adductor magnus, 10, 48, 49, 54, 60, 61, 62, 63, 64, 70, 88, 89
adductor minimus, 62, 64
adductor muscles & fat, 2
adductor pollicis, 50, 51, 52, 53, 60, 61, 64, 77, 87
adductor pollicis (oblique head), 76, 79, 80, 81, 82, 86
adductor pollicis (transverse head), 76, 78, 79, 80, 81,82, 86
adductor pollicis brevis, 52
adductors (grouped), 9
adductor tubercle, 29, 43
afferent arteriole, 198
afferent glomerular arteriole, 199
afferent lymphatic vessels, 189
ala, 22, 27
alaeque nasi, 71
alar cartilage, 6
alar ligament, 36
alveolar capillaries, 152

alveolar duct, 152
alveolar epithelium, 152
alveolar pores of Kohn, 152
alveolar sac, 152
alveoli cluster, 152
alveolus, 152
amnion, 209
ampulla, 137
anal column, 142
anal crypt, 142
anal gland, 142
anal sinus, 142
anal verge, 142
anaphase, 205, 207
anastomosis, 207
anatomical snuffbox, 4, 7, 8
anconeus muscle, 7, 51, 76, 77
angle of mandible, 5, 6
angular artery, 159, 160, 163
angular gyrus, 114, 115
angular vein, 160
annular ligament, 37, 50, 61, 78, 79, 86, 93, 94, 150
annulus fibrosis, 34, 39, 176
anococcygeal nerve, 105
anocutaneous line, 142
anorectal (dentate) line, 142
anterior (superior), semicircular canals & ducts, 128
anterior atlanto-occipital membrane, 34
anterior basal, lower lobe bronchus, 150
anterior cerebral artery, 122, 160, 162, 163
anterior chamber, 127
anterior choroidal artery, 162
anterior ciliary vein, 127
anterior commisure, 116, 121
anterior communicating artery, 122, 162, 163
anterior cruciate ligament, 43
anterior cusp, 176
anterior cutaneous rami, 104
anterior deep temporal nerve, 110
anterior division, 110
anterior ethmoidal nerve, 129, 148
anterior femoral cutaneous nerve, 104
anterior horn, 121
anterior horn of lateral ventricle, 120
anterior inferior cerebellar artery, 122, 162
anterior inferior iliac spine, 27
anterior internodal tract, 175
anterior interosseous artery, 157, 159
anterior interventricular branch (left anterior descending) of left coronary artery, 172, 175, 176
anterior ligaments of the fibular head, 43
anterior lingual glands, 136
anterior longitudinal ligament, 34, 42, 193
anterior margin of tibia, 10
anterior mediastinal nodes, 184
anterior meniscofemoral ligament, 43
anterior notch, 5
anterior parietal artery, 163
anterior part of internal capsule, 120
anterior perforate substance, 119
anterior right atrial branch of right coronary artery, 172
anterior sacrococcygeal ligament, 42
anterior semilunar cusp, 176
anterior septal branch of anterior ethmoidal artery, 129, 148

anterior skull (posterior view), 23
anterior spinal artery, 122, 162
anterior sternoclavicular ligament, 36
anterior superior alveolar nerve, 110
anterior superior iliac spine, 2, 3, 9, 42, 74, 151
anterior superior iliac spine, 16, 27
anterior surface, 24
anterior talofibular ligament, 44
anterior temporal artery, 163
anterior tibial artery, 156
anterior tibial vein, 156
anterior tibiofibular ligament, 44
anterior tibiotalar ligament, deltoid ligament, 44
anterior tubercle, 23
anteromedial basal, lower lobe bronchus, 150
antihelix, 5, 6
antitragus, 5, 6
anus, 134, 135, 140, 194, 196, 202, 203, 209, 210
aorta, 70, 187
aortic arch, 156, 157, 159, 160, 164, 170, 171, 174, 208
aortic arch node, 184
aortic hiatus, 62
aortic valve, 159, 172, 173, 174, 175, 176
apex, heart, 171, 174
apex, lung, 146, 150
apex, tongue, 130, 136
appendicular artery, 141
appendicular nodes, 188
appendicular vein, 165
appendix, 134, 135, 165
appendix epididymis, 192
appendix testis, 192
arachnoid matter, 108
arcuate arteries & veins, 197
arcuate artery, 157, 159
arcuate artery & vein, 199
arcuate line, 42, 61, 70
area cribosa, 194, 196, 197, 199
areola, 2, 3
areolar glands, 8
arm (anterior view), origins & insertions, 50
arm (posterior view), origins & insertions, 51
arm axilla & thorax, lymphatic system, 184
arm & hand, right (lateral & medial views), surface anatomy, 7
arm & hand muscles (anterior & medial views), 76
arm & hand muscles (lateral & posterior arm), 77
arterial system, 159
arteries (blood flows away from heart), 164
arteries, brain, 122
arteries, muscular system, 75, 99
arteriole, 189
artery, 166
articular capsule, 34, 37, 39, 42, 44, 98, 109
articular capsules of zygapophysial joints, 34
articular cartilage, 75
articular cartilage, 20
articular circumfrence, 28
articular disc, 35, 36
articular disc of temporo-mandibular joint, 71
articularis genus, 54, 62
articular tubercle, 22, 35
articular tubercle, 22
arytenoid cartilage, 149

ascending aorta, 157, 159, 170, 173
ascending colon, 140, 141, 165
ascending colon (large intestine), 134, 135
ascending frontal artery, 122, 162
ascending limb, loop of Henle, 199
ascending pharyngeal artery, 161
atlantoaxial ligament, 34
atlas vertebrae (C I), 18, 19, 21, 23, 36, 62, 147
atrioventricular (AV) artery, 175
atrioventricular (AV) node, 175, 178
auditory canal (external acoustic meatus), 5, 22, 128
auditory ossicles, 128
auditory (eustachian) tube, 114, 126, 128, 146, 147
auricle, 128
auricular cartilage (external acoustic meatus cartilage), 128
auricularis anterior, 60, 67, 71
auricularis posterior, 67
auricularis superior, 60, 71
auricular surface, 34
auricular surface (for ilium), 18
auricular (Darwin's) tubercle, 5
auriculotemporal nerve, 110, 111
axilla & breast, surface anatomy, 8
axillary lymph nodes, 182
axillary artery, 156
axillary artery & vein, 157
axillary nerve, 102, 103, 104, 105, 106
axillary tale of mammary gland, 8
axis vertebrae (C II), 18, 19, 21, 23, 36, 104, 105, 106
axon (nerve fiber), 111
axonal terminal, 111
axon hillock, 111
azgomatic, 22
azygos vein, 185

B

bachioradialis muscle, 2
Bachmann's bundle, 175
band of Richer (fascia lata), 2, 10, 60
Bartolin's (greater vestibule) gland & duct, 195
basal anterior, lower lobe bronchus, 150
basal lamina, 99, 152, 166, 204
base of 1st metatarsal, 10
base of 5th metatarsal, 10
base of pyramid, kidney, 192, 194, 195, 196, 197
base of skull, 36
base of skull, origins & insertions, 57
basement membrane, 99, 152, 198
basilar artery, 122, 157, 160, 162, 163
basilar part, occipital bone, 22
basilar part of occipital bone, 23, 62
basilic vein, 7, 8, 156, 157
biceps brachii, 3, 4, 7, 8, 60, 67, 77
biceps brachii (long head, short head), 2, 50, 61
biceps brachii tendon, 37
biceps brachii tendon (long head), 36
biceps femoris, 48, 49, 54, 63, 67, 69
biceps femoris (long head, short head), 54, 61, 64, 70, 76, 88, 89
biceps femoris tendon, 2, 3 4
biceps tendon, 2, 7
bicipital aponeurosis, 2, 7, 60, 76

bicipital groove, 14, 15
bifurcate ligament, 44
bile & pancreatic duct, 138
bladder, 209, 210
blastocyst, 207
bleeding, uterine cycle, 206, 207
blood circuits, circulatory system, 164
blood vessels, 111
blood vessels, circulatory system, 166
blood vessels, uterine cycle, 207
blood vessels within haversian or central canal, 30
blood vessels within Volkmann's or perforating canal, 30
body (sternum), 8
body of clitoris, 195
body of the vertebra, 20
body (corpus) of uterus, 195
bone structure, skeletal system, 30
bony rib, 26
bony trabeculi, 41
Bowman's capsule, 198, 199
brachial artery, 156
brachial artery & vein, 157
brachialis, 2, 7, 50, 51, 61, 64, 67, 76, 77
brachial nodes, 184
brachial plexus, 102, 103
brachial plexus (C IV-T I), 106
brachial vein, 156
brachiocephalic artery, 156, 157
brachiocephalic trunk, 160, 171, 174
brachiocephalic vein, 158, 160, 182
brachioradialis, 4, 7, 50, 51, 60, 67, 76, 77, 78, 79, 80, 81
brain, the, 102, 103, 113–23
brain (anterior view), 117
brain, arteries of, 122, 162
brain (frontal section), 118
brain (horizontal section), 120
brain (inferior view), 119
brain (lateral view), 115
brain (medial/sagittal view), 116
brain, ventricles of, 121
brain & neck, circulatory system, 163
brain, in place, 114
branch to sinoatrial node left coronary artery, 172
bregma, 23
bridge of nose (nasal bone), 5, 6
bronchial tree, 150
bronchiole & lobule, 146
bronchopulmonary (hilar) nodes, 184
brow ridge (frontal sinus), 5, 6
buccal branches, 111
buccal fat, 6
buccal nerve, 110
buccal node, 183
buccinator, 48, 57, 61, 69, 71, 73, 135, 136, 146
Buck's fascia, intercavernous septum of deep, 192
bulb of penis, 192
bulb of vestibule, 195
bulbospongiosus muscle, 195
bulbourethral (Cowper's) gland, 192, 194, 202, 210
bursal openings, 36
buttock fat, 4

C

C I (atlas vertebrae), 18, 19, 21, 23, 36, 62, 147
C I-C VII (cervical vertebrae), 16, 17, 18
C II (axis vertebrae), 18, 19, 21, 23, 36, 104, 105, 106

C III, 19, 21, 23, 104, 105, 106
C IV, 19, 21, 23, 104, 105, 106
C V, 19, 23, 104, 105, 106
C VI, 19, 23, 34, 104, 105, 106
C VII, 19, 23, 34, 49, 62, 66, 104, 105, 106
C VIII, 104, 105, 106
calcaneal fat pad, 3, 10
calcaneal heel fat, 4
calcaneal tendon (triceps crurae), 56
calcaneofibular ligament, 44
calcaneus 29, 89, 98
calcerine sulcus, 119
callosomarginal artery, 163
canaliculi, 30
cancellous bone, 30, 41
capillaries of arms & hands, 164
capillaries of colon, 164
capillaries of head, 164
capillaries of legs & feet, 164
capillaries of liver, 164
capillaries of pelvis, 164
capillary bed, 166, 189
capillary endothelium, 152
capitate, 17, 28, 38, 75
capitulum, 14, 28
capsular branches, 199
capsular ligament, 40
capsular & perirenal arteries, 197
capsule, lens, 127
capsule, muscular system, 83, 84, 85, 90, 91, 92, 95, 96, 97, 98
capsule, nodes & vessels, 189
capsule of atlanto-occipital joint, 34, 36
capsule of lateral atlantoaxial joint, 34, 36
cardiac impressions, 150
cardiac notch, 137, 146
cardiac opening, 137
cardiac ring of nodes, 186
cardiac veins & coronary arteries, 172
carotid canal, 22, 23, 62
carpal bones, 28, 85
carpals, 14, 15, 17
carpal tunnel, 38
carpi ulnaris, 77
carpometacarpal articulation & ligaments, 38
cauda equina, 102, 103
cavity of concha, 5
cecal nodes, 188
cecal veins, 165
cecum, 134, 135, 140, 141, 165
celiac node, 186, 187, 188
celiac trunk, 156
cell body, 111
cell nucleus, 198
central artery, 163
central artery of retina, 163
central artery & vein of retina, 127
central axillary nodes, 184
central band, 87
central canal, 121
central fissure (of Rolando), 115
central fissure, brain, 114
central sulcus, 116, 126
central tendon, 62
central tendon of diaphragm, 70
cephalic vein, 2, 7, 8, 156, 157
cerebellum, 102, 103, 114, 115, 117, 119, 126
cerebral aqueduct, 119, 121
cerebral peduncle, 116, 117, 118, 119
cerivcal vertebrae (VII), 126
cervical canal, 196, 203
cervical canal with palmate folds, 195
cervical lymph nodes, 182
cervical nerve (1st), 102, 103, 117, 119

cervical nerve (2nd), 117
cervical nerve (8th), 102, 103
cervical nerves, 114, 126
cervical nerves (C I-C VIII), 106
cervical plexus, 102, 103
cervical plexus (C I-C IV), 106
cervical spine, 4
cervical vertebra (I) (transverse process), 106
cervical vertebra (VII) (pedicle & transverse process), 106
cervical vertebrae, 14, 15, 26
cervical vertebrae (C I-C VII), 16, 17, 18
cervical vertebrae, 21
cervical vertebrae (anterior view), 23
cervical vertebrae (posterior view), 19, 23
cervical vertebrae (VII), 3, 4
cervical vertebrae (VI), 150
cervicobrachial plexus (posterior view), 106
cervix, 196, 203, 206, 209
cervix (fully dilated), 209
cervix of uterus, 195
cheek bone (zygomatic), 5
chiasma tendinum, 80, 81
choanae, 22
chorda tympani nerve, 110
choroid, 127
choroid plexus, 119, 120
choroid plexus of lateral ventricle, 118
cilia (lashes), 5
ciliary ganglion, 110
ciliary muscle, 127
ciliary process, 127
ciliated columnar epithelium, 152
cingulate gyrus, 116
cingulate sulcus, 116
circle of Willis, 160, 162
circular esophageal muscle, 135
circular fibers, ciliary muscle, 127
circular layer, stomach musculature, 137
circular muscle layer, 142, 146
circular muscle of colon, 141
circular muscle of ileum, 141
circulation, 174
circulatory system, 155–67
circulatory system (anterior view), 156
circulatory system (posterior view), 157
circumferential lamellae, 30
circumflex artery, 157
circumflex branch, 175
circumflex branch of left coronary artery, 171, 172
cisterna chyli, 182, 185, 186, 187, 188
claustrum, 118, 120
clavicle, 3, 4, 8, 14, 15, 16, 17, 25, 26, 36, 68, 76
clavicle (inferior view), origins & insertions, 51
clavicle (superior view), origins & insertions, 50
clavicle (superior & inferior views), skeletal system, 24
clavicular head of sternocleido-mastoid muscle, 3
clavicular segment, 2
clitoris, 196, 210
clitoris (dorsal nerve of), 104, 107
clivus, 23
coccygeal nerve, 107
coccygeus, 48, 62, 64
coccyx, 16, 27, 34, 39, 109, 196, 203, 209
coccyx (III-IV), 17, 18
cochlea (inner ear), 126, 128

cochlea (inner ear) (semicircular canals), 114
cochlear duct, 128
cochlear nerve, 128
collateral ligament, 38, 44, 45, 75, 80, 81, 82, 83, 84, 85, 86, 87, 90, 91, 92, 95, 96, 97, 98
collateral sulcus, 119
collecting duct, 199
colon, 142
columns of fornix, 118, 120
common annular tendon, 72
common AV bundle (of His), 175
common bile duct, 138
common carotid artery, 156, 157, 160, 163
common dorsal digital nerves, 102, 103
common hepatic artery, 159
common hepatic duct, 138
common iliac nodes, 182, 186, 188
common iliac veins, 165
common peroneal nerve, 102, 103, 104, 105
common plantar digital nerves, 102, 103
common sheath of peroneus longus & brevis tendons, 90
communicating branch, 110
compact bone, 30
concentric lamellae, 30
concha, 5, 6
condylar canal & fossa, 22
condyloid depression, 4
conjunctiva, 127
connecting tissue stroma, 206
connective tissue, 131
constrictor pharyngis, 69
contracting atria, 177
contracting ventricles, 178
conus medullaris, 102, 103
coracoacromial ligament, 36
coracobrachialis, 7, 8, 50, 61, 76
coracoclavicular ligament (conoid part), 36
coracohumeral ligament, 36
coracoid cartilage, 126
coracoid process, 14, 15, 25, 26, 68, 76
coracoid process (scapula), 3
coracoid tuberosity, 24
cornea, 5, 127
corniculate cartilage, 149
coronal suture, 21, 23
corona of glans, 192
coronary arteries & cardiac veins, 172
coronary ligament, 138
coronary sinus, 171, 172, 176
coronoid fossa, 28
coronoid process, 14, 15, 22, 28
corpora quadrigemina, 116
corpus albicans, 195, 205, 207
corpus callosum, 116
corpus callosum trunk, 118
corpus cavernosum, 192, 194, 202
corpus luteum, 195, 205, 207, 210
corpus (body) of uterus, 195
corpus spongiosum, 192, 194, 202
corpuscle of Krause (cold), 131
corpuscle of Ruffini (heat), 131
corrugator, 48
corrugator cutis ani muscle, 142
corrugator fibers, 60, 71
corrugator supercilii, 61
cortex, large intestine, 189
cortex, lens, 127
cortex, nephron, 199
costal cartilage, 3, 8, 14, 15, 26, 36
costal facet, 18, 34
costal margin, 8
costal pleural, 150
costal ridge, 2

costal segment, 2
costal tuberosity, 24
costoclavicular ligament, 36
costodiaphragmatic recess of pleura, 146
costoiliac space, 2
costotransverse joint, 34
Cowper's (bulbourethral) gland, 192, 194, 202, 210
craniocervical, joints & ligaments, 36
cremaster, 60, 186, 192
cribiform fascia, 186
cribiform plate (ethmoid bone), 147
cricoid cartilage, 73, 114, 146, 147, 149, 150
cricoid cartilage & thyroid gland, 3
cricothyroid, 73
cricothyroid ligament, 149, 150
crista galli, 23, 148
cruciate ligament, 79, 86, 93, 94
cruciform eminence, 23
cruciform ligament, 36
crura of antihelix, 5
crus of clitoris, 195
crus of helix, 5
crus of penis, 192
cubital lymph nodes, 182
cubital nodes, 184
cuboid, 14, 15, 16, 45
cuboid bone, 29
cuneiform bones (medial, intermediate, lateral), 29
cuneiform ligament, 10
cupola (dome) of pleura, 146
cutaneous innervation: dermatomes & peripheral nerve distributions (anterior view), 104
cutaneous innervation: dermatomes & peripheral nerve distributions (posterior view), 105
cutaneous nerve of forearm, 106
cuticle, hair follicle, 131
cuticle, nail, 75
cymba conchalis, 5
cystic duct, 138
cystic vein, 165

D
dartos, 192
Darwin's tubercle, 5
daughter cells, 205
daughter cells A & B: primary spermatocytes, 204
day 28, uterine cycle, 207
deep abdominal & inguinal nodes, lymphatic system, 186
deep brachial artery, 159
deep branch, 102, 103
deep cervical vein, 160
deeper (accessory) portion of tectorial membrane, 36
deep femoral artery, 157, 159
deep femoral vein, 156
deep inguinal nodes, 186
deep lingual artery, 161
deep muscles (layer II, lateral view), 69
deep muscles (layer III, medial view), 70
deep muscles (layers II & III, posterior view), 64
deep muscles (layers III & IV, anterior view), 61
deep muscles (layers IV & V, posterior view), 65
deep muscles (layers V &VI, anterior view), 62
deep muscles (layers VI & VII, posterior view), 66
deep palmar arch, 157, 159
deep peroneal nerve, 102, 103, 104

deep transverse metacarpal ligament, 38, 78, 79, 80, 81, 82
deep transverse metatarsal ligament, 44, 45, 94, 95, 96, 97
degenerating corpus luteum, 207
deltoid, 4, 8, 50, 51, 60, 63, 67, 76, 77
deltoid ligament, 44
deltoid & pectoralis major muscles, 7
deltoid tuberosity, 25
deltopectoral groove, 2, 8
dens of axis (odontoid process), 19, 23
dental branches, 110
dentate (anorectal) line, 142
deoxygenated red blood cell, 152
depressor anguli oris, 48, 69
depressor labii inferioris, 48, 61
depressor supercilii, 48
dermal papillae, 131
dermatomes & peripheral nerve distributions (anterior & posterior views), 104–5
dermis (temperature), 131
descending abdominal aorta, 164
descending aorta, 170, 208
descending branch of lateral circumflex artery, 157, 159
descending colon, 140, 165, 196, 203
descending colon (large intestine), 135
descending colon (large intestine) behind transparent (small intestine), 134
descending limb (loop of Henle), 199
descending ramus of pubis, 16
descending thoracic aorta, 157, 159, 164
development of ovarian follicles, 210
diaphragm, 74, 146, 150, 151, 187, 193
diaphragm (costal part), 62
diaphragmatic pleural, 150
diaphysis, 30
diastole, beginning of, 177
diastole, end of, 177
diastole, heart in, 176
digastric (anterior belly), 57, 69, 71, 73, 146
digastric (posterior belly), 57, 69, 70, 71, 73, 146
digestive system, 133–43
digestive system (anterior view), 134
digestive system (posterior view), 135
digital artery & vein, 157
digital branches, 104
digital creases, 7
digital fat pad, 2, 3, 4, 7, 8, 10
digital flexor tendons, 86
digital slips of plantar aponeurosis, 93
digital web, 8
digits, 2
disc, 18
distal carpals, 17
distal convoluted tubule, 198
distal convoluted uriniferous tubule, 199
distal epiphysis, 30
distal heel crease, 3, 4, 10
distal interphalangeal crease, 2, 8
distal interphalangeal joint, 8
distal interphalangeal joint, 75
distal palmar crease, 2, 8
distal phalanx, 75, 87, 98
distal phalanx (head, body, base, tuberosity), 28
distal radioulnar joint capsule, 38

213

INDEX

distal wrist crease, 2, 7, 8
dome (fundus), 206
dome (cupola) of pleura, 146
dorsal branch, 102, 103
dorsal calcaneocuboid ligament, 44
dorsal carpometacarpal ligaments, 38
dorsal cuboidonavicular ligament, 44
dorsal cuneocuboid ligaments, 44
dorsal cuneonavicular ligaments, 44, 98
dorsal digital artery, 159
dorsal digital nerves, 102, 103
dorsal expansion, 77
dorsal expansion (hood), 75, 85, 86, 87
dorsal finger creases: distal & proximal, 8
dorsal foot (layer I), muscles, 90
dorsal foot (layer II), muscles, 91
dorsal foot (layer III), muscles, 92
dorsal hand (layer I), muscles, 83
dorsal hand (layer II), muscles, 84
dorsal hand (layer III), muscles, 85
dorsal intercarpal ligaments, 38
dorsal intercuneiform ligament, 44
dorsal interossei, 96, 97
dorsal interosseous, 50
dorsal interosseous (1st), 51, 52, 53, 56
dorsal interosseous (2nd & 3rd), 52
dorsal interosseous (2nd), 51, 53, 56
dorsal interosseous (3rd), 51, 53, 56
dorsal interosseous (4th), 51, 52, 53, 56
dorsal interosseous muscle, 61, 64, 90, 91, 92
dorsal interosseous muscle (1st), 75, 77, 78, 79, 80, 81, 82, 83, 84, 85, 87
dorsal interosseous muscle (2nd), 77, 80, 81, 82, 83, 84, 85
dorsal interosseous muscle (3rd), 77, 80, 81, 82, 83, 84, 85
dorsal interosseous muscle (4th), 77, 80, 81, 82, 83, 84, 85
dorsal interosseus, 54
dorsal interosseus muscle (1st), 4, 7, 8
dorsalis pedis artery, 156, 159
dorsal metacarpal ligaments, 38
dorsal metatarsal ligaments, 44
dorsal nerve of penis (clitoris), 104, 107
dorsal radiocarpal ligament, 38
dorsal root (sensory), 108
dorsal sacral foramina, 37, 109
dorsal talonavicular ligament, 44
dorsal tarsometatarsal ligaments, 44
dorsal vein, 4
dorsum sellae, 23
ductus arteriosus, 208
ductus deferens, 192, 193, 210
ductus deferens & cremaster muscle, 9
ductus reuniens, 128
duodenal longitudinal folds, 138
duodenojejunal flexure, 138, 139
duodenum (small intestine), 135, 137, 138, 139, 165, 187
duodenum (small intestine) (behind transverse colon), 134
dura mater, 108

E

ear, 126
ear, surface anatomy, 5

ear (internal), 147
eardrum (tympanic membrane), 114, 126, 128
earlobe, 5
early spermatides, 204
edge of acetabulum, 27
edge of latissimus dorsi muscle, 3
edge of the liver, 135
edge of the stomach, 135
edge of the transparent heart, 172
edges of annular & cruciate ligaments, 94
efferent arterioles, 198
efferent glomerular arteriole, 199
efferent lymphatic vessels, 189
ejaculatory duct, 194, 202, 210
elastic fibers, 152
elbow (anterior & lateral views), joints & ligaments, 37
elbow (anterior & posterior views), skeletal system, 28
endolymphatic duct, 128
endolymphatic sac, 128
endometrium, 195, 196, 203, 206
endomysium, 99
endoneurium, 111
endosteum, 30
endothelial cell, 152, 166, 189, 198
endothelium, 198
endothelium, tunica intima, 166
epicolic nodes, 188
epicranial aponeurosis (galea aponeurotica), 2, 3, 4, 5, 6, 69, 71
epidermis (touch), 131
epididymis, 192, 194, 202, 210
epiglottis, 126, 130, 136, 147, 149
epimysium, 99
epineurium, 111
epiphysial line, 30, 41
epiphysis, 75
epithelial cells, 152
epithelium, 206
epoöphoron, 195
erector pili muscle, 131
erector spinae, 3, 4, 64, 65, 69, 70
erector spinae muscle (iliocostalis lumborum), 4
erector spinae tendons over multifidous muscle, 4
esophageal foramen, 150
esophageal hiatus, 62, 146
esophageal mucosa, 137
esophagus, 70, 73, 126, 130, 134, 136, 146, 147, 165, 187, 193
ethmoid, 21
ethmoidal cells, 148
ethmoid bone, 129, 148
eustachian (auditory) tube, 114, 126, 128, 146, 147
extensor carpi radialis brevis, 7, 50, 51, 53, 63, 67, 76, 77, 78, 83, 84, 85, 87
extensor carpi radialis brevis tendon, 8
extensor carpi radialis longus, 4, 7, 50, 51, 53, 67, 76, 77, 78, 83, 84, 85, 87
extensor carpi radialis longus tendon, 7
extensor carpi ulnaris, 4, 50, 51, 52, 53, 63, 76, 77, 78, 80, 81, 82, 83, 84, 85, 86
extensor digiti minimi, 4, 7, 51, 53, 76, 77, 78, 83, 84, 85
extensor digiti minimi tendon, 4, 7, 8
extensor digitorum brevis, 2, 3, 4, 10, 54, 56, 60, 61, 67, 88, 89, 90, 91, 92, 98
extensor digitorum brevis tendon, 10, 92
extensor digitorum communis, 4, 7, 51, 63, 67, 76, 77, 78, 83, 84, 85, 86, 87

extensor digitorum communis (central bands), 51, 53, 86
extensor digitorum communis (lateral bands), 53, 86
extensor digitorum communis & extensor indicis tendons, 8
extensor digitorum longus, 2, 3, 54, 56, 60, 61, 67, 69, 77, 88, 89, 98
extensor digitorum longus tendon, 2, 3, 10, 60, 75, 90, 91, 92, 98
extensor expansion, 8, 76, 90, 91, 92, 98
extensor fasciae latae muscle, 2, 3, 9
extensor hallucis brevis, 3, 54, 56, 60, 61, 88, 90, 91, 92, 98
extensor hallucis brevis & extensor digitorum brevis muscles, 10
extensor hallucis brevis tendon, 2, 10, 92
extensor hallucis longus, 54, 56, 60, 61, 88, 90, 91, 92, 98
extensor hallucis longus tendon, 2, 3, 10, 44, 60, 70, 88
extensor indicis, 51, 53, 77, 78, 83, 84, 85, 87, 99
extensor insertions, 75
extensor pollicis brevis, 4, 51, 53, 64, 67, 77, 78, 80, 81, 83, 84, 85, 87
extensor pollicis brevis in synovial sheath, 79
extensor pollicis brevis tendon, 7, 8
extensor pollicis longus, 51, 53, 64, 77, 78, 83, 84, 85, 87
extensor pollicis longus tendon, 4, 7
extensor retinaculum, 2, 7, 8, 10, 63, 76, 77, 86, 87, 88, 89
extensor ulnaris muscle, 7
external acoustic meatus, 5, 22, 23, 128
external anal sphincter, 64
external anal sphincter muscle (deep, superficial, subcutaneous), 140, 142
external capsule, 118, 120
external carotid artery, 161, 163
external iliac artery, 156
external iliac artery & vein, 157
external iliac vein, 156
external iliac vessels, 192
external intercostal, 62, 65, 69, 70, 74, 151
external jugular vein, 156, 157, 160, 174
external occipital crest, 22
external occipital protuberance, 22
external sheath, hair follicle, 131
external spermatic fascia, 192
external urethral meatus, 192, 194, 202
external uretral meatus, 192
external uterine opening, 195
extra cellular space, ovum, 205
eye, 126
eye, surface anatomy, 5
eyebrow, 5
eye muscles (extrinsic eye muscles: right lateral view, anterior view), 72

F

face & head, nerves of the, 111
facial artery, 159, 160
facial artery & vein, 156, 157
facial nerve, 111, 118
facial nerve VII, 117, 119, 128
facial vein, 160
falciform ligament, 138
fallopian (uterine) tube, 206

fascia lata (band of Richer), 2, 10, 60
fascia lata, 186
fascicle, 99, 111
fat, 2, 4, 10, 73, 75, 93, 142, 194
fat, muscular system, 73
fat lobule, 131
fat pads, muscular system, 60
feet (plantar & dorsal views), skeletal system, 29
female hips, surface anatomy, 9
female reproductive system, 203
female skeleton (anterior view), skeletal system, 14
female urogenital system (anterior view), 195
female urogenital system (lateral view), 196
femoral artery, 155, 156, 159, 164
femoral artery & vein, 10
femoral head cartilage, 41
femoral nerve, 102, 103, 107
femoral triangle, 9
femoral vein, 156, 157, 164
femur, 14, 15, 16, 17, 27, 29, 30, 41, 42, 43, 107
femur (medial epicondyle), 4
femur shaft, 41
fenestrations, 152, 166, 198
fertilization, 205, 207, 210
fetal circulation, 208
fibers from flexor retinaculum, 79, 80, 81
fibers of ciliary zonule (suspensory ligament of lens), 127
fibrous capsule, 192, 194, 195, 196, 197, 199
fibrous digital sheath, 75
fibrous loop for intermediate digastric tendon, 70
fibrous septum, 142
fibula (lateral malleolus) 2, 3, 4, 10, 14, 15, 16, 17, 29, 43, 44, 45, 90, 91, 92, 97, 98
fibular (lateral) collateral ligament, 43
fibular head, 2, 4, 89
filament to connective tissue, 189
filiform papillae, 130, 136
filtration slits (slit pores), 198
fimbria, 195, 196, 203, 206, 210
fimbria on infundibulum, 205
finger, components of, 75
finger (medial view), joints & ligaments, 38
flank fat pad, 3
flexor carpal ligament, 60
flexor carpi radialis, 2, 7, 50, 52, 60, 76, 78, 80, 81, 82
flexor carpi radialis in synovial sheath, 79
flexor carpi radialis tendon, 2, 8
flexor carpi ulnaris, 2, 4, 7, 50, 51, 52, 60, 61, 63, 76, 77, 78, 79, 80, 81, 82, 86
flexor carpi ulnaris tendon, 2, 7, 8
flexor digiti minimi, 76
flexor digiti minimi brevis, 50, 52, 55, 69, 76, 78, 79, 80, 81, 82, 86, 94, 95, 96, 97, 98
flexor digitorum, 54
flexor digitorum brevis, 45, 54, 55, 63, 69, 88, 94, 95, 96, 97, 98
flexor digitorum brevis (under plantar aponeurosis), 70
flexor digitorum brevis tendon, 45, 95, 96, 97
flexor digitorum longus, 4, 54, 55, 63, 64, 70, 88, 89, 90, 91, 92, 94, 95, 98
flexor digitorum longus tendon, 45, 96, 97, 98
flexor digitorum profundus, 50, 51, 52, 61, 76, 77, 78, 79, 80, 81, 82, 86, 87

flexor digitorum profundus tendon, 75
flexor digitorum superficialis, 2, 7, 8, 50, 52, 61, 76, 77, 78, 79, 80, 81, 82, 86, 87
flexor digitorum superficialis tendon, 75
flexor hallucis longus, 69
flexor hallucis brevis, 54, 55, 64, 94, 95, 98
flexor hallucis brevis (lateral head), 97
flexor hallucis brevis (medial head), 97
flexor hallucis longus, 3, 54, 55, 64, 67, 70, 88, 89, 90, 91, 92, 94, 95, 96, 97, 98
flexor hallucis longus tendon, 45, 70, 88, 95, 96, 97, 98
flexor knee crease, 4, 10
flexor pollicis brevis, 7, 8, 50, 52, 76, 86
flexor pollicis brevis (deep head), 79, 80, 82
flexor pollicis brevis (superficial head), 78, 79, 80, 81, 82
flexor pollicis longus, 8, 50, 52, 61, 78, 80, 81, 82
flexor pollicis longus in radial bursa, 79
flexor pollicis longus tendon, 7
flexor retinaculum, 4, 38, 45, 60, 61, 63, 70, 76, 78, 79, 80, 81, 82, 86, 87, 88, 89, 94, 95, 96, 98
flexor synovium, 60
flocculus, 117, 118
foliate papillae, 130, 136
follicle, 189, 195
follicles (ovarian), 205, 207, 210
foot, muscular system, 90–97, 98
foot (dorsal view), origins & insertions, 56
foot (plantar view), origins & insertions, 55
foot, right (inferior view), joints & ligaments, 45
foot, right (lateral & medial views), joints & ligaments, 44
foot & leg (anterior & posterior views), origins & insertions, 54
foot & leg (medial, dorsal, plantar views), surface anatomy, 10
foramen caecum, 118
foramen cecum, 130, 136
foramen lacerum, 22
foramen magnum, 22, 23
foramen of inferior vena cava, 70
foramen ovale, 22, 110, 208
foramen rotundum, 23
foramen spinosum, 22
fornix, 116
fornix of vagina, 195
4 cell stage, 205
fovea capitis, 30
fovea central in macula lutea, 127
free nerve endings, 131
free tenia (tenia libera), 135, 140, 141
frenulum of clitoris, 195
frontal, 14, 15, 16, 21, 23
frontal artery, 160
frontal bone, 22
frontal branch of superficial temporal artery, 161
frontal crest, 23
frontal eminence, 5, 6
frontalis, 67, 71
frontalis muscle, 2, 5, 6
frontal lobe, 114, 116, 117, 126
frontal nerve, 110
frontal pole, 117
frontal pole of cerebrum, 119
frontal process (zygomatic bone), 6

frontal sinus (brow ridge), 5, 6, 114, 126, 129, 147, 148
frontal vein, 160
front deltoid muscle, 3, 7
frontopolar artery, 163
full-term baby, 209
full-term baby being delivered, 209
full-term baby prior to delivery, 209
fundic stomach, 137
fundus (dome), 206
fundus of uterus, 195
fungiform papillae, 130, 136
fused basal laminas, 152

G

galea aponeurotica (epicranial aponeurosis), 2, 3, 4, 5, 6, 69, 71
gallbladder, 134, 138, 165
gastric aorta, 164
gastric canal, 137
gastric folds (rugae), 137
gastrocnemius, 60, 63, 67
gastrocnemius (lateral head, medial head), 2, 3, 4, 10, 54, 61, 70, 88, 89
gastrocnemius & soleus via tendo calcaneus (Achilles), 54
genicular arteries, 156, 157, 159
genicular veins, 156
genioglossus muscle, 70, 73, 114, 126, 147
geniohyoid, 57, 70, 114, 147
genitofemoral & ilioinguinal nerve, 104
genitofemoral nerve, 104, 107
genu of corpus callosum, 120
geratocricoid ligament, 149
germinal center in follicle, 189
ghost of the appendix, 139
ghost of the ascending colon, 139
ghost of the descending colon, 139
ghost of the rectum, 139
ghost of the transverse colon, 139
gingiva, 130
gingival branches, 110
glabella, 5
glans of clitoris, 195
glans of penis, 192
glans penis, 194, 202
glassy membrane, 131
glenohumeral ligament, 36
glenoid cavity of scapula, 25
globus pallidus, 118, 120
glomerular capillary, 198
glomerular capsular space, 198
glomerular (Bowman's) capsule, 198, 199
glomerulus, 198, 199
glossopharyngeal nerve IV, 117
glossopharyngeal nerve IX, 119
glossopharyngeal nerve rootlets, 118
glottis, 146
gluteal fat, 3, 10
gluteus maximus, 3, 4, 49, 54, 63, 64, 67, 69, 70, 89
gluteus medius, 3, 4, 9, 49, 54, 60, 63, 64, 69, 88, 89
gluteus minimus, 49, 54, 61, 64, 69
graafian (vesicular) follicle, 205
gracilis, 2, 4, 10, 49, 54, 60, 63, 70, 88, 89
granulosa cells, 205
gray matter, 108
gray & white rami communicantes, 108
great auricular nerve, 104, 105
great cardiac vein, 170, 171, 172
great saphenous vein, 10, 156, 157, 186

great trochanter, 3, 4, 9, 14, 15, 27, 30, 39, 40, 41, 42, 109
greater alar cartilage, 5, 6, 61, 129, 146, 148
greater arterial circle of iris, 127
greater curvature, stomach, 137
greater horns (cornu), 24
greater occipital nerve, 104, 105
greater palatine foramen, 22
greater sciatic foramen, 42
greater sciatic notch, 27, 39, 109
greater tubercle, 14, 15, 25, 68
greater vestibule (Bartolin's) gland & duct, 195
greater wing, sphenoid bone, 22
groove for auditory tube, 22
groove for flexor hallucis longus tendon, 29
groove for occipital artery, 22
groove for peroneus longus, 29
groove for poplíteus, knee, 29
groove for superior sagittal sinus, 23
groove for transverse sinus, 23
grooves for meningeal vessels, 23

H

hair cuticle, 131
hair follicle, 131
hair matrix, 131
hair shaft, 131
hallux (great toe), 2, 10
hamate, 17, 28, 38, 75
hamate & hamulus, 38
hamatometacarpal ligament, 38
hamstrings group muscles, 4
hamulus of hamate, 38, 79, 80, 81, 82
hand, components of (dorsal view), 75
hand, muscular system, 78–83, 86, 87
hand (dorsal view), origins & insertions, 53
hand (palmar view), origins & insertions, 52
hand (dorsal & palmar views), surface anatomy, 8
hand & arm, right (lateral & medial views), surface anatomy, 7
hands (palmar & dorsal views), skeletal system, 28
hard palate, 126, 130, 147
haustra, 140
haversian system (osteon), 30
head: eye, ear, nose & mouth, 126
head, surface anatomy, 5, 6
head & face, nerves of the, 111
head muscles, 71
head & neck, circulatory system, 160
head & neck, lymphatic system, 183
head & trunk (anterior view), origins & insertions, 48
head & trunk (posterior view), origins & insertions, 49
head of 1st metatarsal, 10
head of caudate nucleus, 118, 120
head of condylar process, 22
head of femur, 27, 30, 39, 109
head of fibula, 3, 43
head of humerus, 3, 7, 25
head of mandible, 114, 126
head of pancreas, 187
head of radius, 4
head of rib, 26
head of ulna, 7
hearing, 128
heart, the, 157, 169–79, 208
heart (anterior view), 170
heart (interior view), 173
heart (posterior view), 171

heart (transparent), 184
heart & lungs, lymphatic system, 184
helix, 4, 5, 6
hemiazygos vein, 185
Henle's layer, 131
hepatic artery, 164
hepatic (right colic) flexure, 140
hepatic portal vein, 156, 164
hepatic portal veins, 165
hepatic vein, 157, 158, 165, 186, 208
hiatus of vertebral artery, 34
highest deep inguinal node (Rosenmuller or Cloquet), 186
hilum, 189
hip bone, left (anterior & posterior views), skeletal system, 27
hip capsule, 41
hip ligaments, 40
hip ligaments (opened), 41
hippocampus, 118
hips (female & male views), surface anatomy, 9
horizontal fissure, right lung, 146
horizontal fissure of cerebellum, 115
horizontal plate, palatine bone, 22
horizontal sections, brain, 120
humerus, 14, 15, 16, 17, 25, 28, 36, 37, 68, 76, 106
humulus of carpal bones, 28
Huxley's layer, 131
hyloglossus, 57
hyoepiglottic ligament, 57, 114, 149
hyoglossus, 73
hyoid bone, 6, 16, 35, 70, 73, 114, 126, 146, 147, 149
hyoid bone (superior view), origins & insertions, 57
hyoid bone, skeletal system, 24
hyoid & temporomandibular, joints & ligaments, 35
hypodermis, 131
hypoglossal canal, 22, 23
hypoglossal nerve XII, 117, 119
hypophysis (pituitary), 116, 117, 119
hypothalamus, 116
hypothenar eminence, 2, 7, 8
hypothenar muscles, 60
H zone, muscle microstructure, 99

I

I band, muscle microstructure, 99
ileal & jejunal vein, 165
ileocecal artery, 141
ileocecal fold, 141
ileocecal sphincter & appendix, 141
ileocecal valve & opening, 141
ileocolic nodes, 188
ileocolic veins, 165
ileum, 139, 140, 165
ileum (small intestine), 135
ileum (small intestine) (transparent), 134
iliac crest, 2, 3, 4, 9, 14, 15, 20, 27, 39, 42, 68, 109
iliac tuberosity, 27
iliacus, 48, 62, 70
iliocostalis cervicis, 65
iliocostalis lumborum, 49, 64, 65, 70
iliocostalis lumborum (erector spinae muscle), 4
iliocostalis thoracis, 49, 64, 65
iliofemoral ligament , 40, 42
iliohypogastric nerve, 102, 103, 104, 107
ilioinguinal nerve, 102, 103, 105, 107
iliolumbar ligament, 42

ilionguinal & genitofemoral nerve, 104
iliopsoas, 2, 54, 60, 61, 62, 69, 70, 88
iliotibial band, 61, 67, 88, 89
iliotibial tract, 2, 3, 54
ilium, 14, 15, 16, 17, 20, 27, 39, 109
implanting blastocyst, 205, 207
incisive fossa, 22
incisivus labii inferioris, 48, 61
incisivus labii superioris, 48, 61
incus (anvil), 128
index finger, 8
inferior alveolar nerve, 110
inferior angle of scapula, 4
inferior articular process, 20, 39
inferior articular process & facet, 108
inferior articulating facet, 19, 20, 23
inferior articulating process, 18
inferior buccolabial sulcus, 5, 6
inferior costal facet, 34
inferior deep cervical node, 184
inferior dental branches, 110
inferior dental plexus, 110
inferior epigastric vessels, 192
inferior extensor retinaculum, 44, 60, 70, 88, 90, 98
inferior fascia of urogenital diaphragm, 195
inferior frontal gyrus, 115
inferior frontal sulcus, 115
inferior gemellus, 49, 64, 69
inferior gingival branches, 110
inferior gluteal nerve, 107
inferior horn of lateral ventricle, 120, 121
inferior horn of thyroid cartilage, 149
inferior labial artery, 160, 161
inferior labial frenulum, 130
inferior labial vein, 160
inferior lacrimal papilla & puncta, 5
inferior lateral cutaneous nerve, 105
inferior lip, 130
inferior lip (labium), 6
inferior longitudinal fascicles, deltoid ligament, 36, 44
inferior & medial cluneal nerves, 105
inferior mesenteric artery, 159, 164, 193
inferior mesenteric nodes, 188
inferior mesenteric vein, 165
inferior nasal artery & vein, 127
inferior nasal concha (turbinate), 146, 148
inferior nasal meatus, 114, 126, 147
inferior neck crease, 3
inferior nuchal line, 22
inferior olive, 118
inferior palpebral sulcus, 5
inferior parietal lobe, 115
inferior parietal lobule, 114
inferior peroneal ligament, 95, 96
inferior peroneal retinaculum(a), 44, 45, 90, 98
inferior pharyngeal constrictor, 73, 135, 146
inferior pillar of mouth, 6
inferior pubic ramus, 42
inferior rectal nerve, 107
inferior rectal valve, 142
inferior rib margin, 3
inferior sagittal sinus, 157, 160
inferior segment, kidney, 192, 195
inferior superficial inguinal nodes, 186
inferior temporal artery & vein, 127

inferior temporal gyrus, 115, 119
inferior temporal sulcus, 115, 119
inferior thyroid vein, 170, 174
inferior tracheobronchial nodes, 184
inferior trunk, 106
inferior vena cava, 156, 157, 158, 164, 165, 170, 171, 174, 186, 193, 208
infracardial nodes, 184
infraclavicular fossa, 2
infraglemoid tubercle, 25
infraorbital canal, 21, 22
infraorbital foramen, 110
infraorbital nerve, 110, 111
infrapatellar branch of the saphenous nerve, 104, 105
infrapyloric nodes, 187
infraspinatus, 3, 4, 51, 63, 64, 68, 69, 70, 76, 77
infraspinous fossa, 25
infratrochlear nerve, 111
infundibular recess, 121
infundibulum, 206
inguinal fold, 9
inguinal ligament, 2, 9, 40, 42, 62, 69, 70, 74, 151, 186, 192
inguinal lymph nodes, 182
inner cell mass, ovum, 205
inner hamstring tendons, 63
innermost intercostals, 61, 74, 151
inner zone of medulla, 199
insertion of central band of extensor tendons to base of middle phalanx, 83, 84, 85
insertion of extensor tendon to base of distal phalanx, 83, 84, 85
insertion of lumbrical muscle to extensor tendon, 87
insular gyrus, 118
interarticular ligament, 34
interarticular round ligament, 41
interarytenoid notch, 130, 136
intercapitular veins, 157
intercarpal articulation & ligament, 38
intercavernous septum of deep (Buck's) fascia, 192
intercellular cleft, 166
interchondral part of internal intercostals, 74, 151
interclavicular ligament, 36
intercondylar eminence, knee, 29
intercondylar fossa, 43
intercondylar line, 43
intercondylar notch, knee, 29
intercostal lymph nodes, 185
intercostal nerves, 102, 103
intercostal nerves (anterior & lateral), 104
intercostal nodes, 182
intercostals, 48, 49
intercostobrachial & medial brachial cutaneous nerves, 105
interdigitations, 2
intergluteal crease, 4
interiliac external iliac node, 186
interlobar arteries, 197, 199
interlobar artery & vein, 199
interlobar vein, 199
interlobular arteries, 197
interlobular lymph vessels, 184
intermedial nerve, 117, 119
intermediate bronchus, 150
intermediate cuneiform, 14, 15, 16, 29
intermediate digastric tendon, 57
intermediate external iliac nodes, 186
intermedius & vestibulocochlear nerve, 118
internal acoustic (labyrinthine) artery, 122, 162

215

INDEX

internal anal sphincter muscle, 142
internal capsule, 118, 120
internal carotid artery, 122, 160, 161, 162, 163
internal elastic lamina, tunica intima, 166
internal iliac artery, 156, 164
internal iliac artery & vein, 157
internal iliac nodes, 186
internal iliac vein, 156, 164
internal iliac vessels, 193
internal intercostal, 61, 62, 69, 70, 74, 151
internal jugular nodes, 184
internal jugular vein, 156, 157, 160, 170, 174
internal occipital protuberance, 62
internal pterygoid, 57
internal spermatic fascia, 192
internal venous plexus in submucous space, 142
interosseous membrane, 37, 38, 51, 83, 84
interosseous spaces (body, base), 28
interosseous tendon passing to base of proximal phalanx & joint capsule, 83, 84, 85
interosseus membrane, 90, 91, 92
interosseus talocalcaneal ligament, 44
interpectoral nodes, 184
interpeduncular fossa, 118
interphalangeal crease, 7, 8
interphalangeal joint, 10
interphalangeal joint (1st), 10
interspinalis lumborum, 66
interspinal ligament, 34, 39
intertendinous bands, 75
intertendinous connection, 83, 84, 85, 86, 87
interthalamic adhesion, 116
intertragic incisure, 5
intertransversarii, 62, 66
intertransversarii cervicis, 66
intertransversarii laterales lumborum, 65
intertransversarius cervicis, 66
intertransverse ligament, 34, 39
intertrochanteric crest, 27
intertrochanteric line & capsule attachment, 40
intertubercular groove, 25
intertubular synovial sheath, 36
interventricular foramen of Monroe, 121
intervertebral disc, 20, 23, 34, 39, 42, 108
intervertebral foramen, 18
intestinal nodes, 182
intestinal trunk, 185
iris, 5, 6, 127
ischial bursa, 39, 109
ischial spine, 16, 27
ischial tuberosity, 27, 39, 41, 42, 109
ischiocapsular ligament, 40
ischiofemoral ligament, 39, 42, 109
ischium, 14, 15, 17, 27
isthmus, 206
isthmus of cingulate gyrus, 119

J

jejunal & ileal vein, 165
jejunum, 138, 139
jejunum (small intestine), 135
jejunum (small intestine) (transparent), 134
joint capsule, 35, 38, 40, 45, 75, 80, 81, 82, 86, 87
joint capsule & tendons, 7
joint cavity, 75
joint of head of rib (opened), 34
joints & ligaments, 33–45

jugular foramen, 23
jugular fossa, 22
jugular notch (suprasternal), 8
jugulodigastric nodes, 183
juxta glomerular cells, 198

K

kidney, 192, 194, 195, 196, 208, 210
kidney, right, 197
kidneys, 164
knee ligaments (anterior & posterior views), 43
knees (anterior & posterior views), skeletal system, 29

L

L I, 20, 104, 105, 107
L II, 20, 104, 105, 107
L III, 20, 104, 105, 107
L IV, 20, 39, 42, 104, 105, 107
L V, 20, 34, 39, 42, 104, 105, 107
labial branches, 110
labialis branch, 111
labial (posterior scrotal) nerve, 105
labia majora, 203
labia minora, 203
labium (inferior lip), 6
labium majus, 9, 195, 196
labium minus, 195, 196
labyrinthine (internal acoustic) artery, 122, 162
lacrimal, 21, 22
lacrimal caruncle, 5
lacrimal lake, 5
lacrimal nerve, 110
lambdoid suture, 21, 23, 62
lamina, 18, 39
lamina cribose of retina, 127
lamina of posterior arch, 19
large intestine, 140, 188
laryngeal prominence, 2
larynx, 126, 146, 147, 152
larynx, 149
lateral antebrachial cutaneous nerve, 104, 105
lateral aortic lymph nodes, 185
lateral arch, 10
lateral arcuate ligament, 62
lateral arcuate ligament of diaphragm, 48
lateral atlantoaxial joint, 23
lateral band, muscular system, 75, 83, 84, 85, 87
lateral band of plantar aponeurosis, 93
lateral bands, origins & insertions, 51
lateral basal, lower lobe bronchus, 150
lateral canthus (angle), 5
lateral cartilage, 6
lateral circumflex artery & vein, 157
lateral common iliac nodes, 186
lateral condyle, 17, 29, 30
lateral condyle of the femur, 43
lateral condyle of the tibia, 14, 15, 43
lateral cord, 106
lateral crus, 6
lateral cuneiform, 14, 15, 16, 29
lateral cutaneous rami, 104
lateral cutaneous rami of thoracic spinal nerves, 105
lateral epicondyle, 4, 37, 43
lateral epicondyle, 28, 29, 30
lateral epicondyle of femur, 3, 14
lateral epicondyle of humerus, 14, 15
lateral femoral cutaneous nerve, 104, 105, 107
lateral fissure, brain, 115
lateral foot, muscles, 98
lateral hand, muscles, 87

lateral intertransversarii lumborum, 66
lateral (temporomandibular) ligament, 35
lateral malleolus, 14, 15
lateral malleolus (fibula), 2, 3, 4, 10, 45, 90, 91, 92, 97, 98
lateral malleous artery, 159
lateral margin, kidney, 192, 195, 197
lateral meniscus, 43, 88
lateral nasal cartilage, 146
lateral occipitotemporal gyrus, 119
lateral orbitofrontal artery, 122, 162
lateral patellar retinaculum, 60
lateral plantar fascia over abductor digiti minimi muscle, 93
lateral plantar nerve, 102, 103, 105
lateral plate, pterygoid process, 22
lateral process, calcaneus, 29
lateral pterygoid, 57, 70, 71, 73
lateral pterygoid nerve, 110
lateral recess of 4th ventricle, 121
lateral rectus muscle, 127
lateral sacrococcygeal ligament, 34
lateral semicircular canals & ducts, 128
lateral sulcus, 117, 118, 119, 120
lateral supracondylar crest, 28
lateral suptacondylar line, knee, 29
lateral sural cutaneous nerve, 104, 105
lateral tarsal artery, 159
lateral tubercle, talus, 29
lateral ventricle(s) (1st & 2nd), 116
lateral ventricle central part, 121
late spermatides, 204
latissimus dorsi, 3, 4, 8, 39, 48, 49, 61, 68, 70, 76, 77, 109
latissimus dorsi (small origin slip), 51
lattissimus dorsi (scapular slip), 64
left adrenal gland, 193
left anterior descending (or anterior intraventricular) coronary artery, 170, 175, 176
left anterior tibial artery, 159
left atrium, 156, 159, 164, 170, 171, 173, 208
left auricle, 171
left axillary artery, 159
left brachial artery, 159
left brachiocephalic vein, 170, 174, 184, 185
left branch, 165
left bundle branch, 175
left circumflex artery, 159
left colic (splenic) flexure, 140
left colic nodes, 188
left colic veins, 165
left common carotid, 170
left common carotid artery, 171, 174
left common iliac vein, 193
left coronary artery, 172, 176
left crus, 62
left crus of diaphragm, 48
left cusp, aortic valve, 175
left dome of the diaphragm, 150
left external iliac artery, 159
left external iliac vein, 158
left external jugular vein, 158, 170
left facial vein, 159
left gastric artery, 186
left gastric nodes, 184, 187
left gastric vein, 165
left gastroepiploic (gastro-omental) node, 187
left gastroepiploic vein, 165
left hepatic duct, 138
left inferior labial vein, 158
left inferior nasal concha, 114, 126, 147
left inferior pulmonary vein, 171

left internal iliac artery, 159
left internal iliac vein, 158
left internal jugular vein, 158
left jugular trunk, 185
left kidney, 193
left lateral circumflex artery, 156, 159
left lateral circumflex femoral veins, 158
left lateral supratrochlear nerve, 111
left lobe of liver, 138
left lumbar nodes, 186
left lumbar trunk, 186
left lung, 146
left main bronchus, 150
left medial supratrochlear nerve, 111
left middle cerebral artery, 163
left middle nasal concha, 114, 126
left middle nasal concha (turbinate), 147
left posterior papillary muscle, 174
left profunda femoris, 158
left psoas major, 70
left pulmonary arteries, 158, 164, 170, 171, 174
left pulmonary veins, 159, 164, 170, 174
left radial artery, 159
left renal artery, 193
left renal vein, 158, 193
left semilunar cusp, 176
left subclavian artery, 160, 170, 171
left subclavian lymph trunk, 185
left subclavian vein, 158
left superficial temporal vein, 158
left superior labial vein, 158
left superior pancreatic nodes, 187
left superior pulmonary vein, 171
left superior tracheobronchial nodes, 184
left testicular vein, 193
left thoracic duct, 182
left tibial recurrent artery, 159
left tracheal nodes, 184
left ulnar artery, 159
left ventricle, 156, 159, 164, 170, 171, 173, 174, 208
left zygomaticofacial nerve, 111
leg & foot (anterior & posterior views), origins & insertions, 54
leg & foot (medial, dorsal, plantar views), surface anatomy, 10
leg & foot surface muscles (anterior & medial views), 88
leg & foot surface muscles (lateral & posterior views), 89
lens, 127
lenticulostriate artery, 162
lesser curvature, 137
lesser duodenal papilla, 138
lesser horns (cornu), 24
lesser occipital nerve, 104, 105
lesser palatine foramen, 22
lesser sciatic foramen, 42
lesser sciatic notch, 16
lesser supraclavicular fossa, 2
lesser trochanter, 14, 15, 27, 39, 40, 41, 42, 62, 109
lesser tubercle, 14, 15, 25
levator anguli oris, 48, 57, 61, 69
levator ani, 62, 64, 142, 193
levator ani (pubococcygeus), 49
levator costae, 65
levatores costarum brevis, 65, 66
levatores costarum longus, 65, 66
levator labii, 71
levator labii alaeque nasi, 48, 61, 69
levator labii superioris, 48, 57, 61, 69, 71

levator palpebrae superioris, 72
levator scapulae, 3, 49, 51, 63, 64, 67, 69, 71
levator veli palatini, 57, 128
ligamentum arteriosum, 174
ligamentum flavum, 34, 36, 39, 42
ligamentum nuchae (nuchal ligament), 34
limbus of cornea, 5
linea alba, 2, 9, 60, 61, 70
linea aspera, 27, 29
linea semilunaris, 2, 3, 9, 60
lineaterminalis, 42
lingual artery, 161
lingual gyrus, 119
lingual nerve, 110
lingual tonsil, 130, 136, 147
lingular bronchus (superior, inferior), 150
lip (acetabular labrum), 41
little finger, 8
liver, 135, 208
liver (transparent), 134, 165
lobar artery & vein, 199
lobar (secondary) bronchus, 152
lobe, 5, 6
lobus simplex, 119
long plantar ligament, 44, 45, 95, 96, 97
long & short ciliary nerves, 110
longissimus capitis, 49, 65, 66
longissimus cervicis, 65
longissimus thoracis, 49, 64, 65, 70
longitudinal cerebral fissure, brain, 117, 118, 119
longitudinal esophageal muscle, 135
longitudinal fasciculus of pons, 118
longitudinal layer, stomach musculature, 137
longitudinal muscle layer, 142
longitudinal muscle of colon, 141
longitudinal muscle of ileum, 141
longitudinalis inferior, 73
longus capitis, 57, 62, 64, 69
longus colli, 48, 62, 70
longus tendon, 91
loop of Henle, 199
lower anterior segment, kidney, 192, 195
lower deep cervical nodes, 183
lower eyelid (palpebra inferior), 5
lower lateral brachial cutaneous nerve, 104
lower lobe, left lung, 146
lower lobe, right lung, 146
lower lobe bronchus, 150
lower pole, 197
lumbar III, IV, V, 109
lumbar nerves (L I-L V), 107
lumbar plexus, 102, 103
lumbar plexus (T XII-L IV), 107
lumbar spine, joints & ligaments, 39
lumbar triangle, 4
lumbar vertebra (I) (pedicle), 107
lumbar vertebra (V) (pedicle), 107
lumbar vertebra (superior view), skeletal system, 18
lumbar vertebrae, 14, 15, 26, 27
lumbar vertebrae (I-V), 17, 18
lumbar vertebrae (III), 194, 196
lumbar vertebrae (posterior view), skeletal system, 20
lumbar vertebrae III, IV, V, 39
lumbosacral plexus (posterior view), 107
lumbricales, 60, 61, 76, 77, 94, 95, 98
lumbrical expansion, 4
lumbrical muscle, 79, 80, 87
lumbrical muscle (1st), 78
lumbrical pad, 2, 7, 8

lumen, 166
lumen (cavity) of uterus, 210
lunate, 17, 28, 38
lung, 61, 208
lunula, 8, 10, 75
lymphatic system, 181–89
lymphatic vessels, 182
lymph capillary, 189
lysis of zona pellucida, 207

M

macula densa, 198
main left coronary artery, 175
main pancreatic duct, 138
main renal artery, 197
main renal vein, 197
main right coronary artery, 175
major duodenal papilla, 138
male hips, surface anatomy, 9
male reproductive system, 202
male skeleton (anterior view), 15
male urinary system (anterior view), 193
male urogenital system (anterior view), 192
male urogenital system (lateral view), 194
malleus (hammer), 128
mamillary body, 116, 117, 118, 119
mamillary process, 18, 20
mammary gland & fat, 8
mandible, 14, 15, 16, 21, 22, 35, 71, 114, 126, 146, 147
mandible (transparent), 35
mandibular fossa, 22
mandibular nerve, 111, 119
mandibular nerve (VIII), 104, 110
mandibular nerve, trigeminal nerve, 104
manubriosternal sychondroses, 26
manubrium, 8, 14, 15, 26, 36
marginal branch, 175, 176
marginal zone, 189
marrow cavity, 41
masseter, 3, 5, 6, 48, 57, 60, 61, 71, 136
mastoid air cells, 23
mastoid canaliculus, 22
mastoid emissary vein, 160
mastoid foramen, 22
mastoid groove for digastric muscle, 22
mastoid process, 3, 4, 6, 21, 22, 23, 62
mature spermatozoön, 204
maxilla, 5, 14, 15, 16, 21, 22
maxillary artery, 160, 161
maxillary nerve (VII), 104, 110
maxillary nerve, trigeminal nerve, 104
maxillary nerve, trigeminal nerve V, 119
maxillary sinus, 148
maxillary vein, 160
M band, muscle microstructure, 99
medial angle, 25
medial antebrachial cutaneous nerve, 104, 105
medial arch, 4, 10
medial arcuate ligament, 62
medial arcuate ligament of diaphragm, 48
medial basal, lower lobe bronchus, 150
medial brachial cutaneous & intercostobrachial nerves, 104
medial commisure of lids, 5
medial condyle, 17, 29
medial condyle of the femur, 10, 43
medial condyle of the femur (articular surface), 43

medial condyle of the tibia, 14, 15, 43
medial cord, 106
medial crural cutaneous branches of the saphenous nerve, 104, 105
medial crus, 6
medial cuneiform, 14, 15, 16, 29, 44, 45
medial cusp, tricuspid valve, 176
medial dorsal cutaneous nerve, 102, 103
medial epicondyle, 2, 28, 29, 37, 43
medial epicondyle (femur), 4, 14
medial epicondyle (humerus), 7, 14, 15
medial epicondyle (tibia), 4
medial foot, muscles, 98
medial frontal branches (posterior, middle, anterior), 163
medial hand, muscles, 86
medial intermuscular septum, 2, 7
medial malleolus, 14, 15
medial malleolus (tibia), 2, 4, 10, 45, 90, 91, 92, 94, 95, 97
medial margin, kidney, 192, 195, 197
medial meniscus, 43, 88
medial occipitotemporal gyrus, 119
medial palpebral ligament, 5
medial patellar retinaculum, 60
medial pharyngeal constrictor, 73
medial plantar fascia over abductor hallucis muscle, 93
medial plantar nerve, 102, 103, 105
medial plate, pterygoid process, 22
medial process, calcaneus, 29
medial pterygoid, 49, 70, 71
medial rectus muscle, 127
medial sacral crest, 16
medial striate artery, 122
medial supracondylar crest, 28
medial supracondylar line, knee, 29
medial talocalcaneal ligament, 44
medial thalamic nucleus, 118
medial trochanter, 15
medial tubercle, talus, 29
median aperture, 121
median crease, 10
median furrow, 4
median glossoepiglottic fold, 130, 136
median nerve, 102, 103, 105, 106
median nerve (palmar branch), 104
median palatine suture, 22
median sulcus, 130, 136
median vein, 156
medulla, 189
medulla oblongata, 110, 115, 116, 117, 118, 119
medullary cavity, 30
medullary cords, 189
medullary sinus, 189
Meissner's corpuscle (touch), 131
mental artery, 159
mental branches, 110, 111
mental foramen, 21, 22, 110
mentalis, 48, 61, 71, 147
mental nerve, 110
mental prominence, 5, 6
mentolabial sulcus, 5, 6
meridional fibers, ciliary muscle, 127
Merkel's discs, 131
mesocolic tenia, 135, 140, 141
mesosalpinx, 195
meta-arteriole, 166
metacarpal (2nd), 7, 8, 75, 87

metacarpal (4th), 38
metacarpal (5th), 7, 38, 75
metacarpal ligament (1st), 38
metacarpals, 14, 15, 16, 17, 85
metacarpals (interosseous spaces, body, base), 28
metacarpophalangeal crease, 8
metacarpophalangeal joint (1st), 8
metacarpophalangeal joint (2nd), 8
metacarpophalangeal (MP) joint, 7, 75
metatarsal (1st), 10, 45
metatarsal (5th), 45
metatarsal muscle (1st), 98
metatarsals, 14, 15, 16, 17, 29
metatarsal tuberosity (5th), 3
metatarsophalangeal fat pad, 4, 10
metatarsophalangeal fat pad (1st), 10
metatarsophalangeal joint, 10
microtubules, 204
microvilli, 198
middle cardiac vein, 171, 172, 176
middle cerebellar peduncle, 118
middle cerebral artery, 160, 162
middle cerebral artery & branches, 122, 162
middle colic nodes, 188
middle colic vein, 165
middle finger, 8
middle frontal gyrus, 115
middle infraorbital artery, 122, 162
middle internodal tract, 175
middle & lateral lenticulostriate arteries, 162
middle lobe, right lung, 146
middle lobe bronchus (lateral, medial), 150
middle nasal concha (turbinate), 146, 148
middle phalanx, 75, 87
middle phalanx (body, base, head), 28
middle pharyngeal constrictor, 57, 135
middle rectal valve, 142
middle superior alveolar nerve, 110
middle temporal artery, 161, 162, 163
middle temporal gyrus, 115
middle thyroid vein, 160
middle trunk, 106
mid-palmar crease, 8
midpiece, spermatozoa, 204
minivalve (lymphatic fluid entrance port), 189
minor diameter, spermatozoa, 204
mitochondria, 99, 204
mitral valve, 156, 159, 173, 174, 176
modiolus (node), 5, 6
mons pubis, 2, 3
mons veneris, 9
morula, 207
motor nerve, 99
motor nerve (autonomic), 131
motor nerve fibers, 108
mouth & salivary glands, digestive system, 136
mouth slit (rima oris), 6
mucous, 152
mucous plug, 209
multifidus, 49, 65, 66
muscle fiber, 99
muscle insertion, 99
muscle microstructure, 99
muscles of respiration, 151
muscular branches, 102, 103
muscular interventricular septum, 173, 174
muscular system, 59–99
musculocutaneous nerve, 103, 106
myelin sheath, 111
mylohyoid, 57, 70, 71, 73

mylohyoid muscle, 114, 136, 146, 147
mylohyoid nerve, 110
myofibril(s), 99, 111
myoid bone, 71
myometrium, 195, 196, 203
myosin molecules, 99

N

nail, 2, 7, 8, 10
nail bed, 75
nail body, 75
nail matrix, 75
nail root, 75
nasal, 21, 22
nasal bone, 114, 147
nasal bone (bridge of nose), 5, 6
nasal cavity, 146, 152
nasal conchae, 22
nasalis, 48, 61, 71
nasal & oral cavity, 147
nasal sensory nerves (olfactory bulb), 114
nasal septum, 148
nasociliary nerve, 110
nasofrontal canal, 148
nasolabial furrow, 5, 6
nasolacrimal duct, 148
nasopalatine artery & nerve in the incisive canal, 129, 148
nasopalatine nerve, 129, 148
navel (umbilicus), 2, 3, 9, 68, 70
navicular tuberosity, 4
navicular, 2, 14, 15, 16, 29, 44, 90, 91, 92
navicular bone, 98
navicular fossa, 192, 194, 202
neck, 14, 15, 28
neck (deep) muscle (lateral view), 73
neck of femur, 27, 30, 39, 109
neck of rib, 26
neck of scapula, 25
nephron, 199
nerve fibers, 108
nerves & arteries, heart, 175
nervous system, 101–11
neurilemma (sheath of Schwann), 111
neurovascular bundle
new skin (white line), 5, 6
nissl substance, 111
node of Ranvier (axon), 111
nodes & vessels, lymphatic system, 189
nostril, 6, 152
nuchal furrow, 4
nuchal ligament, 34, 57, 65, 114
nuchal ridge, 3
nucleolus, nerve, 111
nucleolus, ovum, 205
nucleus, 99
nucleus, lens, 127
nucleus, nerve, 111
nucleus, ovum, 205
nucleus, spermatozoa, 204
nucleus of sertoli's cell, 204
nucleus pulposus, 34, 39
nutrient artery, 30
nutrient foramen, 30
nutrient vein, 30

O

oblique cord, 37
oblique fissure, left lung, 146
oblique fissure, right lung, 146
oblique layer, stomach musculature, 137
oblique vein of left atrium, 172
obliquus capitis superior, 66
obliquus externus, 2, 4, 9, 60, 67, 68, 69, 74, 76, 151
obliquus externus abdominis, 48, 49

obliquus externus muscle (area of interdigitation), 3
obliquus externus muscle (flank portion), 2, 3
obliquus inferior, 49, 72
obliquus internus, 61, 64, 69, 74, 151
obliquus internus abdominis, 48, 49
obliquus superior, 57, 72
obturator externus, 48, 54, 61, 62
obturator foramen, 17, 27, 39, 42, 109
obturator internus, 39, 48, 49, 54, 64, 70, 88, 109
obturator membrane, 39, 40, 42, 48, 62, 109
obturator nerve, 104, 105
occipital bone, 16, 17, 21, 22, 23, 34, 62
occipital condyle, 22, 23
occipitalis, 49, 57, 63, 67, 71
occipitalis artery, 161
occipital lobe, 114, 116, 126
occipital nodes, 183
occipital pole of cerebrum, 119
occiput, 4
oculomotor nerve III, 117, 119
odontoid process (dens of axis), 19, 23, 147
olecranon, 16, 77
olecranon creases, 4
olecranon folds, 7
olecranon fossa, 28
olecranon of ulna, 7
olecranon with overlying bursa, 4
olfactory bulb, 126, 129, 147, 148
olfactory bulb (nasal sensory nerves), 114
olfactory bulb I, 115, 117, 119
olfactory nerves, 129, 147, 148
olfactory sulcus, 119
olfactory tract, 117, 119, 129, 148
omental tenia, 135, 140, 141
omohyoid, 2, 3, 8, 50, 57, 60, 67, 68, 69, 70, 71, 73
omohyoid (inferior & superior belly), 61, 71
oöcyte, 207
oöplasm, 205
opening of auditory (eustachian) tube, 148
opening of bulbourethral duct, 192
opening of coronary sinus, 172
opening of ejaculatory duct (seminal colliculus), 192
opening of frontal sinus, 148
opening of greater vestibular gland, 195
opening of maxillary sinus, 148
opening of prostatic ducts, 192
opening of sphenoidal sinus, 148
opening of ureter, 192, 195
opening of vermiform appendix, 141
opercular part, inferior frontal gyrus, 115
ophthalmic artery, 162, 163
ophthalmic nerve, 119
ophthalmic nerve (VI), 104, 110
opponens digiti minimi, 50, 52, 78, 79, 80, 81, 82
opponens pollicis, 7, 50, 52, 77, 78, 79, 80, 81, 82, 87
optic canal, 23
optic chiasma, 117, 121
optic chiasma II, 119
optic nerve, 72, 114, 126, 127
optic nerve II, 117, 119
optic radiation, 120
optic recess, 121
optic tract, 118, 119
oral cavity, 134, 146, 152

INDEX

ora serrata, 127
orbicularis oculi, 48, 60, 71
orbicularis oris, 48, 60, 69, 71, 114, 126, 147
orbit, frontal bone, 22
orbital gyri, 119
orbital part, inferior frontal gyrus, 115
orbital sulci, 119
orifice of vagina, 195
origins & insertions, 47–57
osteon (haversian system), 30
outer hamstring tendon, 67
outer layer, bone, 30
outer zone of medulla, 199
ovarian (right testicular) artery, 159
ovarian follicles, development of, 210
ovarian ligament, 205, 210
ovary, 195, 196, 203, 205, 207, 210
ovulated oöcyte, 207
ovulated secondary oöcyte, 205
ovum, 205
ovum & sperm, stages of, 204, 205
oxygenated red blood cell, 152
oxygenation of alveoli cluster, 152

P

palatine bone, 21, 22
palatine process, 22
palatine raphe, 130
palatine tonsil, 126, 130, 136, 147
palatoglossal arch & muscle, 130, 136
palatopharyngeal arch & muscle, 130, 136
palmar aponeurosis, 60, 76, 78
palmar artery anastomoses, 156
palmar branch, 102, 103
palmar carpal ligament, 78
palmar carpometacarpal ligaments, 38
palmar digital branch, 104
palmar digital nerves, 102, 103
palmar hand (layer I), muscles, 78
palmar hand (layer II), muscles, 79
palmar hand (layer III), muscles, 80
palmar hand (layer IV), muscles, 81
palmar hand (layer V), muscles, 82
palmar interosseous, 50, 51, 64
palmar interosseous (1st), 52, 53, 75, 80, 81, 82
palmar interosseous (2nd), 52, 53, 81, 82
palmar interosseous (3rd), 52, 81, 82
palmaris brevis, 7, 76, 78
palmaris longus, 2, 7, 50, 60, 76, 86
palmaris longus tendon, 2, 7, 8, 78
palmar ligament (palmar plate), 38, 75, 81, 82
palmar metacarpal ligaments, 38
palmar radiocarpal ligament, 38
palmar ulnocarpal ligament, 38
palmar venous network, 156
palmar vessels, 182
palpebra inferior (lower eyelid), 5
palpebra superior (upper eyelid), 5
pampiniform venous plexus, 192
pancreas, 135, 165
pancreas (behind stomach), 134, 187
pancreas (tail, body, neck, head), 138
pancreatic notch, 138
pancreaticoduodenal veins, 165

papilla (nipple), 2, 3, 8
papilla ilealis, 141
papilla of hair follicle, 131
papillary foramina, 197
papillary layer, 131
papillary muscles, 173
para-aortic nodes, 182
paracolic nodes, 188
parahippocampal gyrus, 119
paramammary nodes, 184
paranasal sinuses, 148
parasternal nodes, 184
paraterminal gyrus, 116
parietal, 16, 17, 21, 22
parietal bone, 22, 23, 62
parietal branch of superficial temporal artery, 161
parietal epithelium, 198
parietal layer, 192
parietal lobe, 114, 126
parieto-occipital artery, 163
parieto-occipital sulcus, 114, 115, 116
parotid duct, 71, 134, 135, 136
parotid gland, 6, 134, 135, 136
parotid lymph nodes, 182
patella, 2, 3, 10, 14, 15, 16, 29, 43, 70, 88, 89
patellar groove, 30
patellar ligament, 2, 3, 10, 43, 54, 70, 88, 89
patellar tendon, 60
pathways to opposite breast, 184
pathways to subdiaphragmatic nodes, 184
pectineus, 48, 54, 60, 61, 88
pectoralis major, 2, 3, 8, 48, 50, 51, 60, 67, 68, 69, 70, 76
pectoralis minor, 48, 50, 61, 68, 76
pedicles, 18, 23, 108, 198
pelvic bone muscles, 194, 196
pelvic bowl muscles, 202, 203
pelvis, connective components of the, 39
pelvis (superior & posterior views), joints & ligaments, 42
penis, 9, 210
perforating branches to interventricular septum, 175
perforating (or Sharpey's) fibers, 30
pericallosal artery, 163
perimysium, 99
perineal nerve, 107
perineurium, 111
periosteum, 30
perirenal & capsular arteries, 197
peritoneal reflection, 142
peritoneum, 141
peritubular capillaries, 199
peroneal artery, 157, 159
peroneal retinaculum, 67, 69, 89
peroneal tendons, 3, 10
peroneal trochlea, 29
peroneal vein, 157
peroneus brevis, 2, 3, 4, 54, 56, 61, 64, 67, 69, 88, 89, 90, 91, 92, 95, 98
peroneus brevis tendon, 44, 45, 67, 96, 97
peroneus longus, 2, 3, 54, 60, 64, 67, 88, 89, 95, 98
peroneus longus tendon, 4, 44, 45, 63, 64, 90, 91, 92, 96, 97, 98
peroneus longus tendon & insertion, 55
peroneus tertius, 2, 3, 54, 56, 60, 61, 67, 69, 88, 89, 90, 91, 92, 98
peroneus tertius tendon, 10, 67, 92
pes anserinus, 2, 10, 70, 88
petrotympanic fissure, 22
petrous part, temporal bone, 22
phalangeal bone bodies, 8

phalangeal fat pads, 10
phalanges, 7, 14, 15, 16, 17, 29, 44
phalanges II-IV, 98
pharyngeal tonsil, 114, 126, 147
pharyngeal tubercle, 22
pharyngobasilar membrane, 73
pharynx, 134, 146, 152
philtrum, 5
pia mater, 108
pillar of mouth, 5
pineal body, 116, 120
piriformis, 39, 48, 49, 54, 62, 64, 69, 70, 109
pisiform, 2, 7, 8, 17, 28, 38, 75, 76, 78, 79, 80, 81, 82
pisohamate ligament, 38
pisometacarpal ligament, 38, 52
pituitary (hypophysis), 116, 117, 119
placenta, 208, 209
plantar aponeurosis, 10, 93, 94, 95, 96, 97
plantar artery anastomosis, 156
plantar calcaneocuboid ligament, 44, 45
plantar calcaneonavicular ligament, 45
plantar cuboidonavicular ligament, 45
plantar cuneonavicular ligament, 45
plantar digital artery & vein, 157
plantar fascia, 4
plantar foot (layer I), muscles, 93
plantar foot (layer II), muscles, 94
plantar foot (layer III), muscles, 95
plantar foot (layer IV), muscles, 96
plantar foot (layer V), muscles, 97
plantar interossei, 96, 97
plantar interosseous, 54, 55, 64
plantaris, 54, 64
plantaris tendon, 90, 91, 92
plantar ligaments (plantar plates), 44, 45
plantar metatarsal artery & vein, 157
plantar metatarsal ligaments, 45
plantar plates, 96, 97
plantar tarsometatarsal ligaments, 45
plantar venous network, 156
plantar vessels, 182
plasma, 198
platysma, 57, 60, 67, 71
pleura, 146, 150
plica semilunaris, 5
podocytes (visceral epithelium), 198
polar body, ovum, 205
pons, 115, 117
pontine arteries, 122, 162
popliteal artery, 156, 157, 159
popliteal fossa, 3, 63, 89
popliteal lymph nodes, 182
popliteal surface, 43
popliteal surface, knee, 29
popliteal swelling, 4, 10
popliteal vein, 157
popliteus, 54, 64
pore of sweat gland, 131
portal circulation, 164
portal vein, 165, 208
postcapillary venule, 166
postcentral gyrus, 115
postcentral sulcus, 115
posterior antebrachial cutaneous nerve, 105
posterior atlanto-occipital membrane, 34, 36
posterior auricular artery, 161
posterior auricular sulcus, 5

posterior basal, lower lobe bronchus, 150
posterior brachial cutaneous nerve, 102, 103, 105
posterior branches of occipital nerve, 105
posterior branch of left coronary artery, 171
posterior branch of sphenopalatine artery, 129, 148
posterior cerebral artery, 122, 160, 162, 163
posterior chamber, 127
posterior ciliary artery, 163
posterior communicating artery, 122, 162, 163
posterior cord, 106
posterior cruciate ligament, 43
posterior cusp, 175, 176
posterior cutaneous rami of thoracic spinal nerves, 105
posterior deep temporal nerve, 110
posterior descending artery (interventricular branch of right coronary artery), 171, 172
posterior division, 110
posterior femoral cutaneous nerve, 102, 103, 105, 107
posterior horn, 121
posterior inferior cerebellar artery, 122, 162
posterior inferior iliac spine, 17, 27, 39, 109
posterior intercostal veins, 185
posterior internodal tract, 175
posterior interventricular branch, 175
posterior interventricular branch of right coronary artery (posterior descending artery), 171, 172
posterior layer of rectus sheath, 70
posterior left ventricular branch, 172
posterior ligaments of the fibular head, 43
posterior longitudinal ligament, 34, 39
posterior meniscofemoral ligament, 43
posterior nasal spine, 22
posterior parietal artery, 163
posterior part of internal capsule, 120
posterior perforated substance, 119
posterior process, talus, 29
posterior sacral foramina, 42
posterior sacrococcygeal ligament, 42
posterior sacrococcygeal profundis ligament, 34
posterior sacrococcygeal superficialis ligament, 34
posterior sacroiliac ligaments, 42
posterior scrotal (labial) nerve, 105
posterior semicircular canals & ducts, 128
posterior semilunar cusp, 176
posterior spinal artery, 122, 162
posterior submandibular node, 183
posterior superior alveolar nerve, 110
posterior superior iliac spine, 4, 16, 17, 27, 39, 109
posterior talocalcaneal ligament, 44
posterior talofibular ligament, 44
posterior temporal artery, 163
posterior tibial artery, 157
posterior tibiotalar ligament, deltoid ligament, 44
posterior triangle of neck, 3

posterior tubercle, 23
posterior tubercle of posterior arch of atlas, 19
posterior vein, 176
posterior vein of left ventricle, 171, 172
posterior wall of oropharynx, 130
preaortic nodes, 187, 188
precapillary sphincter, 166, 189
precentral artery, 162
precentral gyrus, 115
prepuce (foreskin), 9, 192, 194, 202
prepuce of clitoris, 195
primary bronchus, 146, 152
primary follicles, 205, 207
primary spermatocytes, 204
primary urine (filtrate), 198
primordial follicles, 205
procerus, 71
promontory, 18, 42
pronator quadratus, 50, 78, 79, 80, 81, 82, 86
pronator teres, 2, 7, 50, 51, 60, 76
proper ovarian ligament, 195
proper palmar digital artery, 159
proper plantar digital nerves, 102, 103
prostate, 193, 194, 202, 210
prostate (sectioned & transparent), 192
prostatic utricle, 192
proximal convoluted uriniferous tubule, 199
proximal digital crease, 2, 8
proximal epiphysis, 30
proximal heel crease, 3, 4, 10
proximal interphalangeal crease, 2, 8
proximal interphalangeal (PIP) joint, 8, 75
proximal palmar crease, 2, 8
proximal phalanx, 75, 98
proximal phalanx (body, base, head), 28
proximal secondary epiphysis, 30
proximal tubule, 198
proximal wrist crease, 2, 7, 8
pseudo fenestrations, 198
psoas major, 48, 62
psoas minor, 48, 62
pterygoid fossa, 22
pterygoid process, 22
pterygomandibular raphe, 71, 73, 146
pterygopalatine branches, 110
pubic arch, 27
pubic bone, 196, 203
pubic crest, 27
pubic symphysis, 2, 9, 14, 15, 27, 42, 209
pubic tubercle, 16, 40, 42
pubis, 14, 15, 27, 39, 109, 194, 202
pubocapsular ligament, 40
pubococcygeus (levator ani), 49
pubofemoral ligament, 40
pudendal nerve, 102, 103, 107
pulmonary artery, 156, 157, 164, 208
pulmonary capillaries, 164
pulmonary nodes, 184
pulmonary trunk, 158, 173, 174
pulmonary valve, 156, 158, 173, 174, 176
pulmonary veins, 157
pupil, 5
Purkinje fibers, 175
putamen, 120
putamen lentiform nucleus, 118
pyloric canal, 137
pyloric opening, 137
pyloric sphincter muscle, 137
pyloris, 187

pyramidal decussation, 118
pyramidalis, 48, 60, 70

Q
quadratus femoris, 48, 49, 54
quadratus lumborum, 48, 49, 62, 65, 66, 74, 151
quadratus plantae, 55, 56, 64, 95, 96, 97, 98
quadriceps femoris, 54
quadriceps femoris tendon, 3

R
radial artery, 156
radial brusa, 86
radial collateral ligament, 37, 38
radialis indicis artery, 159
radial nerve, 102, 103, 106
radial nerve superbrachial branch, 104
radial nerve superficial dorsal digital branches, 105
radial notch of ulna, 28
radial styloid, 2, 8
radial tuberosity, 28
radial vein, 157
radiate carpal ligament, 38
radiate ligament of head of rib, 34
radiate sternocostal ligament, 36
radiocarpal articulation & ligament, 38
radius, 4, 7, 14, 15, 16, 17, 28, 37, 38, 75, 80, 81, 82, 83, 84, 85, 87
ramus, 22
ramus of ischium, 39, 109
rear deltoid muscle, 3, 4, 7
rectal fascia, 142
rectosigmoid junction, 140
rectum, 134, 135, 140, 165, 193, 194, 196, 202, 203, 210
rectum, digestive system, 142
rectus abdominis, 2, 3, 9, 48, 60, 67, 68, 70, 74, 151
rectus capitis anterior, 57
rectus capitis lateralis, 57, 62
rectus capitis posterior major, 49, 57, 66
rectus capitis posterior minor, 49, 57, 66
rectus femoris, 2, 3, 9, 10, 48, 49, 54, 60, 67, 69, 70, 88, 89
rectus inferior, 72
rectus lateralis, 72
rectus medialis, 72
rectus sheath, 60
rectus superior, 72
recurrent artery (of Heubner), 162
recurrent tibial artery, 156
red blood cells, 166, 198
red lip margins, 5, 6
red nucleus, 119, 120
renal arteries, 156, 164
renal artery & vein, 157
renal calices, 192, 197
renal column, 192, 194, 195, 196, 197
renal corpuscle, 198
renal corpuscle, 199
renal cortex, 192, 194, 195, 196, 197
renal hilum, 192, 194, 195, 196
renal impression, 135
renal medulla, 192, 194, 195, 196, 197, 199
renal papillae, 192, 194, 195, 196, 197
renal pelvis, 193, 194, 195, 196, 197
renal pyramid, 192, 194, 195, 196, 197, 199
renal (uriniferous) tubule (nephron), 199
renal vein, 156, 164
renal vessels, 192, 195
reproductive system, 201–11

respiration, muscles of, 74
respiratory system, 145–53
reticular layer, 131
reticulin, 189
retina, 127
retroaortic lymph nodes, 185
retroauricular nodes, 183
retrocardiac nodes, 184
retromandibular vein, 160
rhinal sulcus, 119
rhomboid major (under trapezius), 4
rhomboid minor, 4, 49, 51, 64
rhomboid triangle, 4
rib (1st), 36
rib (2nd), 36
ribs, 3, 4, 14, 15, 16, 26
ridge of philtrum, 5, 6
right adrenal gland, 193
right anterior tibial vein, 158
right arm & hand (lateral & medial views), surface anatomy, 7
right atrium, 156, 158, 164, 170, 173, 174, 208
right basilic vein, 158
right brachial veins, 158
right brachiocephalic vein, 170, 174, 184
right branch, 165
right bundle branch, 175
right cephalic vein, 158
right colic flexure, 165
right colic (hepatic) flexure, 140
right colic nodes, 188
right colic vein, 165
right common carotid, 170
right common carotid artery, 159
right common carotid trunk, 174
right common iliac artery, 159, 193
right common iliac vein, 158
right coronary artery, 170, 171, 172, 176
right crus, 62
right crus of diaphragm, 48
right cusp, aortic valve, 175
right dome of the diaphragm, 150
right dorsal digital veins, 158
right dorsal venous arch of foot, 158
right dorsal venous network of foot, 158
right external jugular vein, 170
right foot (inferior view), joints & ligaments, 45
right foot (lateral & medial views), joints & ligaments, 44
right gastric vein, 165
right gastroepiploic (gastro-omental) nodes, 187
right genicular vein, 158
right great saphenous vein, 158
right hepatic duct, 138
right inferior pulmonary vein, 171
right intercapitular veins, 158
right jugular trunk, 185
right kidney, 193
right kidney, 197
right lobe of liver, 138
right lumbar trunk, 185, 186
right lung, 146, 152
right lymphatic duct, 183, 185
right lymphatic trunk, 182
right main bronchus, 150
right marginal branch of right coronary artery, 172
right median cubital vein, 158
right peroneal vein, 158
right proper palmar digital veins, 158
right psoas major, 70
right pulmonary artery, 164, 171, 174
right pulmonary veins, 164, 174
right radial vein, 158
right renal artery, 159, 193

right renal vein, 193
right semilunar cusp, 176
right subclavian artery, 159, 170
right subclavian lymph trunk, 185
right subclavian vein, 160
right superficial femoral artery, 159
right superficial femoral vein, 158
right superficial palmar venous arch, 158
right superior pancreatic node, 187
right superior pulmonary vein, 171
right superior tracheobronchial nodes, 184
right testicular artery, 193
right testicular (ovarian) artery, 159
right tracheal nodes, 184
right ulnar vein, 158
right ventricle, 156, 164, 173, 174, 208
right vertebral artery, 159
rima oris (mouth slit), 6
ring finger, 5
risorius, 60, 67, 71
root of nose, 5, 6
rotatores cervicis brevis, 66
rotatores cervicis longus, 66
rotatores lumborum brevis, 66
rotatores lumborum longus, 66
rotatores thoracis brevis, 66
rotatores thoracis longus, 66
round ligament, 138, 195
round (cochlear) window, 128
rugae (gastric folds), 137

S
S I, 104, 105, 107
S II, 104, 105, 107
S III, 104, 105, 107
S IV, 105, 107
S V, 105, 107
saccule, 128
sacral dimple, 4
sacral hiatus, 39, 109
sacral nerves, 193
sacral nerves S I-S V, 107
sacral nodes, 186
sacral plexus, 102, 103
sacral plexus L V-S IV, 107
sacral triangle, 4
sacroiliac joint, 17, 39, 109
sacrolumbar ligament, 34
sacrospinous ligament, 39, 42, 49, 70, 109
sacrotuberous ligament, 39, 42, 64, 70, 88, 109
sacrouterine ligament, 195
sacrum, 14, 15, 16, 20, 27, 34, 194, 196, 202, 203, 209, 210
sacrum (5 fused), 17, 18, 107
sagittal suture, 23, 62
saliva, 136
salivary pool, 136
saphenous nerve, 102, 103
sarcolemma, 99
sarcomere, 99
sarcoplasm, 99
sarcoplasmic reticulum, 99
sartorius, 2, 3, 4, 9, 10, 48, 49, 54, 60, 67, 69, 70, 88, 89
satellite cell, 99
scalenes (anterior, middle, posterior), 74, 151
scalenus anterior, 2, 48, 61, 62, 67, 69, 70, 71
scalenus medius, 48, 61, 62, 64, 67, 69, 70, 71
scalenus posterior, 48, 49, 61, 62, 64, 69
scali vestibuli, 128
scaphoid, 38, 75
scaphoid bone, 17, 28
scaphoid fossa, 5, 22

scapula, 14, 15, 16, 17, 25, 26, 36, 76
scapula (acromion), 3, 4, 36, 68, 76
scapula (coracoid process), 3
scapula (anterior & posterior views, anterior view), skeletal system, 25
scapula medial border, 4
scapular spine, 4
Schlemm's canal (sinus venosus of sclera), 127
Schwalbe's line, 127
Schwann cell, 111
sciatic nerve, 39, 102, 103, 107, 109
sciatic nerve, 109
sclera, 5, 6, 127
scrotal branch of perineal nerve, 104
scrotal skin, 192
scrotum, 9, 194, 202, 210
sebaceous glands, 131
secondary bronchi, 146
secondary (lobar) bronchus, 152
secondary follicle, 207
secondary oöcyte, 210
secondary spermatides, 204
second metacarpal head, 8
seeing, 127
segmental artertery & veins, 197
segmental (tertiary) bronchus, 152
sella turcica, 129, 148
semicircular canals & ducts, 128
semilunar fold, 140, 141
semilunar (trigeminal) ganglion, 110
semimembranosus, 3, 4, 10, 49, 54, 63, 64, 70, 88, 89
seminal colliculus (opening of ejaculatory duct), 192
seminal vesicle, 192, 194, 202, 210
seminiferous tubule, cross section of, 204
seminiferous tubules, 210
semispinalis capitis, 4, 49, 57, 63, 64, 65, 66, 67, 70, 71
semispinalis cervicis, 66
semispinalis thoracis, 65, 66
semitendinosus, 2, 4, 10, 48, 49, 54, 63, 70, 88, 89
semitendinosus tendon, 4
senses, the, 125–31
sensory ganglion, 108
sensory nerve, 131
septa, 75
septal branch of posterior ethmoidal artery, 129, 148
septal cartilage, 129, 148
septum pellucidum, 118, 120
serosa, 137
serratus anterior, 2, 3, 48, 49, 50, 60, 61, 64, 67, 68, 69, 76
serratus anterior (under latissimus dorsi muscle), 4
serratus anterior & rib cage, 8
serratus posterior inferior, 49, 64, 69
serratus posterior superior, 49
sertoli's cell, 204
sesamoid bone, 28, 38, 44, 45, 80, 82, 86, 94, 95, 96, 97, 98
sesamoid bones (lateral & medial), 29
sexual intercourse, 210
side deltoid muscle, 3, 7
sigmoid colon, 134, 135, 140, 165, 186, 194, 196, 202, 203
sigmoid nodes, 188
sigmoid veins, 165
sinoatrial (SA) artery, 175
sinoatrial (SA) node, 175, 177
sinus venosus of sclera (Schlemm's canal), 127
skeletal system, 13–31

skeletal system (lateral view), 16
skeletal system (posterior view), 17
skin web, 7
skull (anterior & inferior views), skeletal system, 22
skull (lateral view), skeletal system, 21
skull (posterior view), skeletal system, 23
skull & arteries, circulatory system, 161
slit pores (filtration slits), 198
small cardiac vein, 172, 176
small intestine, 139
small origin slip (latissimus dorsi), 51
smell, 129
smooth muscle, 152, 198
soft palate (uvula), 57, 126, 130, 147, 148
soleus, 2, 3, 4, 10, 54, 61, 63, 64, 67, 69, 70, 88, 89, 90, 91, 92, 98
soleus & gastrocnemius via tendo calcaneus (Achilles), 54
sperm, 207, 210
sperm & ovum, stages of, 204, 205
spermatagonium, 204
spermatic cord, 194, 202
spermatid cytoplasm, 204
spermatogenesis, 204
spermatozoa, 204
sphenoidal sinus, 126, 129, 147, 148
sphenoid bone, 21, 22, 23
sphenoid sinus, 114
sphenomandibular ligament, 35, 71
sphenosquamosal suture, 23
spinal cord, 102, 103, 114, 116, 117, 118, 119, 126, 210
spinal cord, 108
spinalis cervicis, 66
spinalis thoracis, 64, 65
spine, 22, 25
spine, joints & ligaments, 34
spine of ischium, 39, 109
spine of scapula, 3, 17
spinous process, 18, 19, 20, 34, 39, 108
spinous process of the axis (bifid), 19
spiral arteries, 206
spiral ganglion, 128
spleen, 135, 164, 165
spleen (lymphatic system), 134
splenic artery, 159, 164
splenic nodes, 187
splenic vein(s), 157, 165
splenius capitis, 49, 57, 63, 64, 65, 66, 67, 69, 70, 71
splenius capitis muscle, 4
splenius cervicis, 49, 64, 65
squamous suture, 21, 22, 23
stapedius muscle, 128
stapes (stirrup), 128
stellate venules, 199
stemothyroid, 61
sternal angle, 8
sternal articulation, 24
sternal body, 2
sternal end, 24
sternal segment, 2
sternal synchondrosis, 36
sternoclavicular & shoulder, joints & ligaments, 36
sternocleidomastoid, 3, 4, 6, 8, 48, 49, 50, 51, 57, 68, 70, 74, 151
sternocostal interarticular ligament, 36
sternohyoid, 2, 51, 57, 60, 61, 68, 69, 70, 73
sternomastoid, 67, 71

INDEX

sternum, 8, 14, 15, 16, 26
sternum (xiphoid), 8
stomach, 135, 164
stomach (transparent), 134, 165, 187
stomach, digestive system, 137
stomach musculature, 137
stomach & pancreas, lymphatic system, 187
straight arterioles, 199
straight gyrus, 119
straight segments of renal tubules, 199
straight sinus, 160
straight venule, 199
stratum basale, 131
stratum corneum, 131
stratum granulosum, 131
stratum lucidum, 131
stratum spinosum, 131
strengthening band, 90
striated muscle, 111
styloglossus, 57, 73
stylohyoid, 70, 71, 73
stylohyoid ligament, 35, 57, 73
stylohyoid muscle, 146
styloid process, 21, 22, 23, 35, 57, 62, 73, 75
styloid process of ulna, 8
stylomandibular ligament, 35
stylomastoid foramen, 22
stylopharyngeus, 57, 73
subaortic common iliac nodes, 186
subcapsular sinus, 189
subclavian artery, 156, 157, 163
subclavian artery & vein, 174
subclavian chain of nodes, 183, 184
subclavian nodes, 184
subclavian vein, 156
subclavius, 48, 50, 51, 61
subcostal nerve, 102, 103
subcostal vein, 185
subendothelial layer, tunica intima, 166
subgluteal crease, 4
sublingual caruncle with opening of submandibular duct, 136
sublingual ducts, 136
sublingual gland, 136
sublingual salivary gland, 134
submandibular duct, 136
submandibular gland, 6, 135, 136
submandibular nodes, 183
submandibular salivary gland, 134
submandibular triangle, 3
submentalis artery, 161
submental nodes, 183
submucosa, 141
subpleural lymphatic plexus, 184
subpleural lymph plexus, 184
subscapularis, 50, 61, 76
subscapularis tendon, 36
subscapular nodes, 184
substantia nigra, 119
subthalamic nucleus, 120
sulcus for inferior vena cava, 135
sulcus of corpus callosum, 116
sulcus precentral, 116
sulcus terminalis, 171
superficial artery & vein, 157
superficial branch, 102, 103
superficial lymph vessels, 182
superficial palmar arch, 157, 159
superficial parotid nodes, 183
superficial peroneal nerve, 105
superficial peroneal (fibular) nerve, 104
superficial temporal artery, 159, 160
superficial temporal artery & vein, 156
superficial transverse metacarpal, 78
superficial transverse metatarsal ligaments, 93
superior articular facet, 18, 42
superior articular process, 18, 20, 39, 108
superior cerebellar artery, 122, 162
superior cluneal nerves, 105
superior colliculus, 120
superior constrictor, 57
superior costal facet, 34
superior costotransverse ligament, 34
superior division bronchus (apical, posterior, anterior), 150
superior extensor retinaculum, 44, 60, 70, 88, 90, 98
superior frontal sulcus, 115
superior gemellus, 49, 64, 69
superior gluteal nerve, 107
superior horn of thyroid cartilage, 149
superior & inferior articulating facets, 39, 109
superior & inferior gluteal nerves, 102, 103
superior & inferior tarsus, 69
superior labial artery, 160, 161
superior labial frenulum, 130
superior labial tubercle, 6
superior labial vein, 160
superior lacrimal papilla & puncta, 5
superior lip, 130
superior longitudinal fascicle, cruciform ligament, 36
superior lower, lower lobe bronchus, 150
superior lower bronchus, 150
superior mesenteric artery, 156, 159, 164, 187
superior mesenteric nodes, 187, 188
superior mesenteric vein, 156, 165
superior nasal concha (turbinate), 146, 148
superior neck crease, 3
superior notch, 25
superior nuchal line, 22
superior orbital fissure, 23
superior palpebral sulcus, 5, 6
superior palpebral tarsus, 72
superior peroneal retinaculum, 44, 45, 90, 98
superior pharyngeal constrictor, 57, 71, 73, 135, 146
superior pillar of mouth, 6
superior pubic ligament, 42
superior pubic ramus, 42
superior ramus of pubis, 27
superior rectal nodes, 188
superior rectal valve, 142
superior rectal vein, 165
superior sagittal sinus, 157, 160
superior segment, kidney, 192, 195
superior semilunar lobule, 119
superior temporal gyrus, 115
superior thyroid artery, 160, 161
superior thyroid notch, 149
superior transverse scapular ligament, 36
superior ulnar collateral artery, 159
superior vena cava, 156, 157, 158, 164, 170, 171, 174, 185, 208
superior vertebral notch, 18
superolateral superficial inguinal nodes, 186
supinator, 50, 51, 61, 64
supraclavicular fossa, 2, 8
supraclavicular nerve, 104, 105, 106
suprahyoid node, 183
supramarginal gyrus, 114
supraorbital artery, 161, 163
supraorbital foramen, 22
supraorbital nerve, 111
supraorbital ridge, 22
supraparietal lobule, 115
suprapyloric nodes, 187
supraspinatus, 50, 51, 61, 64, 68, 69, 76
supraspinous ligament, 34, 39, 42, 109
supraspinous tendon, 36
suprasternal (jugular notch), 8
supratragic tubercle, 5
supratrochlear artery, 161
supratrochlear vein, 158
sural nerve, 104, 105
surface anatomy, 1–11
surface anatomy (anterior view), 2
surface anatomy (lateral view), 3
surface anatomy (posterior view), 4
surface muscles (layer I, lateral view), 67
surface muscles (layer I, posterior view), 63
surface muscles (layer IA, lateral view), 68
surface muscles (layers I & II, anterior view), 60
suspensory ligament of ovary, 195
sustenaculum tali, 29
sutural (wormian) bone, 23
sweat gland, 131
sympathetic ganglion, 108
sympathetic trunk, 108
symphysis pubis, 40, 210
synapse, 108
synovial flexor tendon sheath, 60, 79, 94
synovial membrane, 75
synovial sheath of extensor digitorum longus tendon, 10
synovial sheath of flexor tendons, 75
systole, 178
systole, end of, 178
systole, heart in, 176

T

T I, 62, 104, 105, 106
T I (transverse costal facet), 34
T II, 104, 105, 106
T III, 104, 105, 106
T IV, 104, 105, 106
T V, 104, 105, 106
T VI, 49, 104, 105, 106
T VII, 104, 105, 106
T VIII, 104, 105, 106
T IX, 104, 105, 106
T X, 104, 105, 106
T XI, 104, 105, 107
T XII, 20, 49, 104, 105, 107
tactile elevation, 7
tail, spermatozoa, 204
tail of caudate nucleus, 120
talus, 14, 15, 16, 29, 44, 90, 91, 92, 98
tarsal bones, 2
tarsometatarsal joint, 29
taste, 130
tectorial membrane, 36, 128
telodendria, 111
temporal, 15, 48, 57
temporal bone, 14, 16, 21, 22, 23, 34, 128
temporal branch, 111
temporal branch of the posterior cerebral artery, 163
temporal fascia, 67, 71
temporalis, 60, 61, 69, 71
temporal lobe, 114, 116, 117, 126, 147
temporal muscle, 2, 3, 4, 5, 6, 128
temporal pole, 119
temporal ridge, 3, 6
temporomandibular & hyoid, joints & ligaments, 35
temporomandibular joint, 35
temporomandibular joint, 6
temporomandibular joint capsule, 71
temporomandibular joint & capsule, 114, 126
temporomandibular (lateral) ligament, 35
temporoparietalis, 67
tendinous arch, 62
tendinous arch of levator ani, 64
tendinous floor of trapezius muscle, 4
tendinous inscription, 60, 68
tendinous intersections, 2, 3
tendo calcaneus (Achilles), 3, 4, 10, 16, 17, 44, 45, 54, 63, 89, 90, 91, 92, 98
tendon, muscular system, 99
tendon levator ani, 203
tendon of extensor indicis, 75
tendon of extensor pollicis longus, 67
tendon sheath of extensor digitorum longus & peroneus tertius, 98
tendon sheath of extensor hallucis longus, 98
tendon sheath of tibialis anterior, 98
tendon sheaths, 86, 87
tendons of extensor digitorum, 75
tendons of quadriceps extensor, 60
tenia libera (free tenia), 135, 140, 141
tensor fasciae latae, 48, 49, 60, 88, 89
tensor tympani muscle, 128
tensor veli palatini, 57
tentorial branch, 110
teres major, 3, 4, 8, 50, 51, 68, 70, 76, 77
teres minor, 3, 4, 51, 63, 64, 68, 77
terminal arteriole, 152, 166
terminal bronchiole, 152
terminal part of ileum, 141
terminal venule, 152
tertiary (segmental) bronchus, 152
testes, 9
testicular artery, 156, 157, 192
testicular tubules, 194, 202
testis, 194, 202, 210
testis (covered by visceral layer of tunica vaginalis), 192
thalamus, 120
thalamus (3rd ventricle), 116
thenar crease, 2, 8
thenar eminence, 2, 7, 8
thenar muscles, 60
thick filament, muscular system, 99
thigh crease, 2, 9
thin filament, muscular system, 99
thoracic bones, skeletal system, 26
thoracic duct, 184, 185, 186, 187, 188
thoracic duct, lymphatic system, 185
thoracic nerve (1st), 102, 103
thoracic nerves T I-T XII, 106, 107
thoracic spine, 4
thoracic vertebra (I) (pedicle), 106
thoracic vertebra (XII) (pedicle), 107
thoracic vertebrae (I-XII), 17, 18
thoracolumbar fasciae, 39, 109
thumb, 8
thumb fat pad, 7
thyrocervical trunk, 160
thyroepiglottic ligament, 149
thyrohyoid, 57, 61, 70, 73
thyrohyoid cartilage, 73, 114
thyrohyoid ligament, 149
thyrohyoid membrane, 73
thyroid cartilage, 3, 6, 126, 135, 146, 147, 149, 150
tibia (medial epicondyle), 4
tibia (medial malleolus), 2, 4, 10, 14, 15, 16, 17, 29, 43, 44, 45, 90, 91, 92, 94, 95, 97, 98
tibia (medial surface), 2
tibial artery & vein, 157
tibial (medial) collateral ligament, 43
tibialis anterior, 2, 3, 10, 45, 54, 55, 56, 60, 67, 69, 70, 88, 89, 95, 98
tibialis anterior tendon, 2, 3, 10, 44, 90, 92, 96, 97, 98
tibialis posterior, 54, 55, 64, 89, 90, 91, 92, 98
tibialis posterior tendon, 2, 4, 10, 45, 70, 88, 94, 95, 96, 97, 98
tibial nerve, 102, 103, 105
tibial plateau, 10
tibial tuberosity, 2, 3, 10, 43
tibia medial surface, 10
tibiocalcaneal ligament, deltoid ligament, 44
tight junction, 152, 166
tight junction between sertoli cells, 204
tongue, 114, 126, 130, 134, 136, 146, 147
tongue, digestive system, 136
tongue surface, 73
touch, 131
trabecula(e), 30, 189, 192
trabecular network, 127
trace of the mandible, 106
trace of the pelvis, 103, 107
trace of the scapula, 103, 106
trace of the spinal column, 103
trachea, 73, 134, 135, 146, 149, 150, 152
tracheal bifurcation, 150
tracheal cartilage, 73, 150, 152
tragus, 5, 6
transverse acetabular ligament, 40
transverse arytenoid muscle, 114
transverse cerebellar fissure, brain, 115
transverse cerebral fissure, brain, 114
transverse cervical nerve, 104, 105
transverse cervical nodes, 183
transverse colon, 140
transverse colon (large intestine), 135
transverse colon (large intestine) (transparent), 134
transverse costal facet (T I), 34
transverse costal facet, 18
transverse facial artery, 157, 160, 161
transverse fasciculi, 78, 93
transverse foramen, 23, 34
transverse ligament of atlas, cruciform ligament, 36
transverse palatine suture, 22
transverse process, 18, 19, 20, 23, 39, 108
transverse sinus, 157, 160
transverse tarsal joint, 29
transverse tubule, 99
transverse vein, 160
transversus abdominis, 48, 61, 62, 64, 65, 69, 70, 74, 151
transversus perinei profundis, 49
transversus thoracis, 61, 70
trapezium, 2, 4, 8, 17, 28, 38, 75
trapezius, 3, 4, 8, 49, 50, 51, 57, 60, 63, 67, 68, 70, 71
trapezius tendon, 4
trapezoid, 8, 17, 28, 38, 75
trapezoid part, 36

triad: cisterns, 99
triangular aponeurosis, 83, 84, 85
triangular fossa, 5, 6
triangular part, inferior frontal gyrus, 115
triceps brachii, 50, 51, 63, 67, 76
triceps brachii muscle (lateral head, long head, medial head,), 2, 4, 7, 8, 64, 76, 77
triceps brachii tendon, 37
triceps crurae (calcaneal tendon), 56
triceps tendon, 7
tricuspid valve, 156, 158, 173, 174, 176
trigeminal (semilunar) ganglion, 110
trigeminal nerve, 110
trigeminal nerve V, 104, 110, 117, 119
trigone of bladder, 195
trigone of urinary bladder, 192
trigonum membrane, 73
triquetral bone, 8, 28
triquetrum, 4, 7, 38, 75
trochanteric bursa, 39, 109
trochanteric line, 27
trochlea, 14, 15, 28, 29, 72, 73, 98
trochlear nerve IV, 117, 119
tropomyosin molecule, 99
troponin molecule, 99
tubal folds, 195
tuber cinereum, 119
tubercle of 5th metatarsal bone, 10
tubercle of navicular bone, 10
tubercle of scaphoid, 28, 38
tubercle of trapezium, 38
tuberosity, 14, 15, 28, 29
tuberosity of calcaneus, 93
tuberosity of navicular bone, 10
tunica adventitia, 166, 189
tunica intima, 166, 189
tunica media, 166, 189
turbinates (superior, middle & inferior conchae), 146, 147, 148
2 cell stage, uterine cycle, 207
tympanic canaliculus, 22
tympanic cavity (middle ear), 128
tympanic membrane (eardrum), 114, 126, 128

U

ulna, 14, 15, 16, 17, 28, 37, 38, 75, 76, 79, 80, 81, 82, 83, 84, 85, 86
ulnar artery, 156, 157
ulnar bursa, 60, 79, 86
ulnar collateral ligament, 37, 38
ulnar head, 7, 8
ulnar nerve, 102, 103, 105, 106
ulnar palmar nerve, 104
ulnar ridge, 4
ulnar styloid process, 4
ulnar vein, 157
umbilical artery, 208
umbilical cord, 209
umbilical vein, 208
umbilicus (navel), 2, 3, 9, 68, 70
unciform processes, 23
uncus, 119
upper anterior segment, kidney, 192, 195
upper deep cervical nodes (jugular), 183
upper eyelid (palpebra superior), 5
upper lobe, left lung, 146
upper lobe, right lung, 146
upper lobe bronchus (apical, anterior, posterior), 150
upper pole, kidney, 197
upper trunk, 106
urachus, 192, 193
ureter, 192, 193, 194, 195, 196, 197, 202, 210

urethra, 193, 194, 195, 196, 202, 203, 209, 210
urethral crest, 192
urethral glands, 192
urethral opening, 195
urinary bladder, 193, 194, 195, 196, 202, 203
urinary bladder (sectioned & transparent), 192
urine, 194, 196, 202, 203, 210
uriniferous (renal) tubule (nephron), 199
urogenital system, 191–99
uterine cycle, 206, 207
uterine gland, 207
uterine tube, 196, 203, 205, 210
uterine (fallopian) tube, 206
uterus, 196, 203, 205, 206, 209, 210
uvula (soft palate), 57, 126, 130, 147, 148
uvula of bladder, 192

V

vagina, 195, 196, 203, 206, 209, 210
vaginal crease, 9
vaginal opening, 196, 203
vagus nerve rootlets, 118
vagus nerve X, 117, 119
vallate papillae, 130, 136
valve, 166, 189
valve of navicular fossa, 192
vasa recta, 199
vascular pattern, 189
vas deferens, 194, 202
vastus intermedius, 54, 61, 64, 69
vastus lateralis, 2, 3, 4, 9, 54, 60, 63, 67, 69, 88, 89
vastus medialis, 2, 10, 54, 60, 70, 88
Vater-Pacini (pacinian) corpuscle (heavy pressure), 131
vein & artery of node, 189
veins (blood flows toward heart), 164
vena caval foramen, 62, 150
venous sinusoid, 207
venous system, 158
ventral root (motor), 108
ventral sacroiliac ligament, 42
ventricle (3rd), 116, 118, 120, 121
ventricle (4th), 116, 121
venule, 189
vermiform appendix, 140, 141
vermis, 119
vermis of cerebellum, 120
vertebral artery, 122, 156, 157, 160, 161, 162, 163
vertebral body, 34
vertebral body (centrum), 108
vertebral column, skeletal system, 18
vertebral foramen, 18
vertebral vein, 160
vertebra prominens (VII), 18
vertebra prominens, 19
vesicula appendix chydatid of Morgagni, 195
vesicular (graafian) follicle, 205
vestibular ganglion, 128
vestibular membrane, 128
vestibule, 130
vestibulocochlear & intermedius nerve, 118
vestibulocochlear nerve VIII, 117, 119
vestivular nerve, 128
vibrissae (external hairs), 6
visceral epithelium (podocytes), 198
vocal fold (cord), 126, 130, 136, 146, 147
vocal ligament, 149

vomer, 21, 22, 129, 148
vulva, 9

W

web, 8
white line (new skin), 5, 6
white matter (motor), 108
white matter (sensory), 108
wing of nose, 5, 6
wing of sphenoid, 21, 22
wormian (sutural) bone, 23
wrist & hand (palmar & dorsal views), joints & ligaments, 38

X

xiphisternal synchondrosis, 26
xiphoid (sternum), 8
xiphoid notch, 2
xiphoid process, 14, 15, 26

Y

yellow marrow, 30
Y ligament of Bigelow, 40

Z

Z band, muscle microstructure, 99
Z line (zigzag) junction of gastric & esophageal mucosa, 137
zona orbicularis, 41, 42
zona pellucida, 205
zygapophysial joints, articular capsules of , 34
zygomatic arch, 5, 6, 14, 15, 21, 71
zygomatic bone, 22
zygomatic branch, 111
zygomatic major, 48, 57, 69, 71
zygomatic minor, 48, 57, 61, 71
zygomatic nerve, 110
zygomaticofacial nerve, 110
zygomaticotemporal nerve, 110
zygomatic process, 22